INTERNATIONAL SERIES IN MATHEMATICS

A JOURNEY INTO PARTIAL DIFFERENTIAL EQUATIONS

The Jones & Bartlett Learning Series in Mathematics

Geometry

Geometry with an Introduction to Cosmic Topology
Hitchman (978-0-7637-5457-0) © 2009

Euclidean and Transformational Geometry: A Deductive Inquiry
Libeskind (978-0-7637-4366-6) © 2008

A Gateway to Modern Geometry: The Poincaré Half-Plane, Second Edition
Stahl (978-0-7637-5381-8) © 2008

Understanding Modern Mathematics
Stahl (978-0-7637-3401-5) © 2007

Lebesgue Integration on Euclidean Space, Revised Edition
Jones (978-0-7637-1708-7) © 2001

Precalculus

Essentials of Precalculus with Calculus Previews, Fifth Edition
Zill/Dewar (978-1-4496-1497-3) © 2012

Algebra and Trigonometry, Third Edition
Zill/Dewar (978-0-7637-5461-7) © 2012

College Algebra, Third Edition
Zill/Dewar (978-1-4496-0602-2) © 2012

Trigonometry, Third Edition
Zill/Dewar (978-1-4496-0604-6) © 2012

Precalculus: A Functional Approach to Graphing and Problem Solving, Sixth Edition
Smith (978-0-7637-5177-7) © 2012

Precalculus with Calculus Previews (Expanded Volume), Fourth Edition
Zill/Dewar (978-0-7637-6631-3) © 2010

Calculus

Single Variable Calculus: Early Transcendentals, Fourth Edition
Zill/Wright (978-0-7637-4965-1) © 2011

Multivariable Calculus, Fourth Edition
Zill/Wright (978-0-7637-4966-8) © 2011

Calculus: Early Transcendentals, Fourth Edition
Zill/Wright (978-0-7637-5995-7) © 2011

Multivariable Calculus
Damiano/Freije (978-0-7637-8247-4) © 2011

Calculus: The Language of Change
Cohen/Henle (978-0-7637-2947-9) © 2005

Applied Calculus for Scientists and Engineers
Blume (978-0-7637-2877-9) © 2005

Calculus: Labs for Mathematica
O'Connor (978-0-7637-3425-1) © 2005

Calculus: Labs for MATLAB®
O'Connor (978-0-7637-3426-8) © 2005

Linear Algebra

Linear Algebra: Theory and Applications, Second Edition
Cheney/Kincaid (978-1-4496-1352-5) © 2012

Linear Algebra with Applications, Seventh Edition
Williams (978-0-7637-8248-1) © 2011

Linear Algebra with Applications, Alternate Seventh Edition
Williams (978-0-7637-8249-8) © 2011

Advanced Engineering Mathematics

A Journey into Partial Differential Equations
Bray (978-0-7637-7256-7) © 2012

Advanced Engineering Mathematics, Fourth Edition
Zill/Wright (978-0-7637-7966-5) © 2011

An Elementary Course in Partial Differential Equations, Second Edition
Amaranath (978-0-7637-6244-5) © 2009

Complex Analysis

Complex Analysis for Mathematics and Engineering, Sixth Edition
Mathews/Howell (978-1-4496-0445-5) © 2012

A First Course in Complex Analysis with Applications, Second Edition
Zill/Shanahan (978-0-7637-5772-4) © 2009

Classical Complex Analysis
Hahn (978-0-8672-0494-0) © 1996

Real Analysis

Elements of Real Analysis
Denlinger (978-0-7637-7947-4) © 2011

An Introduction to Analysis, Second Edition
Bilodeau/Thie/Keough (978-0-7637-7492-9) © 2010

Basic Real Analysis
Howland (978-0-7637-7318-2) © 2010

Closer and Closer: Introducing Real Analysis
Schumacher (978-0-7637-3593-7) © 2008

The Way of Analysis, Revised Edition
Strichartz (978-0-7637-1497-0) © 2000

Topology

Foundations of Topology, Second Edition
Patty (978-0-7637-4234-8) © 2009

Discrete Mathematics and Logic

Essentials of Discrete Mathematics, Second Edition
Hunter (978-1-4496-0442-4) © 2012

Discrete Structures, Logic, and Computability, Third Edition
Hein (978-0-7637-7206-2) © 2010

Logic, Sets, and Recursion, Second Edition
Causey (978-0-7637-3784-9) © 2006

The Jones & Bartlett Learning Series in Mathematics

Numerical Methods

Numerical Mathematics
Grasselli/Pelinovsky (978-0-7637-3767-2) © 2008

Exploring Numerical Methods: An Introduction to Scientific Computing Using MATLAB®
Linz (978-0-7637-1499-4) © 2003

Advanced Mathematics

Mathematical Modeling with Excel®
Albright (978-0-7637-6566-8) © 2010

Clinical Statistics: Introducing Clinical Trials, Survival Analysis, and Longitudinal Data Analysis
Korosteleva (978-0-7637-5850-9) © 2009

Harmonic Analysis: A Gentle Introduction
DeVito (978-0-7637-3893-8) © 2007

Beginning Number Theory, Second Edition
Robbins (978-0-7637-3768-9) © 2006

A Gateway to Higher Mathematics
Goodfriend (978-0-7637-2733-8) © 2006

The Jones & Bartlett Learning International Series in Mathematics

Linear Algebra: Theory and Applications, Second Edition, International Version
Cheney/Kincaid (978-1-4496-2731-7) © 2012

Multivariable Calculus
Damiano/Freije (978-0-7637-8247-4) © 2012

Complex Analysis for Mathematics and Engineering, Sixth Edition, International Version
Mathews/Howell (978-1-4496-2870-3) © 2012

A Journey into Partial Differential Equations
Bray (978-0-7637-7256-7) © 2012

Functions of Mathematics in the Liberal Arts
Johnson (978-0-7637-8116-3) © 2012

Advanced Engineering Mathematics, Fourth Edition, International Version
Zill/Wright (978-0-7637-7994-8) © 2011

Calculus: Early Transcendentals, Fourth Edition, International Version
Zill/Wright (978-0-7637-8652-6) © 2011

Real Analysis
Denlinger (979-0-7637-7947-4) © 2011

Mathematical Modeling for the Scientific Method
Pravica/Spurr (978-0-7637-7946-7) © 2011

Mathematical Modeling with Excel®
Albright (978-0-7637-6566-8) © 2010

An Introduction to Analysis, Second Edition
Bilodeau/Thie/Keough (978-0-7637-7492-9) © 2010

Basic Real Analysis
Howland (978-0-7637-7318-2) © 2010

For more information on this series and its titles, please visit us online at http://www.jblearning.com. Qualified instructors, contact your Publisher's Representative at 1-800-832-0034 or info@jblearning.com to request review copies for course consideration.

INTERNATIONAL SERIES IN MATHEMATICS

A JOURNEY INTO PARTIAL DIFFERENTIAL EQUATIONS

WILLIAM O. BRAY
University of Maine

JONES & BARTLETT LEARNING

World Headquarters
Jones & Bartlett Learning
40 Tall Pine Drive
Sudbury, MA 01776
978-443-5000
info@jblearning.com
www.jblearning.com

Jones & Bartlett Learning
Canada
6339 Ormindale Way
Mississauga, Ontario L5V 1J2
Canada

Jones & Bartlett Learning
International
Barb House, Barb Mews
London W6 7PA
United Kingdom

Jones & Bartlett Learning books and products are available through most bookstores and online booksellers. To contact Jones & Bartlett Learning directly, call 800-832-0034, fax 978-443-8000, or visit our website, www.jblearning.com.

Substantial discounts on bulk quantities of Jones & Bartlett Learning publications are available to corporations, professional associations, and other qualified organizations. For details and specific discount information, contact the special sales department at Jones & Bartlett Learning via the above contact information or send an email to specialsales@jblearning.com.

Production Credits
Publisher: Cathleen Sether
Senior Acquisitions Editor: Timothy Anderson
Associate Editor: Melissa Potter
Production Director: Amy Rose
Associate Production Editor: Tiffany Sliter
Senior Marketing Manager: Andrea DeFronzo
V.P., Manufacturing and Inventory Control: Therese Connell
Cover and Title Page Design: Kristin E. Parker
Composition: Northeast Compositors, Inc.
Cover Image: © Comstock Images/age fotostock
Printing and Binding: Malloy, Inc.
Cover Printing: Malloy, Inc.

Library of Congress Cataloging-in-Publication Data
Bray, William O., 1955–
A journey into partial differential equations / William O. Bray.
p. cm.
Includes bibliographical references and index.
ISBN-13: 978-0-7637-7256-7 (casebound)
ISBN-10: 0-7637-7256-9 (casebound)
1. Differential equations, Partial. 2. Wave equation. 3. Heat equation.
4. Laplace transformation. I. Title.
QA377.B735 2011
515'.353—dc22

2010030536

6048
Printed in the United States of America
14 13 12 11 10 10 9 8 7 6 5 4 3 2 1

This book is dedicated to my wife, Terri, for her patience and support over the years.

Contents

Preface

Historically, differential equations arose as a tool to describe physical laws and processes. Indeed, as the eminent French mathematician Henri Poincaré described [25], the harmonious order of nature seen through observation and experimentation manifests in the qualitative and quantitative explorations of differential equations. This book is an introduction to the latter in the realm of partial differential equations. It has been written with the guiding principle: motivate with physics and then do mathematics. As such, it is a book on applicable mathematics rather than applied mathematics.

The text represents material suitable for a one-semester course in partial differential equations and is aimed at junior- and senior-level undergraduates of mathematics, the physical sciences, and engineering, as well as graduate students in a variety of disciplines. Much of the material has been used for such a course and audience at the University of Maine over the past decade. Prerequisites for reading the book are simple: (i) a working knowledge of calculus in one and several variables; (ii) familiarity with the theory and methods of solution for ordinary differential equations; and (iii) acquaintance with complex arithmetic and the complex exponential function. Some knowledge of linear algebra would be useful, although it appears explicitly only in the discussion of classification of second-order linear partial differential equations. The concept of uniform convergence of sequences of functions is used in many segments of the text, so I have included a brief appendix on the topic.

Although nonlinear equations are mentioned in the first chapter, the entire focus is on linear partial differential equations, in particular on the classical trinity of equations: the heat or diffusion equation, the wave equation, and Laplace's equation. Historically, these stand out as cornerstones and the methodologies developed for these equations carry the fundamental ideas for more advanced treatment. Methodological focal points include: (i) the divergence theorem and its roles; (ii) symmetries of the Laplace operator and its role; (iii) separation of variables and Fourier expansions; and (iv) early introduction to the theory of distributions and their use throughout the text.

An overview of the chapters is as follows. Chapter one provides precisely what the title states "Unfolding Ideas via Examples". It is intended to be a guide to the rest of the book. The classical trinity are introduced with emphasis given to the underlying physics of the heat and wave equations in one spatial dimension. Boundary and initial conditions are motivated and defined and the basic questions regarding partial differential equations are laid out and exemplified. By the end of chapter one, the reader will have a glimpse of key qualitative methodologies as well as an understanding of the behavior of solutions of the heat and wave equations. In chapter two, we overview the theory and solution methods for first-order partial differential equations with emphasis on the connection with the geometry of vector fields and is further applied to classification of linear equations of second order. The third chapter begins with a review of the divergence theorem and its role in describing the physics underlying the classical trinity. The invariance of the Laplacian under translations and rotations is discussed and is used to motivate solution methodologies, in particular separation of variables. Further, the qualitative ideas glimpsed at in chapter one are followed up more thoroughly for the diffusion and wave equations. Distributions are introduced heuristically in chapter one, whereas chapter four presents a guide to the theory and then develops the notions of weak solutions and fundamental solutions of partial differential equations in the context of the heat and wave equations. Chapters five and six go hand in hand motivating and developing the theory of Fourier expansions and applying them to the diffusion and wave equations. Chapter six also introduces multiple Fourier expansions and their application. The qualitative methods and solution methodologies are revisited in chapter seven in the context of Laplace's equation. From a pedagogical perspective, this represents recurrent themes and serves my belief that Laplace's equation, as the basic model of stationary phenomena, deserves to be considered the pinnacle of the classical trinity. Solutions are constructed via Green's function ideas and Fourier methods. Finally, chapter eight introduces spherical harmonic expansions and their role in boundary value problems for Laplace's equation. This chapter represents a partial completion of the ideas of chapter seven connecting Fourier and Green's function methodologies. After chapter one, the core of the text lies in chapters three through seven; chapter two is largely independent of the rest of the text. The text contains over 200 exercises with varying levels of difficulty. Many of these I have devised over the years, others are variations of problems found in textbooks of other authors, in particular those acknowledged on the next page.

As a final remark:

> *Historically, the role of partial differential equations in the development of mathematics and its dual role in understanding the physical world is a fascinating and profound story, a story whose*

> *threads continue in modern day research. The student who takes up the study embarks on a journey through a diversity of mathematical ideas and techniques along with their interpretation. Let the journey begin.*

William O. Bray

Text Supplements

- Complete solutions to all exercises are available to qualified instructors
- Image files for all figures in the textbook using a range of colors
- *Mathematica*® notebook files to supplement course instruction

Acknowledgments

The author of any mathematics text today has been influenced by the style, content, and perspectives of works which came before. I would like to acknowledge the following authors for the impact their work has had on the current text: G. Folland [10], M. A. Pinsky [23], R. Strichartz [30], W. A. Strauss [29], and the work of E. C. Zachmanoglou and D. W. Thoe [35]. Any of these works could be used as supplemental reading for this book.

I would like to thank Sergey Lvin (University of Maine), Wally Madych (University of Connecticut), and Mark A. Pinsky (Northwestern University) for their comments on this work during its early stages. Strong words of appreciation go to Sergey Lvin for his careful reading of the final manuscript.

Stan Wakefield originally connected me to Tim Anderson, Acquisitions Editor, at Jones & Bartlett Learning. I appreciate both of them for their encouragement and support as well as for the efforts of the rest of the Jones & Bartlett Learning staff.

This text was originally typeset using the delightful program called Lyx[1] and then converted to LaTeX for final editing. Most of the graphics in the text were created using *Mathematica*[2].

1. [http://www.lyx.org.]

2. Registered trademark of Wolfram Research, Inc., http://www.wolfram.com.

Historical Prelude

The study of partial differential equations spans the past few centuries with contributions from many mathematicians and scientists. Throughout this text many results, formulas, and ideas carry the name of the originator. Below is a partial list of historical figures which instills a sense of the breadth of time in which partial differential equations have been studied and hints at the broad impact this study has had on mathematics as a whole. By no means is this list complete, but it certainly covers key players from the eighteenth and nineteenth centuries whose names appear elsewhere in the text, and usually more than once.

Jean Le Rond d'Alembert (1717–1783) A pioneer in the application of differential equations to problems of physics. His primary contributions were in equilibrium problems and fluid motion.

Adrien-Marie Legendre (1752–1833) His work on gravitation in the late 1700s introduced what are now known as Legendre functions (chapter eight). He also made contributions to number theory and the theory of elliptic integrals.

Carl Friedrick Gauss (1777–1855) Perhaps the most famous mathematician throughout history, his work embraced a variety of areas in both mathematics and physics including analysis, differential geometry, number theory, astronomy, electromagnetism, geodesy, and optics. A result known as Gauss's theorem, or the Divergence Theorem (chapter three) plays a fundamental role throughout this text.

Pierre-Simon Laplace (1749–1827) Mathematical physicist who gained fame for his application of differential equations to astronomy, in particular demonstrating the stability of the solar system in 1799. The equation which bears his name originated in his investigations in celestial mechanics.

Siméon-Denis Poisson (1781–1855) Mathematician known for his work on differential equations, partial differential equations, potential theory and their application to problems in electrostatics, vibrations, and heat propagation.

Jean-Baptiste-Joseph Fourier (1768–1830) A mathematician and scientist most recognized for his work on heat propagation in solid materials and the mathematical theory he initiated to describe it. Indeed, his mathematical ideas had considerable impact on the mathematical thought of the time period and foreshadowed much of which came after. Later work by many mathematicians became known as an area of mathematics called Fourier analysis, a realm of thought of considerable interest in contemporary research and is a fundamental tool for the study of partial differential equations.

George Green (1793–1841) Mathematician and physicist whose work laid the foundations to the modern day theory of electricity and magnetism. He is most noted for a monograph *An Essay on the Application of Mathematical Analysis to the Theories of Electricity and Magnetism* published in 1828 [13].

Johann Peter Gustav Lejeune Dirichlet (1805–1859) Dirichlet made significant contributions to several areas of mathematics in the nineteenth century. Notably, his contributions in approximation theory (Dirichlet approximation theorem), Fourier analysis (we will see this in chapter five), and number theory are cornerstones in much of modern day mathematical research.

Georg Friedrich Bernhard Riemann (1826–1866) Noted for his significant advances in the areas of complex analysis and geometry, his study of the problem of representibility of a function by a trigonometric series led to the development of what we now call the Riemann integral.

Carl Gottfried Neumann (1832–1925) Known for his work in partial differential equations, in particular on the Dirichlet principle (which we will explore in chapter seven), and the theory of integral equations.

Oliver Heaviside (1850–1925) Mathematical physicist who made contribtutions to vector calculus and electromagnetism. He developed an approach to solving differential equations, known in mathematics as operational calculus, which in many respects foreshadowed the theory of generalized functions.

Jules Henri Poincaré (1854–1912) A mathematician regarded by many as the greatest mathematician of his era. His work embraced the areas of the qualitative theory of differential equations, complex analysis, algebraic topology, celestial mechanics, special relativity, electrodynamics, and fluid mechanics. Poincaré's work on the three-body problem is often cited as the origin of the theory of chaos.

Victor Robin (1855–1897) A mathematician who worked in the realm of thermodynamics.

Jacques Salomon Hadamard (1865–1963) Mathematician known for his significant contributions in number theory, complex analysis, differential geometry, partial differential equations, the calculus of variations, and the foundations of functional analysis.

Henri Léon Lebesgué (1875–1941) Like Riemann earlier, motivated by deep problems regarding the representability of a function by a trigonometric series, Lebesgué constructed a theory of integration extending Riemann's which is a cornerstone in all of modern analysis. He is also known for his work in potential theory and on the Dirichlet problem.

Sergei Lvovich Sobolev (1908–1989) Noted for his fundamental work in the 1930s on weak solutions of partial differential equations. Within mathematics, the concept of a generalized function (chapter four) has origins in his work.

Laurent Schwartz (1915–2002) Well known for laying the foundations of the theory of distributions—a fundamental tool in the study of partial differential equations.

Chapter 1

Unfolding Ideas via Examples

Our primary goal in this chapter is to set the stage for the study of partial differential equations by

- defining and exemplifying the object of our study: partial differential equations.
- relating partial differential equations in concrete physical problems to provide historical backdrop and physical insight.
- putting forth the basic questions of the study of partial differential equations and exploring these questions via examples.

1.1 What Are Partial Differential Equations?

A differential equation is an equation involving the derivatives of an unknown function. The **order** of the differential equation is the order of the highest derivative involved in the equation. Here are a few examples; the first three are second order and the fourth is first order.

$$m\frac{d^2u}{dt^2} + c\frac{du}{dt} + ku = 0 \tag{1.1.1}$$

$$\frac{d^2u}{dt^2} + k\sin u = \sin\omega t \tag{1.1.2}$$

$$\frac{\partial^2 u}{\partial x^2} + \frac{\partial^2 u}{\partial y^2} + ku = 0 \tag{1.1.3}$$

$$\frac{\partial u}{\partial t} + u\frac{\partial u}{\partial x} = 0 \tag{1.1.4}$$

In each case, variables t, x, and y represent independent variables. If u is the unknown function, then structurally any differential equation has form

$$L(u) = F,$$

where the left-hand side represents all dependence of the equation on u and its derivatives and F typically is a function of the independent variable(s). The map $u \to L(u)$ is called an **operator**; L may also depend on the independent variables. By a solution of a differential equation we mean a function u with sufficient smoothness (discussed below) that when substituted into the equation yields an identity.

The equation is said to be **linear** if the operator $u \to L(u)$ is linear in the sense that

$$L(\alpha u_1 + \beta u_2) = \alpha L(u_1) + \beta L(u_2), \tag{1.1.5}$$

where α and β are scalars and u_1 and u_2 are functions. A linear differential equation for which $F \equiv 0$ is said to be **homogeneous**. Equations (1.1.1) and (1.1.3) are homogeneous linear equations whereas (1.1.2) and (1.1.4) are nonlinear. The key principles associated with linear equations are the following.

Proposition 1.1 Let L be a linear differential operator on a region D of the independent variables. Then the following hold on D.

1. Superposition Principle. A linear combination of solutions of the homogeneous linear equation $L(u) = 0$ is also a solution.

2. Any solution of the linear equation $L(u) = F$ has the form $u = u_p + u_h$, where u_p is a particular solution to this equation and u_h is some solution of the corresponding homogeneous equation $L(u) = 0$.

Both of these statements are easily verified on the basis of the defining property (1.1.5).

A differential equation is called an **ordinary differential equation**, abbreviated **ODE**, if the unknown function is dependent on a single variable. A familiar example is the nonhomogeneous form of (1.1.1),

$$m\frac{d^2u}{dt^2} + c\frac{du}{dt} + ku = F(t), \tag{1.1.6}$$

which models a damped spring–mass system. Here $u = u(t)$ represents the displacement of the mass from equilibrium as a function of time t; $m > 0, c \geq 0$, and $k > 0$ are constants representing mass, the damping constant, and the spring constant; and $F(t)$ is an externally applied force. Typically, we would be interested in solving this equation given a pair of initial conditions of the form

$$u(0) = u_0 \quad \text{and} \quad \frac{du}{dt}(0) = u_0', \tag{1.1.7}$$

which specify the initial position and velocity of the mass. Taken together, the equation (1.1.6) and the conditions (1.1.7) are called an **initial value problem**.

A second example of a different flavor is

$$\frac{d^2u}{dx^2} = -\frac{g}{c^2}, \tag{1.1.8}$$

subject to the conditions

$$u(0) = u(L) = 0. \tag{1.1.9}$$

Here $u = u(x)$ is the displacement of an elastic string of length L and x is the spatial variable specifying the position along the string. This linear equation with these auxiliary conditions models the equilibrium position of a string that is clamped at its ends under the influence of gravity (g is the acceleration of gravity and c^2 is a constant dependent on the material properties of the string and the tension). This provides an example of a **boundary value problem**.

A **partial differential equation**, abbreviated **PDE**, is a differential equation where the unknown function is dependent on more than one variable. Two examples were given in equations (1.1.3) and (1.1.4). Following are several more examples, many of which we will study in detail in the coming chapters. In all of these examples, t represents time and the variables x and y are spatial variables.

- The heat equation,

  $$u_t - \alpha^2 u_{xx} = F(x,t), \tag{1.1.10}$$

  models heat flow in one spatial dimension in homogeneous media. Here $L(u) = u_t - \alpha^2 u_{xx}$ is a linear operator. Hence, the equation is a second-order linear PDE.

- The wave equation,

  $$u_{tt} - c^2 u_{xx} = F(x,t), \tag{1.1.11}$$

  governs the motion of an ideal vibrating string. It also is a second-order linear PDE.

- Laplace's equation in two variables is the second-order linear PDE

  $$u_{xx} + u_{yy} = 0. \tag{1.1.12}$$

  This equation appears in a variety of steady-state problems and in complex analysis. Its variant,

  $$u_{xx} + u_{yy} = F(x,y),$$

  is known as Poisson's equation. Both equations are sometimes referred to as potential equations, a term from electrostatics.

- The transport equation,

$$u_t + cu_x = 0,$$

models the transport of a suspended material in a fluid. Here, c represents the velocity of the transporting fluid and u is the concentration of the suspended substance. It is a homogeneous first-order linear equation.

- Schrödinger's equation from quantum mechanics in one spatial variable takes the form

$$iu_t + \alpha u_{xx} + V(x)\, u = 0,$$

where $V(x)$ is a potential function.

The first three equations are sometimes called the **classical trinity** because of their importance in mathematical physics, and most of this text is devoted to their study.

Here are a few examples of nonlinear PDEs.

- Burger's equation from fluid mechanics,

$$u_t + uu_x - \gamma u_{xx} = 0,$$

is a second-order nonlinear equation used as a model of one-dimensional viscous compressible flow.

- The Korteweg–deVries equation,

$$u_t + uu_x + u_{xxx} = 0,$$

is a third-order equation used as a model of lossless propagation of shallow water waves.

- The Sine–Gordon equation,

$$u_{tt} - u_{xx} + \sin u = 0,$$

is found in solid-state electronics.

- The minimal surface equation,

$$(1 + u_y^2)u_{xx} - 2u_x u_y u_{xy} + (1 + u_x^2)u_{yy} = 0,$$

is a fundamental equation found in differential geometry. Physically, it models soap films.

Generally speaking, a partial differential equation may involve a time variable $t > 0$ and one or more spatial variables. Typically, the spatial variables are restricted to some open set. For example, we may be interested in heat transfer in a wall of thickness L, in which case we consider the heat equation (1.1.10) with domain $\{(t, x) \mid t > 0,\ 0 < x < L\}$. Physically it makes sense that information concerning the transfer of heat energy on the boundary is needed; in other words, **boundary conditions**, abbreviated **BCs**, need be specified at $x = 0$ and $x = L$. There are three principal types of boundary conditions that occur in most applications. The condition that specifies the unknown function on the boundary is called a **Dirichlet condition**, named after the French mathematician Johann Dirichlet (1805–1859). In a heat flow problem, this corresponds to specifying the temperature on the boundary. The condition that specifies the normal derivative of the unknown function on the boundary is called a **Neumann condition**, named after the German mathematician Carl Neumann (1832–1925). This corresponds to prescribing the rate of energy transfer on the boundary in a heat flow problem. The third type is called a **Robin condition** after the mathematical physicist Victor Robin (1855–1897). Here, a linear combination of the normal derivative and the function is specified on the boundary and typically represents a linear energy transfer between the domain of the PDE and the exterior region. In equations involving time, one or more **initial conditions** are usually needed in order to completely describe the solution of a physical problem. In a heat flow problem, one condition is needed to specify the initial temperature distribution of the object. For problems involving the vibration of a mechanical object, two initial conditions are needed to specify the initial position and the velocity of the object. The combination of a PDE with associated boundary conditions and initial conditions is called an **initial boundary value problem**, abbreviated **IBVP**. If time is not involved in the PDE, the combination of the equation with associated boundary conditions is called a **boundary value problem**, or **BVP**. This terminology is consistent with that used for ordinary differential equations. All three types of boundary conditions, initial conditions, and the formulation of IBVPs are illustrated in the next section.

Our study of partial differential equations involves two interrelated aspects, qualitative and quantitative. Qualitative aspects concern deriving information about the solution to a given problem from the structure of the PDE and associated boundary and initial conditions. More often than not, qualitative aspects fit physical intuition. For example, for a taut elastic vibrating string with ends clamped and without damping, the total mechanical energy defined as the sum of kinetic energy (energy of motion) and potential energy (energy from the string's tension), is conserved. Hence, from an energy perspective, the motion of the string is an exchange process between the kinetic and potential forms of energy. This can be seen as a basis for Hamilton's principle in mechanics, a topic we pursue later. Quantitative aspects of partial differential equations concern the explicit construction of solution formulas or numerical approximation of such. Almost always, determining solution formulas for PDEs involves clever reduction of the equation to an ODE or finite set of ODEs. The reduction process is always related to an underlying symmetry of the PDE, i.e., a transformation in the underlying space of independent variables that preserves solutions of the equation. This same idea is used in the study of ODEs to reduce the equation to an algebraic one (see Exercise 1.1).

A useful question to ponder is: What does it mean for a function to be a solution of a given PDE? For example, consider the heat equation $u_t = \alpha^2 u_{xx}$, where we assume α is a constant. Since the equation involves partial derivatives of the function $u = u(x,t)$, a natural assumption would be that u possesses at least one derivative in the time variable t and two in the spatial variable x. Then, for $u(x,t)$ to be a solution would mean that when substituted into the equation, an identity results. Mathematically, it is useful to strengthen the notion of what it means to be a solution. A function u is said to be a **classical** or **strong solution** of a PDE if u and its partial derivatives of all orders up to the order of the equation are continuous in the domain of the equation. If the order of the equation is m, notationally we say that u is of class C^m, or write $u \in C^m(D)$, where D is the domain of the independent variables. So in the example of the heat equation, a strong solution would be a function $u(x,t)$ that gives an identity when substituted into the equation and has the property that the following seven functions,

$$u(x,t),\ u_t(x,t),\ u_x(x,t),\ u_{xt}(x,t),\ u_{tx}(x,t),\ u_{tt}(x,t),\ \text{and}\ u_{xx}(x,t),$$

are all continuous. Thus, a strong solution of the heat equation is of class C^2. This may seem like a sticky technical point. However, it has consequences such as equality of mixed partials, a fact that will permit useful manipulations.

Example 1.1 The functions $t + \frac{1}{2}x^2$, $e^{-t}\sin x$, and $e^{-t}\cos x$ are all strong solutions to the heat equation $u_t = u_{xx}$ for any t and x. In fact all of these functions have continuous derivatives of any order. The function

$$\mathcal{S}(x,t) = \frac{1}{2\alpha\sqrt{\pi t}} \exp\left(-\frac{x^2}{4\alpha^2 t}\right)$$

is a solution of the equation $u_t = \alpha^2 u_{xx}$ in the domain where $t > 0$ and $x \in \mathbb{R}$. This can be easily checked by differentiation.

Example 1.2 Consider the transport equation $u_t - u_x = 0$ on the domain $\{(x,t) \mid x \in \mathbb{R}\ \text{and}\ t > 0\}$. Then if $f(x)$ is a continuously differentiable function of one variable on $\mathbb{R}$, a chain rule computation will show that $u(x,t) = f(x+t)$ is a strong solution to the PDE. For an explicit example, let

$$f(x) = \begin{cases} \cos^2 x, & -\pi/2 < x < \pi/2 \\ 0, & \text{elsewhere.} \end{cases}$$

It is easy to check that $f(x)$ is of class C^1. On the other hand, the function

$$g(x) = \begin{cases} \cos x, & -\pi/2 < x < \pi/2 \\ 0, & \text{elsewhere} \end{cases}$$

is continuous on $\mathbb{R}$ but the derivative does not exist at $x = \pm\pi/2$. Hence the function $u(x,t) = g(x+t)$ is not a strong solution of the PDE.

Throughout this text, usually when discussing solutions to PDEs, we will mean strong solution and the adjective *strong* or *classical* will typically be omitted. However, interesting physical situations, particularly related to wave phenomena, arise in which it is useful to relax the smoothness assumption in the definition. One instance is the previous example. Hence it is useful to introduce the notion of a **weak solution** of a PDE, a topic we describe by example in Section 1.3 and follow up in detail in Chapter 4.

Exercises

1.1 Consider the second-order homogeneous linear ODE

$$a_2 y'' + a_1 y' + a_0 y = 0,$$

where the coefficients a_i are constants and $y = y(t)$.

a. Verify that if $y = f(t)$ is a solution to this equation, then so is $y = f(t+\tau)$ for any constant τ.

b. Show that the vector space generated by the single function $e^{\mu t}$, i.e., $\{ce^{\mu t} \mid c \in \mathbb{R}\}$ is also preserved under this transformation. In other words, show that if $f(t)$ is in this space of functions, then so is the function $f(t+\tau)$.

c. Seek solutions of the ODE in this vector space by showing that if $e^{\mu t}$ is a solution of the ODE, then the constant μ must satisfy the algebraic equation $a_2\mu^2 + a_1\mu + a_0 = 0$.

1.2 (Review of ODE Solution Methods) Apply the ideas of the previous exercise to find fundamental sets for the following ODEs.

a. $y'' + 3y' + 2y = 0$

b. $y'' + 4y = 0$

c. $y'' + 2y' + y = 0$

1.3 Verify the conclusions of Example 1.1.

1.4 Verify the conclusions of Example 1.2.

1.2 The Classical Trinity

In this section we will explore in simple terms the physics behind the wave, heat, and Laplace equations and formulate explicitly associated initial and boundary conditions.

Derivation of the Classical Trinity

The Vibrating String Vibrations of an elastic string occur in three dimensions and are completely modeled using vector methods.[1] Herein we provide a simplified view that captures the essential ideas. Imagine a string of length L stretched between two points; the string will vibrate only if sufficient tension is present. Assume that the string will vibrate only in a plane, the xu-plane (see Figure 1.1). A particle of the string in vibration could have both horizontal and vertical components of its motion. A key assumption is that the slope of the string is small, in which case it can be shown that the horizontal component of motion can be neglected. Under these assumptions, the motion of the string is described by a function $u(x,t)$ that represents the vertical displacement of the string as a function of position x and time t.

Suppose the string has density $\rho = \rho(x)$ measured in mass units per unit length. Newton's second law will be applied to a small segment of the string $[x, x+\Delta x]$. The string is assumed to be perfectly flexible, i.e., offers no resistance to bending. Hence the force exerted on the ends of the segment by the rest of the string act in a tangential direction. This tangential force is called **tension** and is denoted $\mathbf{T}(x,t)$, a vector quantity. Under this assumption, the vectors $\mathbf{T}(x,t)$ and $\mathbf{T}(x+\Delta x,t)$ lie in the same plane and we may introduce the angle $\theta = \theta(x,t)$ that these vectors make with the horizontal. The slope of the string is given by $\frac{\partial u}{\partial x} = \tan\theta(x,t)$.

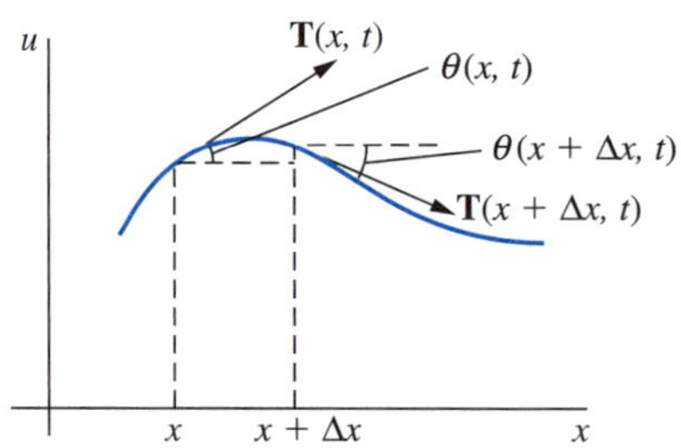

Figure 1.1 The Vibrating String

1. A detailed derivation of the wave equation in one dimension is given in the book by Weinberger [33]. Further analysis, in particular when the small slope assumption is not assumed, is given in the paper by Antman [1].

Applying Newton's second law in the vertical direction, we have

$$\rho(x)\Delta x\frac{\partial^2 u}{\partial t^2} = |\mathbf{T}(x+\Delta x,t)|\sin\theta(x+\Delta x,t) - |\mathbf{T}(x,t)|\sin\theta(x,t) + \rho(x)\Delta x Q(x,t),$$

where Q represents vertical external forces measured in force per unit mass on the string, e.g., gravity. Dividing by Δx and letting Δx tend to zero gives

$$\rho(x)\frac{\partial^2 u}{\partial t^2} = \frac{\partial}{\partial x}\left[|\mathbf{T}(x,t)|\sin\theta(x,t)\right] + \rho(x)Q(x,t). \tag{1.2.1}$$

This can be simplified using

$$\sin(\theta,t) = \frac{\tan\theta(x,t)}{\sec\theta(x,t)} = \frac{u_x(x,t)}{\sqrt{1+u_x^2(x,t)}}.$$

Consequently, the small-slope hypothesis ($|u_x| \ll 1$) implies $\sin\theta(x,t) \approx u_x(x,t)$. Substituting in (1.2.1) produces the linearized wave equation

$$\rho\frac{\partial^2 u}{\partial t^2} = \frac{\partial}{\partial x}(T\frac{\partial u}{\partial x}) + \rho Q, \tag{1.2.2}$$

where T denotes the magnitude of the tension $\mathbf{T}$.

Assuming that the string is perfectly elastic, meaning that the tension $\mathbf{T}(x,t)$ only depends on the local stretching of the string, the small-slope assumption then implies that the local stretching is essentially the same as for the string in equilibrium. Under this assumption, the magnitude of $\mathbf{T}$ is constant and equation (1.2.2) becomes

$$u_{tt} = c^2 u_{xx} + Q, \tag{1.2.3}$$

where $c^2 = T/\rho$, an equation we refer to as the wave equation with external force.

If the only external force is due to gravity, $Q = -g$, then often this is small in comparison with the tension term. Then the equation then becomes the **one-dimensional wave equation**

$$u_{tt} = c^2 u_{xx}. \tag{1.2.4}$$

Notice that c^2 has units of velocity squared; the role of the velocity c is very important in the sequel.

We derived the wave equation as a model for the vibrating string. It is also used as a model for compression waves (sound) traveling in a thin tube (an acoustic wave guide). Here the equation is written $p_{tt} = c^2 p_{xx}$, where $p = p(x,t)$ is the sound pressure at position x along the tube and time t. The wave equation derived for this model is based on the laws of gas dynamics and is useful for investigating the physics of a number of musical instruments, e.g., the flute, clarinet, and organ pipes; see [31] for an overview of the applications of the wave equation in this realm.

The Heat Equation Consider a thin rod of cross-sectional area A and length L that is insulated laterally (see Figure 1.2) and contains no heat sources. By "thin" we mean that heat can only flow along the length of the rod. (One can also think in terms of an infinite slab or wall of thickness L where heat can flow only in one direction.) Let $E = E(x,t)$ be the heat energy per unit volume in the rod. **Heat flux**, $\phi = \phi(x,t)$, is defined as the rate of change of heat energy per unit area per unit time; this is taken to be positive in the direction of the positive x-axis. The time rate of change of energy in the slice $[x, x+\Delta x]$ is given by

$$\frac{d}{dt}\left(A\int_x^{x+\Delta x} E(s,t)ds\right) = A\int_x^{x+\Delta x} E_t(s,t)ds.$$

Conservation of heat energy in the slice $[x, x+\Delta x]$ states that the rate of change of energy is proportional to the flux difference at the endpoints. Mathematically, this is

$$A\int_x^{x+\Delta x} E_t(s,t)ds = A(\phi(x,t) - \phi(x+\Delta x,t)).$$

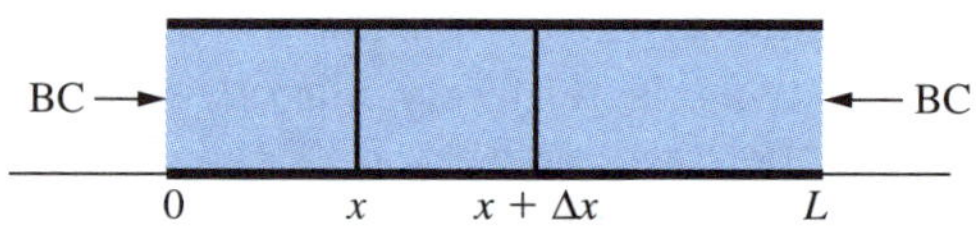

Figure 1.2 The Thin Rod

Dividing through by Δx, applying the Fundamental Theorem of Calculus on the left, and applying the definition of partial derivative on the right results in

$$\frac{\partial E}{\partial t} = -\frac{\partial \phi}{\partial x}.$$

This equation represents an infinitesimal version of conservation of energy and involves two unknown functions E and ϕ, neither of which is physically measurable. However, energy is related to the temperature, $u = u(x,t)$, through the specific heat c, which is defined as the heat energy required to raise one mass unit one degree (all in appropriate units). If we let ρ denote the density (mass/unit volume) of the rod, then $E = c\rho u$ is the symbolic expression defining specific heat. In general, c and ρ can depend on x and u, which for simplicity here are assumed to be constants. The heat flux is related to temperature through Fourier's law: Heat flux is proportional to the rate of change of temperature in the spatial variable, i.e., $\phi = -ku_x$. Here k is the thermal conductivity (assumed constant for the material). Substituting the relations into the energy conservation relation gives the **homogeneous heat equation**,

$$u_t = \alpha^2 u_{xx}, \quad 0 < x < L, \quad t > 0. \tag{1.2.5}$$

All constants have been absorbed into one positive constant $\alpha^2 = k/c\rho$.

Under similar assumptions, a thin rod with internal heat source $Q(x,t)$ is governed by the nonhomogeneous equation

$$u_t = \alpha^2 u_{xx} + Q(x,t). \tag{1.2.6}$$

As any physicist knows, the phrase "heat flow" is not really accurate—the spread of heat energy is more of a diffusive process much like what occurs when a drop of dye is placed in water. In fact, the heat equation also provides a simple model for the latter, as we will see later on. In this manner, the heat equation is often called the **diffusion equation**, and we will use the terminology interchangeably.

Laplace's Equation Under appropriate assumptions, a heat flow problem will have a steady state. To determine this, we seek time-independent solutions $U = U(x)$ of the heat equation (1.2.5). These time-independent solutions will satisfy the simple ODE

$$U'' = 0, \qquad 0 < x < L,$$

the analog of Laplace's equation in one dimension. It follows by integration that all steady-state solutions are linear functions, i.e.,

$$U(x) = c_1 x + c_2,$$

where c_1 and c_2 are constants. The latter can be determined from the boundary conditions, as we will be see shortly.

The nonhomogeneous heat equation (1.2.6) may also possess steady-state solutions provided the heat source is time independent, $Q = Q(x)$, or has limit as $t \to \infty$ (which we also denote as $Q(x)$). In either case, the time-independent solutions would satisfy the equation

$$U'' = \frac{-Q(x)}{\alpha^2},$$

which can be solved by integration.

In two dimensions, Laplace's equation has form

$$u_{xx} + u_{yy} = 0$$

and has close connections with complex analysis. Let D be a bounded open subset of the complex plane $\mathbb{C}$. By open we mean that for each point $z \in D$, there is an $r > 0$ such that the set $\{w : |w-z| < r\}$ is contained in D. Examples of open sets include the interior of a circle or square and a punctured disk, i.e., $\{z : 0 < |z - z_0| < 1\}$ (the unit disk about z_0 excluding z_0). A complex-valued function $f : D \to \mathbb{C}$ is said to be *analytic* on D if about each point $z_0 \in D$, f has a power series

$$f(z) = \sum_{n=0}^{\infty} a_n (z - z_0)^n$$

convergent in some open disk about z_0. (This should remind you of Taylor series in calculus.) Power series can be differentiated term by term inside their disk of convergence. It follows that analytic functions are infinitely differentiable, the derivative being defined in the usual way:

$$f'(z) = \lim_{h \to 0} \frac{f(z+h) - f(z)}{h}.$$

Break f into its real and imaginary parts: $f(z) = u(x,y) + iv(x,y)$, where $z = x + iy$. We compute the limit by letting h tend to zero through real values and through imaginary values, i.e.,

$$\begin{aligned} f'(z) &= \lim_{h \to 0} \frac{u(x+h,y) + iv(x+h,y) - (u(x,y) + iv(x,y))}{h} \\ &= u_x(x,y) + iv_x(x,y) \\ f'(z) &= \lim_{h \to 0} \frac{u(x,y+h) + iv(x,y+h) - (u(x,y) + iv(x,y))}{ih} \\ &= -iu_y(x,y) + v_y(x,y). \end{aligned}$$

Comparing the two limits, the real and imaginary parts of an analytic function satisfy the Cauchy–Riemann equations, i.e.,

$$\begin{aligned} u_x &= v_y \\ u_y &= -v_x. \end{aligned}$$

Differentiating the first with respect to x and the second with respect to y and using equality of mixed partials, it follows that the real part u satisfies Laplace's equation. Similarly, the imaginary part v also satisfies Laplace's equation.

Boundary and Initial Conditions

The physics of a vibrating string or heat flow problem can be structurally represented in a space–time picture as in Figure 1.3. The important point is that the PDE represents the physical laws governing the process inside the vertical shaded region. Physically, conditions are needed on the boundary of the slab to completely describe the state of the physical process.

Along the line segment $t = 0$, initial conditions are specified to prescribe the initial state of the physical system. For a heat flow problem this would take the form of specifying the initial temperature distribution, i.e.,

$$u(x,0) = f(x), \quad 0 < x < L.$$

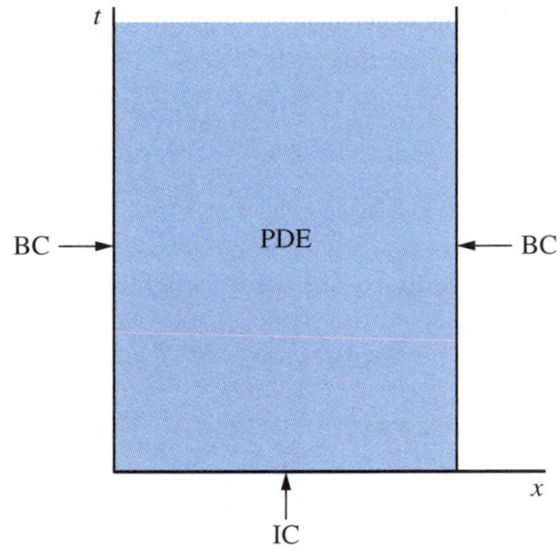

Figure 1.3 Space–Time Picture

In higher dimensions, the interval $[0, L]$ would be replaced by a suitable multidimensional domain (set). In the case of the vibrating string, there would be a pair of conditions

$$u(x, 0) = f(x)$$
$$u_t(x, 0) = g(x),$$

specifying the initial position and velocity of the string. This is natural from the perspective of mechanics: The state of a mechanical process is prescribed by knowing the position and velocity at any point in space–time.

In terms of conditions on the spatial boundaries of a PDE problem, we formulate the three key types described in Section 1.1.

- **Dirichlet Conditions.** The function $u(x, t)$ is specified at the boundary point.

In a one-dimensional problem this might take the form $u(0, t) = g(t)$, where g is a known function. For the vibrating string, g specifies explicit motion of the left endpoint; if $g(t) = 0$, the string is said to be clamped. The same boundary condition in a heat flow problem specifies the temperature of the left boundary as a function of time. These conditions can also be placed at the right endpoint.

- **Neumann Conditions.** The normal derivative of u is specified at the boundary point.

The normal derivative is denoted generally by $\frac{du}{dn}$, where $\mathbf{n}$ is the outward normal vector to the boundary. In one dimension, this normal derivative is

$-u_x(0,t)$ at the left endpoint and $u_x(L,t)$ at the right. For example, suppose the inward heat flux $\phi(t)$ is specified at $x = L$. The inward heat flux in terms of temperature is given by $ku_x(L,t)$, hence the boundary condition takes the form

$$ku_x(L,t) = \phi(t).$$

The same reasoning would lead to the boundary condition

$$ku_x(0,t) = -\phi(t)$$

at the left endpoint. In either case, if $\phi(t) = 0$, the boundary point is said to be insulated; there is no heat flow. In general, the Neumann condition in a heat flow problem specifies the rate of energy flow across the boundary. For a vibrating string, the boundary condition $u_x(0,t) = 0$ states that the vertical component of tension is zero, e.g., the endpoint is attached to a ring that slides on a frictionless rod. Alternatively, for an acoustical wave guide, a homogeneous Neumann condition models the case of an open end, e.g., open-ended organ pipes or a flute.

- **Robin Conditions.** A linear combination of u and its normal derivative is specified at the boundary point.

In general this would look like

$$a\frac{du}{dn} + bu = \phi,$$

where a, b, and ϕ are known, possibly functions of t. For a one-dimensional heat flow problem, suppose the right end convects heat into a medium with temperature $T(t)$. Newton's law of cooling states that the inward heat flux is proportional to the difference between the medium temperature and the temperature of the end of the rod. The boundary condition then takes the form

$$ku_x(L,t) = H(T(t) - u(L,t)),$$

where $H > 0$ is the proportionality constant.

The mechanical analog for a vibrating string can be derived as follows. Imagine the left end of a perfectly elastic string attached to a spring–mass device, as depicted in Figure 1.4. Assume that the mass can move only in the vertical direction and the device obeys Hooke's law with spring constant $k > 0$. Let the natural spring length be l and assume the support mechanism for the spring is such that the equilibrium position of the mass is the equilibrium position of the string. In such a case, the position of the mass as a function of time is $u(0,t)$. In the absence of friction, the only vertical external forces on the mass stem from its weight mg and vertical component of string tension

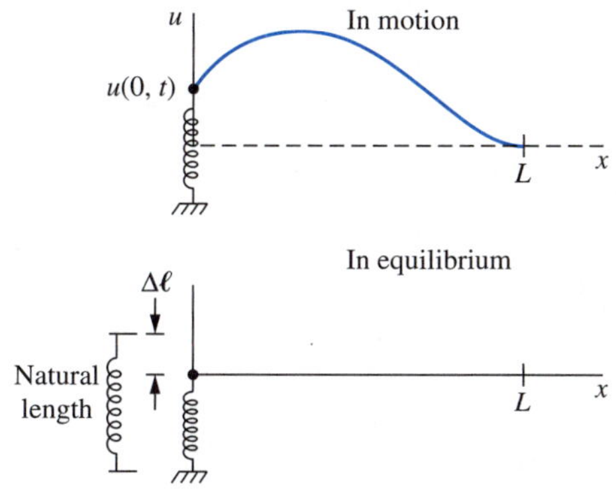

Figure 1.4 Spring–Mass Device

$T\sin(\theta(0,t))$. In the equilibrium state, Hooke's law gives $k\Delta l = mg$, where Δl is the magnitude of the change in length of the spring. Applying Newton's law of motion,

$$m\frac{d^2u}{dt^2}(0,t) = -k\left[u(0,t) - \Delta l\right] + T\sin\theta(0,t) - mg.$$

As before, $T\sin\theta(0,t) \approx T\frac{\partial u}{\partial x}(0,t)$; consequently, applying the equilibrium condition, the formula reads

$$m\frac{d^2u}{dt^2}(0,t) = -ku(0,t) + T\frac{\partial u}{\partial x}(0,t). \tag{1.2.7}$$

In a situation where the mass m is negligible, the boundary condition is a Robin condition of the form

$$T\frac{\partial u}{\partial x}(0,t) = ku(0,t). \tag{1.2.8}$$

This is a mechanical analog of Newton's law of cooling at the left end of a thin rod. Note the graphical implication of (1.2.8): If $u(0,t) > 0$, then $u_x(0,t) > 0$.

All three types of boundary conditions can be concisely written in the form

$$\begin{aligned} a_{11}u_x(0,t) + a_{12}u(0,t) &= g_1(t) \\ a_{21}u_x(L,t) + a_{22}u(L,t) &= g_2(t). \end{aligned}$$

Here the coefficients a_{ij} (possibly dependent on t) and the functions g_i are known. When $g_i(t)$ is the zero function, the respective condition is called a **homogeneous boundary condition**. We note that in the preceding physical problems, the coefficients are not completely arbitrary, e.g., $a_{21}a_{22} \geq 0$ and $a_{11}a_{12} \leq 0$ in the cases of heat flow and vibrating strings. In the case where the boundary conditions are different on different parts of the boundary, we say the problem has **mixed boundary conditions**.

Putting It All Together: IBVP and BVP

The complete formulation of an initial boundary value problem for one-dimensional heat flow or for a vibrating string takes the following form.

$$\text{PDE:}\begin{cases} u_t = \alpha^2 u_{xx}, \quad 0 < x < L, \quad t > 0 \,(\text{HEAT}) \\ \text{OR} \\ u_{tt} = c^2 u_{xx}, \quad 0 < x < L, \quad t > 0 \,(\text{WAVE}) \end{cases}$$

$$\text{BC:}\begin{cases} a_{11}u_x(0,t) + a_{12}u(0,t) = g_1(t) \\ a_{21}u_x(L,t) + a_{22}u(L,t) = g_2(t) \end{cases}, \quad t > 0 \tag{1.2.9}$$

$$\text{IC:}\begin{cases} u(x,0) = f(x), \quad 0 < x < L \,(\text{for HEAT}) \\ \text{OR} \\ \begin{cases} u(x,0) = f(x) \\ u_t(x,0) = g(x) \end{cases}, 0 < x < L \quad (\text{for WAVE}) \end{cases}$$

From a physical perspective, this gives a complete description associated with Figure 1.3.

These IBVPs are typical of all such from a physical perspective: *The partial differential equation governs the function u in some domain in space–time, the boundary conditions govern the solution on the spacial boundaries, and the initial condition(s) specify the state of the solution at $t = 0$. The structure of the PDE and the form of the BCs and ICs are usually physically motivated.*

Steady-state solutions for the IBVP for the heat equation are defined as time-independent solutions. These exist under suitable conditions, e.g., both g_1 and g_2 and the coefficients a_{ij} are constants. This leads to a BVP for an ordinary differential equation:

$$\text{ODE: } U'' = 0, \quad 0 < x < L$$

$$\text{BC:}\begin{cases} a_{11}U'(0) + a_{12}U(0) = g_1 \\ a_{21}U'(L) + a_{22}U(L) = g_2. \end{cases}$$

Example 1.3 Formulate the IBVP for one-dimensional heat flow without sources where the left end is held at temperature T_1 and the right end convects heat into a medium with constant temperature T_2. Assume the initial temperature is given by $f(x)$. Determine the boundary value problem for steady state, and solve for the steady-state temperature distribution. Assume uniform material and thermal properties.

The formulation of this IBVP will involve the homogeneous heat equation, a Dirichlet boundary condition at $x = 0$, and a Robin condition at $x = L$. Hence, its form is

$$\text{PDE}: u_t = \alpha^2 u_{xx}, \quad 0 < x < L, \quad t > 0$$
$$\text{BC}: \begin{cases} u(0,t) = T_1 \\ ku_x(L,t) + Hu(L,t) = HT_2 \end{cases}$$
$$\text{IC}: u(x,0) = f(x).$$

For the steady state the temperature depends only on x, $U = U(x)$, and satisfies the following ODE boundary value problem:

$$\text{ODE}: U'' = 0, \quad 0 < x < L$$
$$\text{B}: \begin{cases} U(0) = T_1 \\ kU'(L) + HU(L) = HT_2. \end{cases}$$

The steady state must be a linear function, $U(x) = ax + b$. The constants a and b can be determined from the boundary conditions. Indeed, the first boundary condition yields $b = T_1$ and the second provides the equation

$$ka + H(aL + b) = HT_2.$$

Solving these equations, the steady-state solution is

$$U(x) = \frac{H(T_2 - T_1)}{k + HL}x + T_1.$$

Example 1.4 Reconsider Example 1.3, but assume the rod is perfectly insulated. Here the IBVP has the same formulation but with the boundary conditions replaced by

$$\text{BC}: \begin{cases} u_x(0,t) = 0 \\ u_x(L,t) = 0. \end{cases}$$

Likewise, the steady state satisfies the equation $U'' = 0$ with boundary conditions $U'(0) = U'(L) = 0$. Proceeding as in Example 1.3, it follows that $U(x) = b$, where b is a constant that cannot be determined from the given conditions. However, returning to first principles yields the solution as follows. Since the rod is completely insulated, total heat energy is constant and given by $E = c\rho \int_0^L f(x)dx$. By the same token, the total heat energy in steady state is given by $E = c\rho \int_0^L b\,dx = bc\rho L$. Comparing the two expressions for energy gives the steady state,

$$U(x) = b = \frac{1}{L}\int_0^L f(x)dx.$$

Notice that the right-hand side of this equation is just the average value of f on the interval $[0, L]$. The physical argument that the energy is constant can be made rigorous as follows. Suppose u is a C^2 solution of the IBVP. Then

$$\begin{aligned}\frac{\partial}{\partial t}\int_0^L u(x,t)dx = \int_0^L u_t(x,t)dx &= \alpha^2 \int_0^L u_{xx}(x,t)dx \\ &= \alpha^2[u_x(L,t) - u_x(0,t)] = 0.\end{aligned}$$

Now by continuity, $\int_0^L u(x,t)dx = \int_0^L f(x)dx$, as claimed.

Exercises

1.5 What is the formulation of the Robin condition at $x = 0$ for the heat equation?

1.6 Consider IBVP (1.2.9). What sign must $a_{11}a_{12}$ and $a_{21}a_{22}$ have for the boundary conditions to be physically realizable in a heat flow problem? Explain.

1.7 Consider a heat flow problem on the spacial interval $[0, L]$. For each of the following:

- **a.** Write the corresponding boundary conditions.
- **b.** Determine the steady-state solution.
- **c.** Compute the outward flux at the boundary points for the steady-state solutions.

  - **(i)** The left end is held at temperature T_1 and the right end at T_2.
  - **(ii)** There is perfect insulation at the left endpoint; the right end of the rod is held at T_2 degrees.
  - **(iii)** The left end convects into a medium with constant temperature T_0 and the right end has constant inward heat flux equal to ϕ_L, a constant.

1.8 What sign must $a_{11}a_{12}$ and $a_{21}a_{22}$ have for the boundary conditions to be physically realizable in the IBVP for the vibrating string? Explain.

1.9 Modify the derivation of the vibrating string equation to incorporate the case where the string is in a medium that provides resistance proportional to velocity.

1.10 Derive the analog of (1.2.8) for the right end of the string.

1.3 The Basic Questions

Given an IBVP or a BVP, the following questions are fundamental.

- **Existence.** Does there exist a solution to the problem?

Clearly, in a description of a physical process leading to a model in the form of a partial differential equation with auxiliary conditions, the absence of a solution would point to incorrect model assumptions. Incorrectness may be in the formulation of the PDE or in the form of boundary conditions. Fortunately, guided by sound physical principles arrived at through careful experimental studies, the formulation of a problem for which a solution exists is usually not the difficulty. From a mathematical perspective, proving that a particular problem has a solution is often an intricate task, particularly in higher-dimensional settings where the geometry of the underlying domain plays a significant role.

- **Uniqueness.** Is the solution unique?

Echoing the previous paragraph, an IBVP formulated under the guidance of sound physical principles should have a unique solution. However, there are physically interesting situations where uniqueness does break down in the formulation. An example is the equation for steady-state heat flow $U'' = 0$ with Neumann boundary conditions. In the sequel, mathematically addressing the question of uniqueness is often guided by the underlying physical principles.

- **Stability.** Is the solution stable under small changes in boundary or initial data?

In the measurement of any physical process we are hindered by experimental error. Hence, in the mathematical formulation the coefficients in the partial differential equation and the functional representation of boundary and/or initial data or the equation source term we cannot have total precision relative to measurements. So this question boils down to: Does the solution to the IBVP change in a small way if one perturbs the data of the equation in some small way? This question clearly needs an explanation of the meaning of *small*; we need to describe a way of determining when two functions are close. There is no universal answer to this question, and the variety of ways of measuring closeness for functions depend on the type of process involved (e.g., diffusion or wave) and whether the solution is in the strong or weak sense.

A partial differential equation problem for which a unique solution exists and is stable in suitable sense is said to be **well posed**, a term introduced by Jacques Hadamard in the early part of the 20th century.

Three rather more pragmatic questions are also apparent:

- **Construction of Solutions.** Can analytic formulas for the solution of a given IBVP be found?

This question is of principal concern throughout the sequel. A key theme in all methodologies for solving PDEs is reduction in the number of independent variables; usually the methods ultimately reduce solving the PDE to solving a class of ordinary differential equations.

- **Regularity of Solutions.** Given regularity (such as smoothness) or lack thereof in the boundary or initial data, how is this reflected in the solution?

For example, it will be seen that the homogeneous heat equation has smooth solutions, no matter what irregularity the initial data possesses. In contrast, the wave equation propagates singular behavior of the initial data. (See the examples below.)

- **Approximation of Solutions.** Where appropriate, how can solutions to an IBVP be approximated for graphical or analytic use?

Construction of explicit solution formulas is only possible for the simplest of equations and/or in higher dimensions, where the boundary has sufficient symmetry. Solution formulas often are expressed using infinite series and integrals. Being able to suitably approximate the latter for plotting purposes or to study asymptotics is an important topic. More complicated equations (e.g., the heat equation where the domain has a nonhomogeneous nature) require the use of numerical approximation of solutions. Explicit formulas for the more basic equations provide ways to benchmark effectiveness of numerical schemes. Stability of the IBVP is critical in order to produce convergent schemes of approximation.

Qualitative and Quantitative Aspects for the Heat Equation

Maximum Principle for the Heat Equation

Consider the following IBVP for one-dimensional heat flow:

$$\begin{aligned} \text{PDE}\colon & u_t = \alpha^2 u_{xx}, \quad 0 < x < L, \quad t > 0 \\ \text{BC}\colon & \begin{cases} u(0,t) = 0 \\ u(L,t) = 0 \end{cases} \\ \text{IC}\colon & u(x,0) = f(x). \end{aligned}$$

Setting aside the question of the existence of a solution, let's argue uniqueness on physical grounds and determine the notion of stability in regards to changes in the initial condition. Let $T > 0$ and consider the following heuristic claim:

$$\max_{0 \le x \le L, t \le T} u(x,t) \le \max\left\{0, \max_{0 \le x \le L} f(x)\right\}.$$

This is called the **maximum principle**. If the left-hand side were strictly larger than the right—i.e., $u(x,t)$ has a maximum at a point (x_0, t_0) where $0 < x_0 < L$ and $t_0 > 0$—then the physical principle of heat energy flowing from hotter regions to colder would be violated. Likewise, there is a **minimum principle**:

$$\min_{0 \le x \le L, t \le T} u(x,t) \ge \min\left\{0, \min_{0 \le x \le L} f(x)\right\}.$$

Suppose there were two solutions u_1 and u_2 of our IBVP. Then the function $u = u_1 - u_2$ would satisfy the PDE, the boundary conditions, and the initial condition $u(x,0) = 0$. In this case, for the difference between the two solutions, the maximum and minimum are both zero. Hence, $u_1 - u_2 = 0$ by the maximum principle and we conclude that the solution of the original problem is unique. In regards to stability, suppose u_1 and u_2 are the unique solutions to the IBVP where $f = f_1(x)$ and $f_2(x)$, respectively. Then the difference $u_1 - u_2$ satisfies the IBVP where $f(x) = f_1(x) - f_2(x)$. Again based on our maximum–minimum argument,

$$|u_1(x,t) - u_2(x,t)| \le \max_{0 \le x \le L} |f_1(x) - f_2(x)|,$$

for all x and t. Therefore,

$$\max |u_1(x,t) - u_2(x,t)| \le \max_{0 \le x \le L} |f_1(x) - f_2(x)|,$$

providing a gauge of closeness between solutions in terms of a uniform closeness between the initial data. This is called **uniform stability**. Another gauge for stability is presented in the next example.

Bounds on Energy for the Heat Equation We now give a brief introduction to the energy method for the heat equation. The following identity is obtained by multiplying the homogeneous heat equation by u and then using standard differentiation formulas:

$$\begin{aligned} 0 &= u(u_t - \alpha^2 u_{xx}) \\ &= \frac{1}{2}\frac{\partial}{\partial t}(u^2) - \alpha^2 \frac{\partial}{\partial x}(u_x u) + \alpha^2 u_x^2. \end{aligned} \tag{1.3.1}$$

Recall that u is proportional to the heat energy, hence a gauge on the total heat energy in the interval $0 \le x \le L$ is given by

$$E(t) = \frac{1}{2}\int_0^L u^2(x,t)dx.$$

Integrating identity (1.3.1) over this interval gives

$$\frac{1}{2}\frac{\partial}{\partial t}\int_0^L u^2(x,t)dx = \alpha^2 \left(u_x(x,t)u(x,t)\right|_{x=0}^L - \alpha^2 \int_0^L u_x^2(x,t)dx.$$

Now, suppose u is a solution of the homogeneous heat equation on the interval $0 < x < L$ and is continuous with continuous first partials on $[0, L] \times [0, \infty)$. The latter condition justifies our integration identity. If u satisfies a homogeneous Dirichlet BC (dbc), a homogeneous Neumann BC (nbc), *or* a mixed Dirichlet/Neumann BC (mbc), then the boundary terms of the identity vanish. We can then write

$$\frac{dE}{dt} = -\alpha^2 \int_0^L u_x^2(x,t)dx \le 0. \tag{1.3.2}$$

Therefore, the total energy of the system is a nonincreasing function of t, i.e.,

$$E(t) \le E(0) = \frac{1}{2}\int_0^L f^2(x)dx, \tag{1.3.3}$$

where $f = f(x)$ is the initial data as usual. At this point similar arguments could be used to deduce that the solution to this IBVP is unique. Further, we have a different way of measuring stability. Namely, with u_1 and u_2 being the solutions corresponding to the initial data f_1 and f_2, then

$$\int_0^L \left[u_1(x,t) - u_2(x,t)\right]^2 dx \le \int_0^L \left[f_1(x) - f_2(x)\right]^2 dx.$$

This is known as **mean square stability**.

The Cauchy Problem for the Heat Equation on $\mathbb{R}$ Recall from Example 1.1 the Gaussian solution

$$\mathcal{S}(x,t) = \frac{1}{2\alpha\sqrt{\pi t}}e^{-\frac{x^2}{4\alpha^2 t}}$$

of the heat equation for $x \in \mathbb{R}$ and $t > 0$. The constant in front of the expression was chosen so that

$$\int_{\mathbb{R}} \mathcal{S}(x,t) = 1 \tag{1.3.4}$$

for all $t > 0$. A second important property of this function is its rapid decay as $|x| \to \infty$. Since the heat equation has constant coefficients, translations of a solution will also be solutions. In this case, $\mathcal{S}(x-y,t)$ is also a solution for any $y \in \mathbb{R}$. Let f be a bounded continuous function on $\mathbb{R}$ and consider

$$u(x,t) = \int_{\mathbb{R}} f(y)\mathcal{S}(x-y,t)dy. \tag{1.3.5}$$

You can think of this as a continuous superposition of solutions of the heat equation using coefficients $f(y)$. In this context the solution $\mathcal{S}(x,t)$ is often called the **heat kernel** for heat equation on the real line. The following theorem unfolds the key properties of the function $u(x,t)$.

Theorem 1.2 Let $f \in C(\mathbb{R})$ and bounded. Then $u(x,t)$ as defined in equation (1.3.5) is a smooth function on $\mathbb{R} \times (0,\infty)$, is a solution of the heat equation, and satisfies the condition

$$\lim_{t \to 0^+} u(x,t) = f(x). \tag{1.3.6}$$

In the realm of partial differential equations, a paraphrase of this theorem is that the function $u(x,t)$ gives the solution to the Cauchy problem on $\mathbb{R}$ given by

$$\begin{aligned} \text{PDE: } & u_t = \alpha^2 u_{xx}, \quad x \in \mathbb{R}, \quad t > 0 \\ \text{IC: } & u(x,0) = f(x). \end{aligned}$$

A heuristic argument for these claims is as follows. The function $\mathcal{S}(x-y,t)$ is infinitely differentiable in either variable and is a solution to the heat equation. To show that u is a smooth function and a solution to the heat equation, it is natural to simply differentiate the integral defining u and interchange the operation of differentiation with the operation of integration. Both of these operations are limits, and interchanging limit operations can be tricky. What justifies it in this case is the rapid decrease of the Gaussian for large $|y|$; this gives control over the improper integral. For the limit conclusion, because of (1.3.4) the integral defining u can be thought of as an average of the function f. In Figure 1.5, the Gaussian is plotted for small and large values of t. The point is that because of the rapid decrease in this function at infinity, it becomes more localized as t gets small. Hence, $u(x,t)$ is a localized average of the function f near the point x. From this perspective, the limit conclusion has intuitive appeal. A rigorous proof of this result is given at the end of this chapter. The following example explores the conclusions, in fact, for a function f with jump discontinuities.

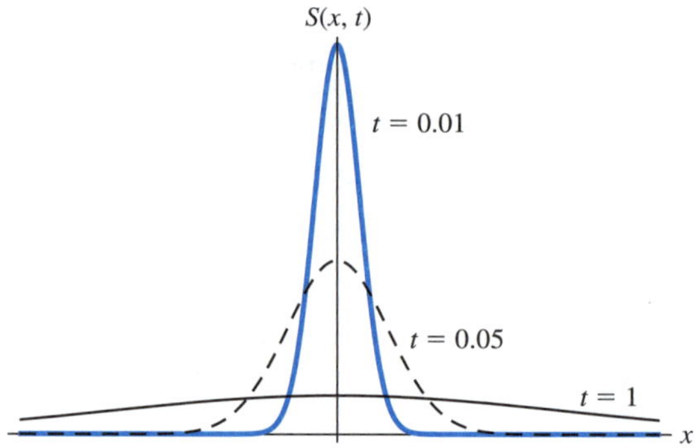

Figure 1.5 The Gaussian Solution

Example 1.5 We take an initial condition of the form $f(x) = 1$ for $-1 \leq x \leq 1$ and $f(x) = 0$ otherwise. Substituting into (1.3.5) we need to compute

$$u(x,t) = \frac{1}{2\alpha\sqrt{\pi t}} \int_{-1}^{1} \exp\left(-\frac{(x-y)^2}{4\alpha^2 t}\right) dy.$$

Making the substitution $s = \frac{x-y}{2\alpha\sqrt{t}}$ this becomes

$$u(x,t) = \frac{1}{\sqrt{\pi}} \int_{\frac{x-1}{2\alpha\sqrt{t}}}^{\frac{s+1}{2\alpha\sqrt{t}}} \exp(-s^2) ds.$$

At this point it is convenient to introduce the **error function**,

$$\operatorname{erf}(x) = \frac{2}{\sqrt{\pi}} \int_{0}^{x} \exp(-s^2) ds.$$

Key properties are that the error function is a smooth odd function and $\lim_{x\to\infty} \operatorname{erf}(x) = 1$. Now we can rewrite the expression for u as

$$u(x,t) = \frac{1}{2}\left[\operatorname{erf}\left(\frac{x+1}{2\alpha\sqrt{t}}\right) - \operatorname{erf}\left(\frac{x-1}{2\alpha\sqrt{t}}\right)\right].$$

Because the error function is commonly available in computer algebra systems, this form is convenient for plotting. It is easy to check that $u(x,t)$ is smooth in both variables and is a solution to the heat equation. Since the error function is odd, it is easy to see that $u(x,t)$ is even in the x-variable. If $x > 1$, then the argument of both error functions defining u are positive and hence $\lim_{t\to 0} u(x,t) = 0$. For $0 \le x < 1$, the arguments of the error functions are positive and negative, respectively. In this case,

$$\lim_{t\to 0^+} u(x,t) = \lim_{t\to 0^+} \frac{1}{2}\left[\operatorname{erf}\left(\frac{x+1}{2\alpha\sqrt{t}}\right) + \operatorname{erf}\left(\frac{1-x}{2\alpha\sqrt{t}}\right)\right] = \frac{1}{2}(1+1) = 1.$$

We also note that

$$\lim_{t\to 0} u(1,t) = \frac{1}{2}\lim_{t\to 0} \operatorname{erf}\left(\frac{1}{\alpha\sqrt{t}}\right) = \frac{1}{2} = \lim_{t\to 0} u(-1,t).$$

Figure 1.6 shows a plot of $u(x,t)$ for several values of t, providing a graphical view of the solution and its properties.

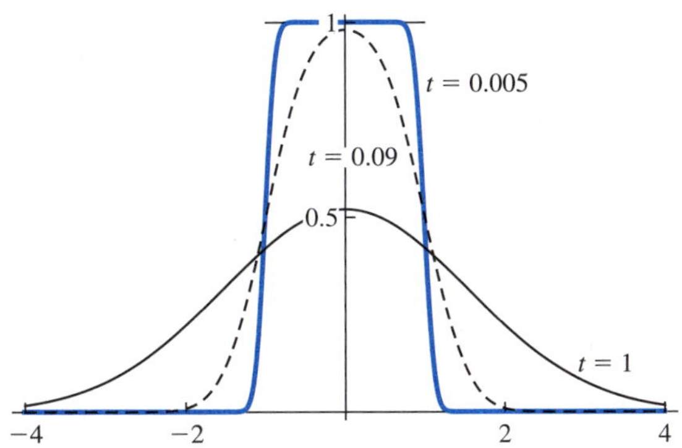

Figure 1.6 Heat Equation Solution for Example 1.5

The function $f(x)$ given in Example 1.5 is a piecewise continuous function. Intuitively, piecewise continuous functions are continuous everywhere except at certain points where they have jump discontinuities. We formalize the definition here as it plays an important role in the sequel.

Definition 1.3 A function $f : [a,b] \to \mathbb{R}$ (or $\mathbb{C}$) is said to be **piecewise continuous** if f is continuous except perhaps at a finite number of points $c \in (a,b)$; the left and right limits of f exist and are finite at those points—i.e., $\lim_{x\to c^+} f(x) \overset{def}{=} f(c+)$ and $\lim_{x\to c^-} f(x) \overset{def}{=} f(c-)$ are both finite; and

$$\lim_{x\to a^+} f(x) \text{ and } \lim_{x\to b^-} f(x)$$

both exist and are finite. The class of such functions is denoted $\mathrm{PC}[a,b]$. A function defined on the real line is said to be piecewise continuous if it is piecewise continuous on any subinterval $[a,b]$ of the real line. On the real line, the class of piecewise continuous functions is denoted $\mathrm{PC}(\mathbb{R})$.

Motivated by Example 1.5, the following result extends Theorem 1.2 for piecewise continuous initial data.

Theorem 1.4 Let $f \in \mathrm{PC}(\mathbb{R})$ be bounded. Then $u(x,t)$ defined by (1.3.5) is a smooth function on $\mathbb{R} \times (0,\infty)$, is a solution of the heat equation, and satisfies

$$\lim_{t\to 0^+} u(x,t) = \frac{1}{2}[f(x+) + f(x-)]$$

for all $x \in \mathbb{R}$.

If x is a point of continuity of the function f, then the right-hand side of the preceding limit is just $f(x)$. The intuition underlying this result (and its proof) is similar to that when $f \in C(\mathbb{R})$. The averaging given in this limit relation stems from the fact that the kernel $\mathcal{S}(x,t)$ is an even function of x.

Qualitative and Quantitative Aspects of Waves

Conservation of Energy for the Wave Equation

The IBVP for the clamped string of length L is

$$\begin{aligned} \text{PDE} &: u_{tt} = c^2 u_{xx}, \quad 0 < x < L, \quad t > 0 \\ \text{BC} &: u(0,t) = u(L,t) = 0 \\ \text{IC} &: \begin{cases} u(x,0) = f(x) \\ u_t(x,0) = g(x), \end{cases} \end{aligned} \tag{1.3.7}$$

where we will assume that $f \in C^1[0,L]$ and $g \in C[0,L]$. Multiplying the wave equation by u_t and applying differentiation formulas gives the identity

$$\begin{aligned} 0 = u_t\left(u_{tt} - c^2 u_{xx}\right) &= \frac{\partial}{\partial t}\left(\frac{1}{2}u_t^2 + \frac{c^2}{2}u_x^2\right) - c^2 u_t u_{xx} - c^2 u_x u_{tx} \\ &= \frac{\partial}{\partial t}\left(\frac{1}{2}u_t^2 + \frac{c^2}{2}u_x^2\right) + \frac{\partial}{\partial x}\left(-c^2 u_x u_t\right). \end{aligned} \tag{1.3.8}$$

This is analogous to the identity (1.3.1) obtained for the heat equation. If $u(x,t)$ is the solution to the IBVP that is continuous on $[0,L] \times [0,\infty)$, then integrating the identity over $[0,L]$ gives

$$0 = \frac{\partial}{\partial t}\frac{1}{2}\int_0^L \left[u_t^2(x,t) + c^2 u_x^2(x,t)\right] dx - c^2 u_t(x,t) u_x(x,t)\Big|_0^L .$$

The boundary terms are zero since the string is clamped ($u_t(0,t) = u_t(L,t) = 0$). Hence, introducing

$$E(t) = \frac{1}{2}\int_0^L \left[u_t^2(x,t) + c^2 u_x^2(x,t)\right] dx, \tag{1.3.9}$$

it follows that that $\frac{dE}{dt} = 0$. Therefore $E(t)$ is a constant function and our continuity hypothesis implies

$$E(t) = E(0) = \frac{1}{2}\int_0^L \left[[g(x)]^2 + c^2\,[f'(x)]^2\right] dx. \tag{1.3.10}$$

From a physical perspective, the function $E(t)$ is the total energy of the vibrating string, the first term in (1.3.9) corresponds to the kinetic energy, and the second is the potential energy (from tension). Thus the principle of conservation of energy holds. It is left to the reader to show uniqueness of solutions for the IBVP (1.3.7).

The Cauchy Problem for the Wave Equation on $\mathbb{R}$ The mathematical formulation for an infinitely long string displaced from equilibrium and released from rest is

$$\begin{aligned} &\text{PDE}: u_{tt} = c^2 u_{xx}, \quad -\infty < x < \infty, \quad t > 0 \\ &\text{IC}: \begin{cases} u(x,0) = f(x) \\ u_t(x,0) = 0 \end{cases}, \quad -\infty < x < \infty. \end{aligned} \tag{1.3.11}$$

On physical grounds, we expect the one-dimensional wave motion to have form of a sum of a left and a right moving wave (pluck a long taut string). More precisely, the homogeneous wave equation has solutions of the form

$$u(x,t) = F(x+ct) + G(x-ct), \tag{1.3.12}$$

where F and G are twice continuously differentiable functions of one variable. This can be verified using the chain rule. Applying the initial conditions leads to the equations $F(x)+G(x) = f(x)$ and $F'(x) - G'(x) = 0$. The second equation implies that $G(x) = F(x) + C$, where C is a constant. Using this with the first relation leads to the solution of the IVP (1.3.11):

$$u(x,t) = \frac{1}{2}\left[f(x+ct) + f(x-ct)\right]. \tag{1.3.13}$$

This represents the sum of a left and a right moving wave, the height of which is defined by the initial state. For example, if $f(x) = e^{-x^2}$, then

$$u(x,t) = \frac{1}{2}[e^{-(x+t)^2} + e^{-(x-t)^2}].$$

The solution is plotted in space–time with $c = 1$ in Figure 1.7(a). Alternatively, if $f(x) = \cos 2x$, then

$$u(x,t) = \frac{1}{2}[\cos 2(x+ct) + \cos 2(x-ct)] = \cos 2x \cos 2ct.$$

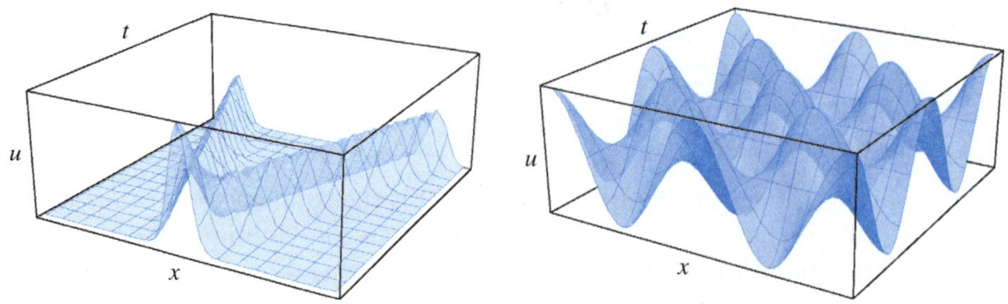

(a) $f(x) = e^{-x^2}$ (b) $f(x) = \cos 2x$

Figure 1.7 Solutions of the Wave Equation in One Dimension

The resultant wave in space–time is plotted in Figure 1.7(b) taking $c = 1$. Here the resulting wave is called a standing wave.

However, what if the initial pluck has the following form?

$$f(x) = \begin{cases} 1 - |x|, & \text{if } |x| \leq 1 \\ 0, & \text{otherwise} \end{cases}$$

In this case, the function $u(x,t)$ given by (1.3.13) in not differentiable on the lines $x \pm ct = -1, 0$, or 1. Figure 1.8 depicts snapshots in time of $u(x,t)$ in space–time for this initial data (again with $c = 1$) and clearly shows the propagation of the points of nondifferentiability. How can it be claimed that (1.3.13) gives the solution? On physical grounds, the argument is that the resultant pluck will cause the formation of two moving waves each with half the height of the

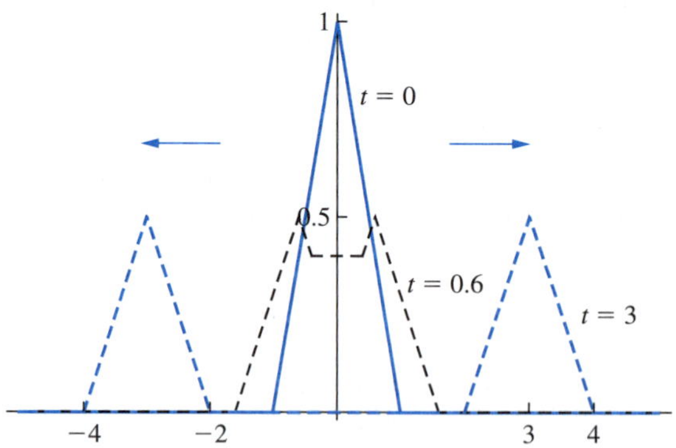

Figure 1.8 Wave Solution: Nonsmooth Initial Data

original pluck, so let's ignore differentiability at these points while realizing that the corners of the pluck will be preserved in the wave motion. *This is a reasonable physical argument but does not resolve the question mathematically.*

A heuristic look at the question can be given by the following thought experiment into the process of measurement. To measure the displacement of the string at some point x_0 along the string and at some time t_0, a measuring rod and a clock are placed at the point x_0. When the clock registers the desired time, a measurement of displacement is taken. Even with inaccuracy in the measurement of time and in the positioning of the measuring rod along the string, the measuring apparatus provides an *average value of displacement relative to both the spatial and time variables. The better the positioning of the measuring rod, the taking of the measurement, and the reading of the clock, the better the measurement of the displacement at* (x_0, t_0). Mathematically, this averaged measurement process can be represented the following way. Let $\phi = \phi(x,t)$ be a positive C^∞ function on $\mathbb{R} \times (0,\infty)$ that is zero outside a given rectangle in space–time (called the support of the function ϕ) and has integral one. Then the act of measuring the string's position in space–time, say at the center of the rectangle, could be represented as the averaging integral

$$\int_0^\infty \int_{\mathbb{R}} u(x,t)\phi(x,t)dx\,dt. \tag{1.3.14}$$

This integral is actually not an improper integral because ϕ vanishes outside a rectangle. The smaller the support rectangle of ϕ, the closer we should actually be to the value of $u(x,t)$ at the center. Derivatives of the function can also be "measured" in this manner, e.g., the following integral would gauge u_x at the center of the rectangle:

$$\int_0^\infty \int_{\mathbb{R}} u_x(x,t)\phi(x,t)dx\,dt.$$

If $u_x(x,t)$ is in fact a continuous function, then this last integral can be integrated by parts and equals

$$-\int_0^\infty \int_{\mathbb{R}} u(x,t)\phi_x(x,t)dx\,dt.$$

In other words, we can measure the derivative of u by measuring u using the derivative of ϕ with respect to x, and by similar argument, measure u_{xx} by measuring u using the second derivative of ϕ. The important point here is that the functions ϕ, thought of as measuring devices, can be pretty much arbitrary as long as they are smooth and vanish outside a rectangle. In this way, the integrals representing measurement (1.3.14) completely define the function $u(x,t)$. The bottom line of our heuristic thought experiment is that a function

$u(x,t)$ could be considered a solution of the wave equation if for every measuring device $\phi(x,t)$,

$$\int_0^\infty \int_{\mathbb{R}} u(x,t)[\phi_{tt}(x,t) - c^2\phi_{xx}(x,t)]dx\,dt = 0.$$

Herein lies the concept of a weak solution of the wave equation. It turns out that (1.3.13) defines a solution of the wave equation in this sense if we merely assume that f is a continuous function. Mathematical details underlying this heuristic argument are provided in Chapter 4.

Remark 1.5 The Cauchy problem for the wave equation

$$\text{PDE: } u_{tt} = c^2 u_{xx}, \quad -\infty < x < \infty, \quad t > 0$$

$$\text{IC: } \begin{cases} u(x,0) = f(x) \\ u_t(x,0) = g(x) \end{cases}, \quad -\infty < x < \infty$$

has solution formula

$$u(x,t) = \frac{1}{2}[f(x+ct) + f(x-ct)] + \frac{1}{2c}\int_{x-ct}^{x+ct} g(s)\,ds. \tag{1.3.15}$$

This is known as d'Alembert's formula; see Exercise 1.19.

Exercises

1.11 Consider the wave equation $u_{tt} = u_{xx}$, where $0 < x < L$ and $0 < t < T$. Impose the Dirichlet boundary conditions: $u(0,t) = u(L,t) = 0$ and $u(x,0) = u(x,T) = 0$. Assume that the ratio T/L is a rational number, say $T/L = m/n$ where m and n are integers. Show that

$$u(x,t) = C \sin\frac{n\pi x}{L} \sin\frac{m\pi t}{T}$$

is a solution to this BVP for any constant C. In other words, the problem is not well posed.

1.12 Use (1.3.3) to demonstrate the uniqueness of solutions for the heat equation with homogeneous Dirichlet, Neumann, or mixed Dirichlet/Neumann boundary conditions.

1.13 Modify the energy method for the heat equation to obtain a uniqueness result for the case of a Robin boundary condition at $x = L$ with either a Dirichlet or Neumann condition at $x = 0$.

1.14 Let

$$f(x) = \begin{cases} 1, & x \geq 0 \\ 0, & x < 0. \end{cases}$$

Compute $u(x,t)$ in terms of the error function using (1.3.5). Plot $u(x,t)$ for representative values of t (take $\alpha = 1$ for this). Verify that $u(x,t)$ is infinitely differentiable in either variable and determine $\lim_{t \to 0^+} u(x,t)$. Find the steady state.

1.15 Let

$$f(x) = \begin{cases} 1 + x, & |x| < 1 \\ 0, & \text{otherwise.} \end{cases}$$

Compute $u(x,t)$ in terms of error functions using (1.3.5). Plot $u(x,t)$ for representative values of t. Verify that $u(x,t)$ is an infinitely differentiable solution to the heat equation $u_t = \alpha^2 u_{xx}$ and

$$\lim_{t \to 0^+} u(x,t) = \begin{cases} f(x), & x \neq 1 \\ 1, & x = 1. \end{cases}$$

HINT: A computer algebra system is handy to compute the integral. It can be done by hand by carefully manipulating the integrand—write the polynomial part of the integrand as $1+y = 1-(x-y)+x$ and split this into three integrals.

1.16 For the string problem (1.3.7) if the left end is clamped and the right end point has a homogeneous Robin condition, energy as written before is not conserved (the string and spring exchange energy). How can one modify the total energy expression in this case to arrive at a conservation law and uniqueness? Show your work.

1.17 Verify by differentiation that (1.3.13) gives the solution to (1.3.11) under the assumption that f is a twice–differentiable function of one variable.

1.18 Let $f(x) = \cos x$ if $|x| < \pi/2$ and $f(x) = 0$ otherwise. Compute $u(x,t)$ as given by (1.3.13) for all x and $t > 0$. Plot this function in space–time (take $c = 1$). Is $u(x,t)$ a strong solution to the wave equation? Explain.

HINT: To compute u, take cases $x + ct < -\pi/2$, $x - ct < -\pi/2 < x + ct < \pi/2$, $-\pi/2 < x - ct < x + ct < \pi/2$, etc.

1.19 Consider the following IVP.

$$\text{PDE}: u_{tt} = c^2 u_{xx}, \quad -\infty < x < \infty, \quad t > 0$$
$$\text{IC}: \begin{cases} u(x,0) = 0, \\ u_t(x,0) = g(x) \end{cases}$$

Show that

$$u(x,t) = \frac{1}{2c}\int_{x-ct}^{x+ct} g(s)ds$$

is a solution to this problem given that g is a differentiable function of one variable by applying the initial conditions to (1.3.12).

1.20 For the following functions $g(x)$, compute $u(x,t)$ using the formula given in Exercise 1.19. Plot $u(x,t)$ in space-time with $c = 1$. Does the function $u(x,t)$ define a strong solution of the wave equation?

a. $g(x) = \cos x$

b. $g(x) = h$ if $|x| < a$, and $g(x) = 0$ otherwise.

HINT: For part (b), show that the solution is given by $u(x,t) = \frac{h}{2c}\text{length}\{(x-ct, x+ct) \cap (-a,a)\}$ using d'Alembert's formula.

Proof of Theorem 1.2

There are three things to be shown: (1) $u(x,t)$ is a smooth function; (2) $u(x,t)$ satisfies the heat equation; (3) is the limit relation (1.3.6). To show the first two, we use Theorem A.7; we must show that the integral defining u and the integrals of partial derivatives are all uniformly convergent. Considering the integral defining u, we have

$$u(x,t) = \int_{\mathbb{R}} f(y)\mathcal{S}(x-y,t)dy = \int_{\mathbb{R}} f(x-y)\mathcal{S}(y,t)dy.$$

Now we let $t_0 > 0$ and assume $|f(y)| \leq M$. Then $|f(x-y)\mathcal{S}(y,t)| \leq M\mathcal{S}(y,t_o)$; this latter function is integrable on the line. This shows that the integral defining u is uniformly convergent for $t \in [t_0, \infty)$. We have the structural formula

$$\frac{\partial^k}{\partial x^k}\mathcal{S}(x-y,t) = \frac{p(x-y)}{t^\beta}\mathcal{S}(x-y,t),$$

where p is a polynomial and β a positive constant (both dependent on k). Hence,

$$\int_{\mathbb{R}} f(y)\frac{\partial^k \mathcal{S}}{\partial x^k}(x-y,t)dy = \int_{\mathbb{R}} f(x-y)\frac{p(y)}{t^\beta}\mathcal{S}(y,t)dy, \tag{1.3.16}$$

and

$$\left| f(x-y)\frac{p(y)}{t^\beta}\mathcal{S}(y,t)\right| \le M\frac{p(y)}{t_0^\beta}\mathcal{S}(y,t_0).$$

Since this latter function is integrable on the line, the integral (1.3.16) is uniformly convergent for $t \in [t_0, \infty)$. We can now apply Theorem A.7 to conclude that partial derivatives of u in the x-variable exist to all orders and that

$$\frac{\partial^k u}{\partial x^k}(x,t) = \int_{\mathbb{R}} f(y)\frac{\partial^k \mathcal{S}}{\partial x^k}(x-y,t)dy.$$

The proof for the partials of u with respect to t is similar. Since $\mathcal{S}(x-y,t)$ is a solution to the heat equation, it follows that $u(x,t)$ is also a solution. To verify the limit relation, we write the integral defining u in the form

$$u(x,t) = \int_{\mathbb{R}} f(x+y)\mathcal{S}(y,t)dt.$$

Then since $\int_{\mathbb{R}} \mathcal{S}(y,t)dy = 1$ for all $t > 0$, we obtain

$$u(x,t) - f(x) = \int_{\mathbb{R}} [f(x+y) - f(x)]\,\mathcal{S}(y,t)dt.$$

Let $\delta > 0$ be a small number (to be chosen shortly) and split the integral up into two pieces, I_δ and R_δ, where the intervals of integration are $|y| \le \delta$ and $|y| > \delta$, respectively. For I_δ, we proceed as follows:

$$|I_\delta| = \left| \int_{|y|\le\delta} [f(x+y) - f(x)]\,\mathcal{S}(y,t)dt\right| \le \int_{|y|\le\delta} |f(x+y) - f(x)|\,\mathcal{S}(y,t)dt.$$

We then introduce the gauge[2]

$$\omega(x;\delta) = \max_{|y|\le\delta} |f(x+y) - f(x)|;$$

this maximum exists as a continuous function achieves its maximum on a closed interval. We now see that

$$|I_\delta| \le \omega(x;\delta)\int_{|y|\le\delta} \mathcal{S}(y,t)\,dy \le \omega(x;\delta)\int_{\mathbb{R}} \mathcal{S}(y,t)\,dy = \omega(x;\delta).$$

Since f is continuous at x, $\omega(x;\delta) \to 0$ as $\delta \to 0$, i.e., given $\varepsilon > 0$, there is a $\delta > 0$ such that $\omega(x;\delta) < \varepsilon/2$ (in general, the choice of δ depends on x and ε).

---

2. $\omega(x;\delta) = \omega_f(x;\delta)$ is called the modulus of continuity of f at x. It can be used as a gauge on the continuity of a function and plays a variety of roles in approximation theory.

In particular, for this choice of δ, $|I_\delta| < \varepsilon/2$. Having chosen δ, we now estimate R_δ. Now, $|f(x)| \leq M$ for some constant M and all $x \in \mathbb{R}$. Hence

$$|R_\delta| = \left| \int_{|y|\geq\delta} [f(x+y) - f(x)]\, \mathcal{S}(y,t)dt \right| \leq \int_{|y|\geq\delta} |f(x+y) - f(x)|\, \mathcal{S}(y,t)dt$$
$$\leq 2M \int_{|y|\geq\delta} \mathcal{S}(y,t)dt = \frac{M}{\alpha\sqrt{\pi t}} \int_{|y|\geq\delta} \exp\left(\frac{-y^2}{4\alpha^2 t}\right) dy.$$

Making the change of variable $s = y/2\alpha\sqrt{t}$, we then have

$$|R_\delta| \leq \frac{2M}{\sqrt{\pi}} \int_{|s|\geq\delta/2\alpha\sqrt{t}} \exp(-s^2)ds.$$

Since the integral $\int_{\mathbb{R}} \exp(-s^2)ds$ is convergent, the right-hand side of this estimate can be made as small as we wish provided t is sufficiently small. Formally, there is a $t_0 > 0$ such that if $t < t_0$, $|R_\delta| < \varepsilon/2$. Putting the estimates together, if $t < t_0$, then

$$|u(x,t) - f(x)| \leq |I_\delta| + |R_\delta| < \frac{\varepsilon}{2} + \frac{\varepsilon}{2} = \varepsilon,$$

concluding the proof. □

Chapter 2

First-Order Linear Partial Differential Equations

In the first chapter, we introduced the transport equation and explored its solution. In this chapter we follow up by developing methods for solving general first-order linear partial differential equations. These equations are directly connected with the geometry of vector fields via integral curves. The methodology lends itself to the problem of classifying second-order linear partial differential equations as well as the notion of characteristics of a partial differential equation.

2.1 Ideas from Vector Calculus

Partial Derivatives and the Gradient

The set of all d-tuples of real numbers is denoted $\mathbb{R}^d$. Points in $\mathbb{R}^d$ will be denoted x, y, z, etc. The Cartesian coordinates of such points are written $x = (x_1, x_2, \ldots, x_d)$, $y = (y_1, y_2, \ldots, y_d)$, etc. Defining vector operations of addition and scalar multiplication componentwise, $\mathbb{R}^d$ carries a vector space structure. It is natural to identify points $x \in \mathbb{R}^d$ with their position vector relative to the origin, also denoted by x. (Boldface for position vectors, $\mathbf{x}$, will be used when emphasis on the vector structure is needed.) In particular, a function $f(x)$ on $\mathbb{R}^d$ can also be thought of as a function defined on vectors x, permitting vector space operations. An example would be the translation of a function $f(x+y)$. The Euclidean inner or dot product on $\mathbb{R}^d$ is defined via

$$x \cdot y = x_1 y_1 + x_2 y_2 + \cdots + x_d y_d$$

and the length or norm of a vector is defined by $|x| = (x \cdot x)^{1/2}$. The distance formula between points in $\mathbb{R}^d$ is given by

$$\rho(x, y) = |x - y| = \sqrt{(x_1 - y_1)^2 + (x_2 - y_2)^2 + \cdots + (x_d - y_d)^2}.$$

The partial derivative of a function $f(x)$ with respect to x_i, $1 \leq i \leq d$, is denoted in one of three forms:

$$\partial_i f = \frac{\partial f}{\partial x_i} = f_{x_i}.$$

Likewise, higher-order derivatives have similar notation, e.g.,

$$\partial_{ij} f = \frac{\partial^2 f}{\partial x_j \partial x_i} = f_{x_i x_j}.$$

Notation. Throughout this text, the dimension $d = 1, 2,$ or 3 and points x will typically carry the physical meaning of spatial location. The use of subscripted variables is very useful in describing conceptual notions that are independent of dimension. Furthermore, most of what is developed in this text can be carried out in an arbitrary number of dimensions where the subscript notation is necessary. In the sequel and in dealing with concrete classical problems in one, two, or three dimensions, we will often revert to more familiar notation, e.g., (x, y, z) as coordinates of a point in three dimensions, u_{xy} as a second mixed partial. The meaning of the notation should be clear from the context.

An **open ball** centered at x of radius r is the set $B_r(x) = \{y | \, \rho(y, x) < r\}$. An **open set** is a set $U \subset \mathbb{R}^d$ with the property that for every point of the set there is an open ball centered at the point contained in the set; i.e. if $x \in U$, then there exists an $r > 0$ such that $B_r(x) \subset U$. Given any set $V \subset \mathbb{R}^d$, a point x is called a **boundary point** if $B(x, r)$ has nonempty intersection with V and its complement for any $r > 0$. *Necessarily, an open set does not contain points of its boundary.* Simple examples of open sets are open balls, the interior of a box, the region between concentric spheres, etc. Other examples would be an open ball with center removed or the interior of a cylinder whose axis of revolution is removed. A set E is said to be a **closed set** if it is the complement of an open set. Necessarily, closed sets contain their boundary. Given any subset E, its **closure**, denoted $\overline{E}$, is the smallest closed set containing E. It is easy to see that $\overline{E} = E \cup \partial E$, where ∂E is the boundary of E.

Often in the hypothesis of results, certain regularity/smoothness of the functions involved is needed. A function f defined on an open set D (or its closure $\overline{D}$) is said to be of class $C^m(D)$ (or $C^m(\overline{D})$) if f and all partial derivatives up to order m are continuous on D. The mathematical notation for this is $f \in C^m(D)$ (or $f \in C^m(\overline{D})$).

Let $f = f(x)$ be a function defined on some open set $D \subset \mathbb{R}^d$. Let $\mathbf{n}$ be a unit vector in $\mathbb{R}^d$. Motivated by the definition of a derivative and keeping in mind our notation convention for position vectors, let's consider the difference quotient

$$\frac{f(x + tn) - f(x)}{t}, \quad t > 0$$

and the associated limit (provided it exists):

$$D_{\mathbf{n}}f(x) = \frac{df}{dn} = \lim_{t\to 0} \frac{f(x+tn) - f(\mathbf{x})}{t}.$$

This is known as the **directional derivative** of f in the direction of $\mathbf{n}$ and gives a measure of the rate of change of f in the direction of the unit vector $\mathbf{n}$. The standard unit basis vectors are denoted $\mathbf{e}_1, \mathbf{e}_2 \ldots, \mathbf{e}_d$, where $\mathbf{e}_i$ has ith component 1 and all other components 0. Notice that for the choice $\mathbf{n} = \mathbf{e}_i$,

$$D_{\mathbf{e}_i} f = \partial_i f.$$

Applying a multivariate chain rule gives the following useful computational formula for the directional derivative:

$$D_{\mathbf{n}} f = \boldsymbol{\nabla} f \cdot \mathbf{n},$$

where

$$\boldsymbol{\nabla} f = (\partial_1 f, \partial_2 f, \ldots, \partial_n f)\,. \tag{2.1.1}$$

The vector quantity $\boldsymbol{\nabla} f$ is called the **gradient** of f. (*Vector-valued functions and vector fields are denoted using bold notation.*) Since, for any two vectors $\mathbf{a}$ and $\mathbf{b}$, $\mathbf{a}\cdot\mathbf{b} = |\mathbf{a}|\,|\mathbf{b}|\cos\theta$, where θ is the angle between the two vectors, we get the basic estimate

$$|D_{\mathbf{n}} f| \le |\boldsymbol{\nabla} f|\,.$$

Further, *the directional derivative is maximized when the unit vector* $\mathbf{n}$ *has the same direction as* $\boldsymbol{\nabla} f$.

A useful geometric property of the gradient is that $\boldsymbol{\nabla} f$ is orthogonal to any level surface, $f(x) = C$, at any point on this surface. This can be seen as follows. Denote the level surface by S and let $x \in S$ be a fixed point. Let $\gamma : [-1, 1] \to S$ be a parametric curve on S passing through x at $t = 0$ (see Figure 2.1), i.e.,

$$f(\gamma(t)) = C \quad \text{with} \quad \gamma(0) = x.$$

Then

$$0 = \frac{d}{dt} f(\gamma(t)) = \boldsymbol{\nabla} f(\gamma(t)) \cdot \boldsymbol{\gamma}'(t),$$

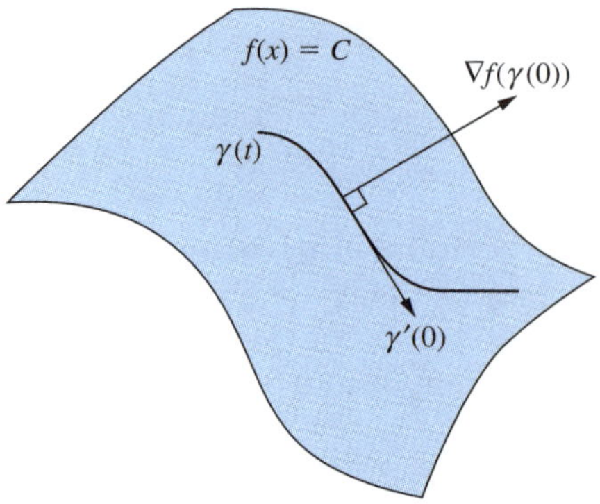

Figure 2.1 Gradient Illustration

in particular, $\nabla f(x) \cdot \gamma'(0) = 0$. This tells us that $\nabla f(x)$ is orthogonal to the tangent vector $\gamma'(0)$ to the surface at the point x. Since this is valid for any such curve with the described properties, the orthogonality property is proved.

Vector Fields and Integral Curves

A **vector field** is simply a vector-valued function defined on a subset of $\mathbb{R}^d$, i.e., $\mathbf{V} : D \to \mathbb{R}^d$, where $D \subset \mathbb{R}^d$. In component form we can write the vector field as

$$\mathbf{V}(x) = (V_1(x), V_2(x), \ldots, V_d(x)),$$

where each V_i is a scalar-valued function. Analogous to the case of functions, we write $\mathbf{V} \in C^m(D)$ (or $\mathbf{V} \in C^m(\overline{D})$), if each of its component functions are in $C^m(D)$ (or $C^m(\overline{D})$).

Intuitively, this is an assignment of a vector at each point x. For example, consider a fluid moving in three dimensions. At each point we can assign a vector representing the fluid's velocity at that point, the so-called velocity field of the fluid. Alternatively, consider a particle of mass M located at the origin in $\mathbb{R}^3$. Then the vector field

$$\mathbf{F}(x) = -GM\frac{\mathbf{x}}{|x|^3},$$

where G is the gravitational constant, represents the Newtonian gravitational field. The units would be force per unit mass. A similar formula holds in electrostatics. Place a charge q at the origin, then the vector field $\mathbf{E}(x) = q\frac{\mathbf{x}}{|x|^3}$ represents the electric field of the particle in free space. The difference in sign between the formulas for the gravitational and electric field is easily explained by the fact that the former is always attractive, whereas the latter is attractive or repulsive depending on whether it is applied to a negative or positive charge, respectively, at the point x.

Throughout the following discussion, we consider vector fields that are $C^1(D)$ for some open set D and $\mathbf{V} \neq \mathbf{0}$ at all points of D.

Definition 2.1 Given a vector field $\mathbf{V}$ defined on U, a curve C in U is called an **integral curve** of the vector field if $\mathbf{V}$ is tangent to C at each of its points.

Other names for integral curves are *field lines* and *flow curves.* In physics, the former comes from the terminology "lines of force" and the latter from fluid mechanics. Notice in the case of the gravitational or electric field, the integral curves are straight lines through the origin.

In principle, it seems easy to find the integral curves of a given field. Indeed, let $\gamma(t)$ be a parametrized curve in the set U. Then for this curve to be an integral curve of $\mathbf{F}$, it is necessary and sufficient that γ satisfy the equation

$$\gamma'(t) = \mathbf{V}(\gamma(t)). \tag{2.1.2}$$

This vector equation, written out in component form, would be a system of d first-order coupled ODEs that are, furthermore, possibly nonlinear. In general form, it is impossible to solve this system of equations analytically; the most we can do is prove the existence of solutions under a suitable regularity hypothesis on the vector field. Rather than pursue the latter (an advanced topic in the theory of ODE, e.g. [4]), we use the rest of this section to give an methodological overview in two and three dimensions and explore examples.

Let $\mathbf{V}(x,y,z) = (P(x,y,z), Q(x,y,z), R(x,y,z))$ be a vector field in three dimensions (note we are dispensing with subscript notation). In component form, $\gamma(t) = (x(t), y(t), z(t))$, and hence the vector equation (2.1.2) becomes the system

$$\frac{dx}{dt} = P(x,y,z), \quad \frac{dy}{dt} = Q(x,y,z), \quad \frac{dz}{dt} = R(x,y,z). \tag{2.1.3}$$

This is an autonomous system, meaning that the right-hand side does not depend on t explicitly and allows eliminating t from the equations. For example, if P is not identically zero, then we can write the system equivalently as

$$\frac{dy}{dx} = \frac{Q}{P}, \quad \frac{dz}{dx} = \frac{R}{P}.$$

Similar formulas hold in the cases where Q or R are not identically zero. We can dispense with these cases by rewriting (2.1.3) in the form

$$\frac{dx}{P} = \frac{dy}{Q} = \frac{dz}{R}, \tag{2.1.4}$$

with the understanding that if one of the denominators is identically zero, then the integral curves lie in planes where the corresponding coordinate is constant. For example, if $R \equiv 0$, then the integral curves lie in planes of the form $z =$ constant.

In the case of a planar vector field ($R \equiv 0$ and P and Q depend only on x and y), the above considerations reduce to a single ODE:

$$\frac{dx}{P} = \frac{dy}{Q}.$$

If we express the general solution of this equation in implicit form, $\xi(x, y) = c$, where c is a constant, then this family of curves describes the integral curves of the field. Analogously, for the three-dimensional case, we seek integral curves that are the intersection between two surfaces implicitly given by

$$\xi(x, y, z) = c_1, \quad \eta(x, y, z) = c_2.$$

A technical condition is needed here, namely,

$$\boldsymbol{\nabla}\xi \times \boldsymbol{\nabla}\eta \neq \mathbf{0} \tag{2.1.5}$$

throughout D or a suitable subset. This ensures that the tangent planes at a point of intersection of the surfaces are not parallel and a curve of intersection will exist. Figure 2.2 depicts the geometry of these surfaces. Since $\boldsymbol{\nabla}\xi$ is perpendicular to level surfaces of $\xi(x, y, z)$ and $\mathbf{V}$ is tangent to these surfaces, it follows that

$$0 = \boldsymbol{\nabla}\xi \cdot \mathbf{V} = P\xi_x + Q\xi_y + R\xi_z,$$

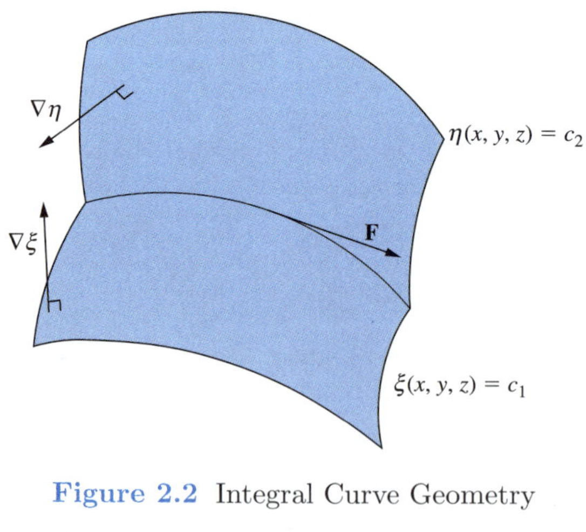

Figure 2.2 Integral Curve Geometry

and similarly for $\eta(x, y, z)$. In other words, both functions are solutions to the first-order linear PDE:

$$Pu_x + Qu_y + Ru_z = 0.$$

A C^1 solution to this PDE is called a **first integral** of the vector field $\mathbf{V}$ (or of the associated system (2.1.4)). Notice that if $\mathbf{V}$ is a C^1 vector field, then any first integral is C^2.

The following examples should make these ideas clear.

Example 2.1 Consider the planar vector field $\mathbf{V} = (-y, x)$ on $\mathbb{R}^2$ with the origin removed (the punctured plane). The equation for the integral curves can be written $x\,dx + y\,dy = 0$, which can be immediately integrated to give $x^2 + y^2 = c$, circles in the plane with center at the origin. A plot of this vector field and typical integral curves is given in Figure 2.3(a). Let's lift this problem to three dimensions by considering $\mathbf{V} = (-y, x, 0)$ on $\mathbb{R}^3$ with the z-axis removed. The equations for the integral curves are then

$$\frac{dx}{-y} = \frac{dy}{x} = \frac{dz}{0}.$$

As in the planar case, the first pair yields the implicit equation $x^2 + y^2 = c_1$, which is a family of cylinders. The last ratio indicates that $z = c_2$, representing a family of planes parallel to the $xy-$plane. Hence the integral curves can be described as the family of all circles that lie in planes parallel to the $xy-$plane.

Example 2.2 Here we examine the planar field $\mathbf{V} = (x+y, x-y)$. The ODE for the integral curves can be written

$$(x - y)dx - (x + y)dy = 0.$$

This is an exact[1] differential equation. As a consequence, the general solution in implicit form, $\xi(x, y) = c$, satisfies $\xi_x = x - y$ and $\xi_y = -(x+y)$. Integrating, the general solution is given by the family $\frac{1}{2}x^2 - xy - \frac{1}{2}y^2 = c$. See Figure 2.3(b) for a plot of the vector field and some typical integral curves.

1. An ODE $M(x, y)dx + N(x, y)dy = 0$ is said to be exact in a region D if there is a function $F(x, y)$ such that $\frac{\partial F}{\partial x} = M$ and $\frac{\partial F}{\partial y} = N$, in which case the general solution of the equation is $F(x, y) = c$, c being a constant.

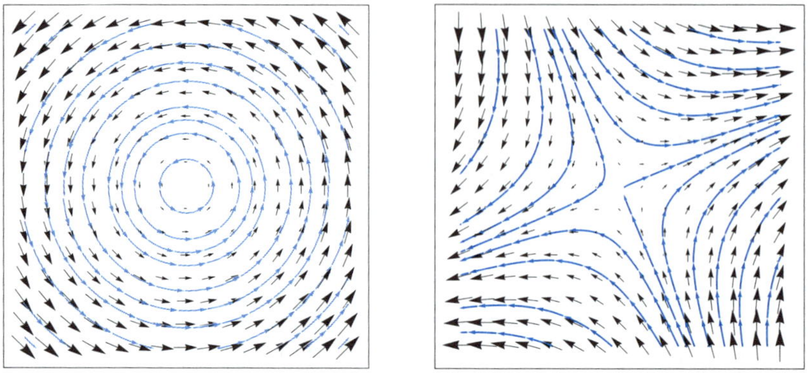

(a) (b)

Figure 2.3 Field and Integral Curves for Examples 2.1 and 2.2

Example 2.3 Let $\mathbf{V} = (1, 1, 1)$, a constant vector field (take $D = \mathbb{R}^3$). The equations for the integral curves can be written $\frac{dy}{dx} = 1$ and $\frac{dz}{dx} = 1$ with solutions $y - x = c_1$ and $z - x = c_2$, respectively. These are two planes with respective normal vectors $(-1, 1, 0)$ and $(-1, 0, 1)$ and that intersect in a line for any choice of constants c_1 and c_2. Setting $t = x$, the parametric equation for the lines is $\boldsymbol{r}(t) = (t, t + c_1, t + c_2)$. Notice the geometry of the solution. The vector field is constant, so it determines a family of planes with normal vector $(1, 1, 1)$. The integral curves are then clearly lines perpendicular to this family of planes, i.e., lines with direction vector $(1, 1, 1)$.

Example 2.4 Let D be $\mathbb{R}^3$ with the origin removed and consider the vector field $\mathbf{V} = (x, y, z)$. The equation for the integral curves is

$$\frac{dx}{x} = \frac{dy}{y} = \frac{dz}{z}.$$

The first pair of ratios is an ODE with general solution $\xi = \frac{y}{x} = c_1$. Equating the first and third ratios is a similar ODE with general solution $\eta = \frac{z}{x} = c_2$. Taken together, these are a family of planes passing through the origin that intersect in a line. It is easily checked that the gradient condition (2.1.5) is satisfied. Setting $t = x$, the integral curves form a two-parameter family with equation $\gamma(t) = (t, c_1 t, c_2 t)$. Choosing parameterizations $y = t$ or $z = t$, it follows that any line through the origin is an integral curve (which should be geometrically clear).

Exercises

2.1 Review the multivariate chain rule and verify (2.1.1).

2.2 Find the integral curves for the following vector fields.

a. $\mathbf{V} = (1, 1, 0)$
b. $\mathbf{V} = (0, z, -y)$
c. $\mathbf{V} = (y^3, x^3, 0)$

2.3 For the following planar fields, determine the integral curves and plot the vector field and some typical integral curves.

a. $\mathbf{V} = (1, 2xy)$
b. $\mathbf{V} = (x^2, y^2)$

2.2 First-Order Linear Equations

Consider a first-order linear equation with two independent variables x and y of the form

$$a(x,y)\,u_x + b(x,y)\,u_y = F(x,y). \tag{2.2.1}$$

The underlying geometric meaning to this equation and a methodology for solution can be found by expressing the equation in gradient form:

$$\nabla u \cdot (a(x,y), b(x,y)) = F(x,y).$$

The left-hand side is a multiple of the directional derivative of u in the direction of the vector $(a(x,y), b(x,y))$. As such, the equation can be interpreted as prescribing the directional derivative of u along the integral curves of the vector field $\mathbf{V} = (a(x,y), b(x,y))$.

Example 2.5 (The Transport Equation)
Imagine a long channel containing a fluid moving with velocity c along the length of the channel. Introduce the coordinate x as position along the channel and let A be the cross-sectional area of the channel. A substance is suspended in the fluid with concentration given by $u(x,t)$. Assume that no diffusion occurs and the fluid moves only in the channel direction. Then in the slice of the channel, $[x, x+\Delta x]$, conservation of mass of the substance in the slice can be written

$$u_t(x,t)\,A\Delta x = cA[u(x,t) - u(x+\Delta x, t)].$$

Dividing by Δx and taking the limit as this tends to zero yields the transport equation

$$u_t + cu_x = 0,$$

an equation representing an infinitesimal version of conservation of mass. If the initial concentration of the substance, $f(x)$, is known, then an initial value problem describing the motion of the substance with the fluid can be written as

$$\begin{aligned} \text{PDE}\colon\ & u_t + cu_x = 0 \\ \text{IC}\colon\ & u(x,0) = f(x). \end{aligned} \tag{2.2.2}$$

This assumes no other boundaries on the channel so the mathematical domain for this problem is $-\infty < x < \infty$ and $t > 0$. If we write the gradient of u as $\nabla u = (u_t, u_x)$, then the equation can be written as $\nabla u \cdot (1, c) = 0$. The interpretation is that the directional derivative of u in the direction of the vector $(1, c)$ is zero, i.e., the function u must be constant along lines with direction vector $(1, c)$. These lines have slope c, and equivalently are given as solutions to the ODE $\frac{dx}{dt} = c$. Hence the integral curves are given by the equation $x - ct = C$, where C is a constant, and u must be a function of $x - ct$. In other words, $u(x,t) = g(x - ct)$ for some continuously differentiable function g of one variable. Applying the initial condition, the solution to initial value problem is

$$u(x,t) = f(x - ct).$$

Notice the form of the solution: a traveling wave moving to the right or left depending on the sign of c. Based on the discussion in the previous chapter, this is a strong solution to the problem if f is continuously differentiable. If f is merely continuous or differentiable, then this defines a weak solution to the problem.

The method of solution given in Example 2.5 can be generalized to the homogeneous equation

$$\nabla u \cdot (a(x,y), b(x,y)) = 0. \tag{2.2.3}$$

The geometric interpretation is that the solution u will be constant along the integral curves of the vector field $(a(x,y), b(x,y))$. The latter can be determined using the method of the previous section—i.e., assuming $a(x,y)$ is continuous and nonzero in some region, then the integral curves are found by solving the ordinary differential equation

$$\frac{dy}{dx} = \frac{b(x,y)}{a(x,y)}.$$

If we write the general solution to this equation in implicit form $\xi(x, y) = C$, then the solution to (2.2.3) is given by

$$u(x, y) = g(\xi(x, y)), \tag{2.2.4}$$

for some differentiable function g.

Example 2.6 Find the general solution to the equation $yu_x + u_y = 0$. Here the solution will be constant along integral curves of the vector field $(y, 1)$. Such curves satisfy the differential equation $\frac{dy}{dx} = \frac{1}{y}$, which has the general solution $y^2 - 2x = C$. Hence $u(x, y) = g(y^2 - 2x)$ is a solution to the PDE for any differentiable function g. This can easily be checked by differentiation.

There is more than meets the eye in the preceding two examples. In both cases, formula (2.2.4) provides a solution for *any* continuously differentiable function g. The fact is that any solution of the PDE can be written in this form for a suitable choice of g. In this sense, we say that (2.2.4) is the **general solution** to the first-order homogeneous PDE (2.2.3). We give the essential ideas of the proof of this by using the integral curves (defined by $\xi(x, y)$) to define a change of variables. Introduce new independent variables

$$\xi = \xi(x, y),\ \eta = \eta(x, y),$$

so that the Jacobian determinant

$$J = \det\begin{pmatrix} \xi_x & \xi_y \\ \eta_x & \eta_y \end{pmatrix} \neq 0. \tag{2.2.5}$$

(The function η is chosen to satisfy this condition.) This latter condition guarantees that the transformation $(x, y) \to (\xi, \eta)$ is invertible. Then the chain rule gives

$$u_x = u_\xi \xi_x + u_\eta \eta_x, \quad u_y = u_\xi \xi_y + u_\eta \eta_y. \tag{2.2.6}$$

Substituting into (2.2.3) and rearranging terms gives

$$[a\xi_x + b\xi_y]u_\xi + [a\eta_x + b\eta_y]u_\eta = 0.$$

The first bracketed term is zero since ξ defines the integral curves; the second bracketed term cannot be zero because of the Jacobian condition (2.2.5). Hence, the change of variables reduces the equation to an ordinary differential equation $u_\eta = 0$. The general solution of the ODE is $u(\xi, \eta) = g(\xi)$, for some differentiable function g. Alas, the general solution to (2.2.3) has form $u(x, y) = g(\xi(x, y))$. The following example illustrates this method and its application to the general nonhomogeneous equation (2.2.1).

Example 2.7 Find the general solution to $yu_x + u_y = x$. In Example 2.6 we found the integral curves, so we set $\xi = y^2 - 2x$. We also choose $\eta(x, y) = y$. The Jacobian determinant is

$$J = \begin{pmatrix} -2 & 2y \\ 0 & 1 \end{pmatrix} = -2$$

and is nonzero. Note that the inverse transformations are given by $x = (\eta^2 - \xi)/2$ and $y = \eta$. From these considerations, the partial differential equation can be written as

$$u_\eta = \frac{\eta^2 - \xi}{2}.$$

Integrating this equation, we have

$$u(\xi, \eta) = \frac{1}{6}\eta^3 - \frac{1}{2}\xi\eta + g(\xi),$$

for some differentiable function g. Hence, the general solution being sought is

$$u(x, y) = \frac{1}{6}y^3 - \frac{1}{2}y(y^2 - 2x) + g(y^2 - 2x).$$

Remark 2.2 Note the form of the general solution in Example 2.7: the sum of a specific function and the general solution to the corresponding homogeneous linear PDE. The specific function is in fact a particular solution to the non-homogeneous equation and hence the structure of the solution is completely analogous to that of linear ordinary differential equations. *A word of warning: This solution structure is a rarity in the realm of PDEs.* Once we enter the realm of second-order equations, typically other factors such as the geometry of the domain and the form of the boundary conditions greatly affect the form of solutions, so the notion of a general solution loses its meaning.

Remark 2.3 The change of variables methodology of Example 2.7 can be applied to a general first-order linear PDE of the form

$$a(x, y)u_x + b(x, y)u_y + c(x, y)u = f(x, y).$$

In this case the transformed equation becomes the first order ODE

$$b'(\xi, \eta)u_\eta + c(\xi, \eta)u = f(\xi, \eta),$$

where $b'(\xi, \eta) = a\eta_x + b\eta_y$. Examples will be seen in the exercises set.

In most applied problems, the interest is in finding the solution of the PDE that satisfies an initial condition that prescribes the values of $u(x, y)$ along a particular curve. One example of this was seen in the case of the transport equation, Example 2.5. Care must be taken in the choice of the initial curve as the following example illustrates.

Example 2.8 Consider again the equation $yu_x + u_y = 0$. Let's find the solution satisfying the condition that $u = x$ on the curve $y = 0$. From Example 2.6, the general solution of the equation is $u(x, y) = g(y^2 - 2x)$. Hence, we need to choose the function g such that $g(-2x) = x$. By inspection, $g(x) = -x/2$ and the desired solution of the problem is $u(x, y) = \frac{1}{2}(2x - y^2)$. On the other hand, suppose the solution were to satisfy the condition that $u = x$ on the curve $y^2 = 2x + 6$. The same line of reasoning would yield $g(6) = x$, a condition satisfied by no function g. The idea here is that one needs to avoid putting initial conditions on integral curves of the vector field.

The preceding ideas can be applied to first-order linear PDEs in any number of variables. We illustrate for the case of three variables and the equation

$$a(x, y, z)u_x + b(x, y, z)u_y + c(x, y, z)u_z = 0, \tag{2.2.7}$$

where we assume the coefficient functions are C^1 in some open set $D \subset \mathbb{R}^3$ and that the associated vector field $\mathbf{V} = (a, b, c)$ is nonzero throughout D. As in the two-variable case, we interpret the equation as saying that the directional derivative of the unknown function u in the direction of $\mathbf{V}$ is zero and expect to construct solutions of the equation by knowing the integral curves. From the previous section, we know that these integral curves are a two-parameter family determined by the intersection of a pair of surfaces of form

$$\xi(x, y, z) = c_1 \quad \text{and} \quad \eta(x, y, z) = c_2,$$

where $\nabla\xi \times \nabla\eta \neq \mathbf{0}$. In analogy with the two-variable case, the function

$$u(x, y, z) = F(\xi(x, y, z), \eta(x, y, z)) \tag{2.2.8}$$

will be a solution of (2.2.7) for any function $F(\xi, \eta)$ of two variables for which first partials exist. This can be verified using the chain rule and the fact noted earlier that both functions $\xi(x, y, z)$ and $\eta(x, y, z)$ satisfy the PDE. The following theorem makes these ideas precise and, with appropriate rephrasing, does the same for the two-variable linear equation.

Theorem 2.4 Let $\xi = \xi(x, y, z)$ and $\eta = \eta(x, y, z)$ be two solutions of (2.2.7) in a region D that satisfy the gradient condition (2.1.5) there, and let $u = u(x, y, z)$ be any solution of (2.2.7) in D. Let $P \in D$. Then there exists a neighborhood $D' \subset D$ containing P and a C^1 function $F(\xi, \eta)$ such that

$$u(x, y, z) = F(\xi(x, y, z), \eta(x, y, z))$$

throughout D'.

Part of Theorem 2.4 can be proved using the change-of-variables scheme used in the two-variable case. The fact that the result typically holds on a subset D' and the conclusion of smoothness of the function F require deeper ideas (see [35]). We now turn to examples.

Example 2.9 Consider the equation $au_x + bu_y + cu_z = 0$, where the coefficients are constants and all are nonzero. The constant vector field (a, b, c) has integral curves given by the intersection of the planes $ay - bx = c_1$ and $az - cx = c_2$. We conclude that the function

$$u(x, y, z) = F(ay - bx, az - cx),$$

where F is a function of two variables is a solution to the PDE, a fact that can be checked by differentiation. Here we will demonstrate that all solutions have this form. We first introduce the invertible change of variables:

$$\xi = ay - bx, \quad \eta = az - cx, \quad \zeta = z.$$

Applying the chain rule to any function u gives

$$\begin{aligned} u_x &= -bu_\xi - cu_\eta \\ u_y &= au_\xi \\ u_z &= au_\eta + u_\zeta, \end{aligned}$$

and substituting these into the PDE leads to the equation $cu_\zeta = 0$. This is an ordinary differential equation! Its general solution is $u = F(\xi, \eta)$, as desired.

Example 2.10 Consider the equation $xu_x + yu_y + zu_z = 0$ in the region where $x > 0$. The integral curves for the vector field (x, y, z) in this region were determined in Example 2.4 to be the intersection of the planes $\frac{y}{x} = c_1$ and $\frac{z}{x} = c_2$. Hence, the function

$$u(x, y, z) = F(\frac{y}{x}, \frac{z}{x})$$

is a solution of the PDE in the domain $x > 0$.

Exercises

2.4 Find the general solution of the following equations.

a. $u_x = x \sin y$

b. $u_{xy} = x \sin y$

2.5 Determine the integral curves, sketch them, and find the general solution of the following equations.

a. $3u_x + 2u_y = 0$

b. $(1 + x^2)u_x + u_y = 0$

c. $u_x + 2xyu_y = 0$

d. $yu_x + xu_y = 0$

2.6 Find the general solution of the following equations.

a. $au_x + bu_y + cu = 0$

b. $u_x + 2u_y + (2x - y)u = 2x^2 + 3xy - 2y^2$

2.7 Solve the following initial value problems.

a. $xu_x - yu_y = 0$ satisfying $u(x, x) = x^2$

b. $xu_x + yu_y = u$ satisfying $u(x, x^2) = 1$

2.8 Verify the computations and details of Example 2.9.

2.9 Construct the general solution of the PDE in Example 2.10 in the domain $y > 0$.

2.10 Find the general solution of the equation $yu_x + xu_y + xu_z = 0$. Where is the solution valid?

2.3 Classification of Second-Order PDEs in Two Variables

In this section we will apply the ideas of the previous section to homogeneous second-order linear PDEs of the form

$$au_{xx} + 2bu_{xy} + cu_{yy} + G(x, y, u, u_x, u_y) = 0, \tag{2.3.1}$$

where the coefficients a, b, and c are functions of x and y, not all zero, and G is a linear function of u, u_x, and u_y. The focus of the section is to show that equations of this form have three canonical forms, depending on the sign of the **discriminant**,

$$\Delta = \Delta(x, y) = b^2 - ac.$$

A simple observation is that given a smooth change of variables $\xi = \xi(x, y)$ and $\eta = \eta(x, y)$ satisfying the Jacobian condition (2.2.5), the equation (2.3.1) transforms into one of the same form:

$$Au_{\xi\xi} + 2Bu_{\xi\eta} + Cu_{\eta\eta} + \tilde{G}(\xi, \eta, u, , u_\xi, u_\eta) = 0. \tag{2.3.2}$$

The following lemma tells us that the sign of the determinant does not change under such transformations. (We use $\tilde{\Delta}$ for the discriminant of the transformed equation.)

Lemma 2.5 Under a smooth change of variables satisfying the Jacobian condition (2.2.5),

$$\tilde{\Delta} = J^2\Delta.$$

Proof This is an elaboration of previous chain rule computations. The fact that a second-order linear equation goes over to a PDE of the same form under a change of variables is apparent. The constants A, B, and C are computed by applying the chain rule to (2.2.6),

$$\begin{aligned} u_{xx} &= (u_{\xi\xi}\xi_x + u_{\xi\eta}\eta_x)\xi_x + u_\xi\xi_{xx} + (u_{\eta\eta}\eta_x + u_{\xi\eta}\xi_x)\eta_x + u_\eta\eta_{xx} \\ &= u_{\xi\xi}\xi_x^2 + 2u_{\xi\eta}\xi_x\eta_x + u_{\eta\eta}\eta_x^2 + \text{lower-order terms,} \end{aligned}$$

where the lower-order terms will be absorbed into $\tilde{G}$. Similarly,

$$\begin{aligned} u_{xy} &= u_{\xi\xi}\xi_x\xi_y + u_{\xi\eta}(\xi_x\eta_y + \xi_y\eta_x) + u_{\eta\eta}\eta_x\eta_y + \text{lower-order terms} \\ u_{yy} &= u_{\xi\xi}\xi_y^2 + 2u_{\xi\eta}\xi_y\eta_y + u_{\eta\eta}\eta_y^2 + \text{lower-order terms.} \end{aligned}$$

Substituting into (2.3.1) and grouping terms, we have

$$\begin{aligned} A &= a\xi_x^2 + 2b\xi_x\xi_y + c\xi_y^2 \\ B &= a\xi_x\eta_x + b(\xi_x\eta_y + \xi_y\eta_x) + c\xi_y\eta_y \\ C &= a\eta_x^2 + 2b\eta_x\eta_y + c\eta_y^2. \end{aligned} \tag{2.3.3}$$

Computing the discriminant for the transformed equation is a tedious algebraic exercise. The result is

$$\tilde{\Delta} = B^2 - AC = (\xi_x\eta_y - \xi_y\eta_x)^2(b^2 - ac) = J^2\Delta.$$

□

This result motivates a definition in analogy with the analytic geometry of quadratic forms.

Definition 2.6 The second-order linear PDE (2.3.1) is said to be *hyperbolic, parabolic,* or *elliptic* at a point (x_0, y_0) if $\Delta > 0$, $\Delta = 0$, or $\Delta < 0$, respectively, at that point. The equation is said to be hyperbolic, parabolic, or elliptic in a region D in the xy–plane if it is of that type at each point of D.

With this definition Lemma 2.5 may be stated thus: *The type of a second-order linear PDE is preserved under invertible changes of variables.*

Example 2.11 The wave equation, heat equation, and Laplace's equation are hyperbolic, parabolic, and elliptic, respectively.

Example 2.12 For the equation

$$yu_{xx} + 2u_{xy} + xu_{yy} = 0,$$

$\Delta = 1 - xy$. Hence, the equation is parabolic on the hyperbola $xy = 1$, hyperbolic in the region $xy < 1$, and elliptic in the region $xy > 1$.

The following theorem spells out the importance of the classification of second-order equations.

Theorem 2.7 Consider the second-order equation (2.3.1) where a, b, and c are constants, not all zero. Then there is a change of variables $\xi = \xi(x, y)$, $\eta = \eta(x, y)$ satisfying the Jacobian condition (2.2.5) and a function $\tilde{G}$ such that:
(1) If $\Delta > 0$, the transformed equation has the form

$$u_{\xi\eta} + \tilde{G}(\xi, \eta, u, u_\xi, u_\eta) = 0, \tag{2.3.4}$$

called the first canonical form of a hyperbolic equation. Alternatively, setting $\xi' = \xi + \eta$ and $\eta' = \xi - \eta$, the equation has the form

$$u_{\xi'\xi'} - u_{\eta'\eta'} + \tilde{G}(\xi, \eta, u, u_{\xi'}, u_{\eta'}) = 0, \tag{2.3.5}$$

known as the second canonical form of a hyperbolic equation.
(2) If $\Delta = 0$, the transformed equation has the form

$$u_{\eta\eta} + \tilde{G}(\xi, \eta, u, u_\xi, u_\eta) = 0, \tag{2.3.6}$$

known as the canonical form of a parabolic equation.
(3) If $\Delta < 0$, the transformed equation has the form

$$u_{\xi\xi} + u_{\eta\eta} + \tilde{G}(\xi, \eta, u, u_\xi, u_\eta) = 0, \tag{2.3.7}$$

known as the canonical form of an elliptic equation.

Proof The proof is constructive and makes use of the formulas (2.3.3).

Case $\Delta > 0$: Notice that if $a = c = 0$, there is nothing to prove. Further, if $a = 0$ or $c = 0$, then $b \neq 0$. Assume $a \neq 0$, the other case being similar. In this case a transformation is determined so that $A = C = 0$ and $B \neq 0$. Completing the square in the formula for A gives

$$A = a\left(\xi_x + \frac{b}{a}\xi_y\right)^2 - \frac{\Delta}{a}\xi_y^2.$$

Since $\Delta > 0$, setting this to zero and taking square roots leads to the pair of first-order PDEs,

$$\xi_x + \frac{b \pm \sqrt{\Delta}}{a}\xi_y = 0. \tag{2.3.8}$$

Set $\gamma_\pm = \frac{b\pm\sqrt{\Delta}}{a}$ and $\xi(x,y) = y - \gamma_+ x$, $\eta(x,y) = y - \gamma_- x$ (so that $\xi(x,y) =$ constant and $\eta(x,y) =$ constant are integral curves for the preceding equations). Then

$$J = \det\begin{pmatrix} -\gamma_+ & 1 \\ -\gamma_- & 1 \end{pmatrix} = \gamma_- - \gamma_+ = -\frac{2\sqrt{\Delta}}{a} \neq 0,$$

so that the Jacobian condition (2.2.5) is satisfied. Moreover, by the choice of ξ and η, $A = C = 0$ and

$$\begin{aligned} B &= a\gamma_+\gamma_- - b(\gamma_+ + \gamma_-) + c \\ &= -\frac{2\Delta}{a} \neq 0. \end{aligned}$$

The transformed equation is $-\frac{2\Delta}{a}u_{\xi\eta} + \widetilde{G} = 0$ and the desired form is obtained by multiplying by $-\frac{a}{2\Delta}$. Obtaining the second form for a hyperbolic equation is a straightforward chain rule computation.

Case $\Delta = 0$: In this case, there is only one equation in (2.3.8). Following the previous case, let $\xi(x,y) = y - \frac{b}{a}x$. If $a > 0$, set $\eta(x,y) = x/\sqrt{a}$, then

$$J = \det\begin{pmatrix} -\frac{b}{a} & 1 \\ \frac{1}{\sqrt{a}} & 0 \end{pmatrix} = -\frac{1}{\sqrt{a}} \neq 0.$$

Furthermore, direct computation gives $A = 0$, $C = 1$, and $B = -\frac{\Delta}{a} = 0$.

Case $\Delta < 0$. Again we assume $a > 0$. Here, the idea is to choose ξ and η so as to force $B = 0$. Choosing $\xi = x$ then to force $B = 0$, η must satisfy $a\eta_x + b\eta_y = 0$. The integral curves of the latter equation are defined by $\eta = ay - bx$. With these choices, $A = a$, $B = 0$, and $C = -a\Delta$. This suggests rescaling the choices for ξ and η. It is left to the reader to show that $\xi = x/\sqrt{a}$ and $\eta = (ay - bx)/\sqrt{a}\sqrt{-\Delta}$ lead to $A = C = 1$. □

This result can be restated the following way: In the case of second-order constant coefficient linear PDEs, the canonical form is either a wave equation, the heat equation, or Laplace's equation (each possibly with first-order terms). *It is a remarkable fact that the physically important second-order equations essentially embody all second-order equations in terms of solution and solution properties.*

Remark 2.8 Theorem 2.7 can be generalized to the case where a, b, and c are functions of x and y and the discriminant condition holds on an open set in the xy–plane. Indeed, in the hyperbolic and parabolic cases, the methodology of the proof will work (see Example 2.15). The difficulty lies in the elliptic case where $\Delta < 0$ and equations (2.3.8) have complex coefficients.

Example 2.13 The equation $u_{xx} + 5u_{xy} + 4u_{yy} = 0$ is hyperbolic in the plane. Equations (2.3.8) become

$$\xi_x + \frac{5 \pm 3}{2}\xi_y = 0.$$

Solving these, we set $\xi(x, y) = y - 4x$ and $\eta(x, y) = y - x$. In this case the Jacobian determinant is $J = -3$. Since ξ and η are linear functions, no terms below second order will appear in the canonical form and formulas (2.3.3) can be used to determine the coefficients: $A = 0$, $B = -9/2$, and $C = 0$. Hence the canonical form is $u_{\xi\eta} = 0$.

Example 2.14 Consider the equation $u_{xx} + 2u_{xy} + u_{yy} - u_y = 0$. In this case $\Delta = 0$ and the equation is parabolic. Then (2.3.8) becomes the single equation $\xi_x + \xi_y = 0$. Hence take $\xi(x, y) = y - x$. Choosing $\eta(x, y) = x$ gives the Jacobian determinant as $J = -1$. Again using formulas (2.3.3): $A = 0$, $B = 0$, and $C = 1$. Furthermore, the second-order computations do not contribute to lower-order terms. Since $u_y = u_\xi$, the canonical form is $u_{\eta\eta} - u_\xi = 0$.

Example 2.15 Consider the equation $yu_{xx} - xu_{yy} = 0$, which is hyperbolic in the first quadrant. Indeed, $\Delta = xy$. In this case equations (2.3.8) can be written $x^{-1/2}\xi_x \pm y^{-1/2}\xi_y = 0$. To determine the integral curves, we must solve the separable ODEs

$$x^{1/2}dx = \pm y^{1/2}dy.$$

The general solutions of the these are $x^{3/2} \mp y^{3/2} =$ constant. With the choice $\xi = x^{3/2} - y^{3/2}$ and $\eta = x^{3/2} + y^{3/2}$, formulas (2.3.3) yield $A = C = 0$ and $B = \frac{9}{2}xy$. Due to the variable coefficients in the original PDE, this change of variable produces lower-order terms. Collecting these, the transformed equation is

$$\frac{9}{2}xyu_{\xi\eta} + \frac{3}{4}\frac{x^{3/2} - y^{3/2}}{(xy)^{1/2}}(u_\xi - u_\eta) = 0.$$

Expressing everything in terms of ξ and η, the desired canonical form is

$$u_{\xi\eta} + \frac{2}{3}\frac{\xi}{\eta^2 - \xi^2}(u_\xi - u_\eta) = 0.$$

Reformulation and Characteristics Borrowing some ideas from linear algebra, we reformulate the classification theorem in terms of a quadratic form naturally associated with a second-order PDE. For concise notational presentation, we write the general form of a second-order linear equation using subscripted variables, i.e., we write $x = (x_1, x_2)$ and consider the equation

$$a(x)\partial_{11}u + 2b(x)\partial_{12}u + c(x)\partial_{22}u + \text{LOT} = 0,$$

where LOT stands for lower-order terms. Associated with this equation is a quadratic form in the $\xi = (\xi_1, \xi_2)$−variables given by

$$Q(x, \xi) = a(x)\xi_1^2 + 2b(x)\xi_1\xi_2 + c(x)\xi_2^2, \tag{2.3.9}$$

called the **principal symbol** of the equation. In matrix form this can be written as

$$Q(x, \xi) = [\xi_1 \quad \xi_2]\begin{bmatrix} a(x) & b(x) \\ b(x) & c(x) \end{bmatrix}\begin{bmatrix} \xi_1 \\ \xi_2 \end{bmatrix} = \xi^T A(x)\xi,$$

where the standard matrix notation $\xi = \begin{bmatrix} \xi_1 \\ \xi_2 \end{bmatrix}$ is used. Notice that $A = A(x)$ is a symmetric matrix and $\Delta = -\det(A)$. Let I be the identity matrix. Recall that the eigenvalues of a matrix are scalars λ such that the matrix $A - \lambda I$ is singular, i.e., does not have an inverse. This occurs precisely when $\det(A - \lambda I) = 0$, or when λ is a solution to the quadratic equation

$$\lambda^2 - \text{Tr}(A)\lambda + \det(A) = 0,$$

where $\text{Tr}(A) = a + c$ is the trace of the matrix. Using the quadratic formula, the eigenvalues are given by

$$\lambda = \lambda(x) = \frac{\text{Tr}(A) \pm \sqrt{\text{Tr}(A)^2 - 4\det(A)}}{2}.$$

This formula demonstrates that the eigenvalues of a symmetric matrix are real and, in this case, the sign of the eigenvalues is controlled by that of $\det(A) = -\Delta$. Since the sign of Δ determines the type of the PDE, we arrive at the following characterization result.

Proposition 2.9 The second-order linear equation (2.3.1) is hyperbolic, parabolic, or elliptic at a point x precisely when the eigenvalues of the matrix of the principal symbol have the following properties, respectively.

1. The eigenvalues are nonzero and have opposite sign.
2. One eigenvalue is zero.
3. The eigenvalues are nonzero and have the same sign.

Remark 2.10 The alternative classification has fundamental use in classifying second-order partial differential equations in higher dimensions; see page 71.

Example 2.16 The matrices of the principal symbol for the wave, heat, and Laplace equations are as follows.

$$\text{Wave equation: } \partial_{11}u - c^2\partial_{22}u = 0, \quad \begin{bmatrix} 1 & 0 \\ 0 & c^2 \end{bmatrix}$$

$$\text{Heat equation: } \partial_1 u - \alpha^2\partial_{22}u = 0, \quad \begin{bmatrix} 0 & 0 \\ 0 & -\alpha^2 \end{bmatrix}$$

$$\text{Laplace equation: } \partial_{11}u + \partial_{22}u = 0, \quad \begin{bmatrix} 1 & 0 \\ 0 & 1 \end{bmatrix}$$

Example 2.17 The equation considered in Example 2.13 has the associated matrix

$$\begin{bmatrix} 1 & 5/2 \\ 5/2 & 4 \end{bmatrix},$$

whose trace is 5 and determinant is $-9/4$. The eigenvalues are $\lambda = \frac{5\pm\sqrt{34}}{2}$, from which we conclude that the equation is hyperbolic (consistent with Example 2.13).

The moral of this section is that the principal symbol determines the type of equation and hence its canonical form. Different canonical forms have solutions with distinctive qualitative and quantitative properties, as we will see in later chapters. A geometric quantity of interest is defined using the principal symbol as follows.

Definition 2.11 A vector $\xi = (\xi_1, \xi_2)$ is said to be a **characteristic direction** at x if $Q(x, \xi) = 0$. A **characteristic curve** is a curve in the x-plane that is normal to the characteristic direction at all points x. Characteristic curves are also referred to as **characteristics.**

Example 2.18 Here we return to Example 2.16 and determine the characteristics for the classical trinity. In the case of the wave equation, $Q(x, \xi) = Q(\xi) = \xi_1^2 - c^2\xi_2^2$. Hence, the characteristic directions are multiples of $(\pm c, 1)$. For $(c, 1)$ to be normal to a curve given by $v(x_1, x_2) =$ constant, the vector $(1, -c)$ will be tangent to the curve and, hence,

$$\partial_1 v - c\partial_2 v = 0.$$

This first-order linear PDE has integral curves given by $x_2 - cx_1 =$ constant that are characteristic for the wave equation. Likewise, corresponding to the characteristic direction $(-c, 1)$, the characteristics are $x_2 + cx_1 =$ constant. Thus, for the wave equation, at each point in the plane, there are a pair of lines as characteristics. These lines play a fundamental role in the qualitative study of the wave equation, as will be seen later on. For the heat equation, $Q(\xi) = -\alpha^2\xi_2^2$ and the characteristic directions are multiples of $(1, 0)$. The characteristic curves are then $x_2 =$ constant. Finally, in the case of Laplace's equation, $Q(\xi) = \xi_1^2 + \xi_2^2$, and the only characteristic direction is $(0, 0)$. There are no characteristic curves in this case.

In the case of variable coefficient equations, determining the characteristics is more labor intensive and involves solving a first-order PDE. The following example is illustrative.

Example 2.19 The PDE $x_2\partial_{11}u + \partial_{22}u = 0$ is called Tricomi's equation and appears in hydrodynamics. The principal symbol is $Q(x, \xi) = x_2\xi_1^2 + \xi_2^2$. In the region $x_2 > 0$, the equation is elliptic and there are no characteristic curves. However, in the region $x_2 < 0$, the equation is hyperbolic and the characteristic directions are multiples of $(1, \pm\sqrt{-x_2})$. As for the wave equation, if the curve $v(x_1, x_2) = c$ is to be characteristic, then

$$\mp\sqrt{-x_2}\partial_1 v + \partial_2 v = 0.$$

Solving these, we find that the characteristic curves are given by two one-parameter families of curves: $x_1 \pm \frac{2}{3}(-x_2)^{3/2} = c$.

Characteristics can also be defined for first-order linear equations,

$$a(x)\partial_1 u + b(x)\partial_2 u + c(x)u = 0.$$

In this case the principal symbol is given by the linear form

$$Q(x, \xi) = a(x)\xi_1 + b(x)\xi_2.$$

It is left to the reader to show that characteristic curves in this case are the integral curves of the PDE.

Remark 2.12 The role of characteristics in the study of partial differential equations is essentially twofold. First, we have seen that solutions of the one-dimensional wave equation propagate singularities in the initial data along the characteristic lines. This propagation of singularities is a general phenomena for hyperbolic equations and the study of their characteristics is fundamental in the analysis as well as other qualitative properties of the equations. Second, the characteristics play a fundamental role in the mathematical existence results for first-order initial value problems. Classically this culminates in the Cauchy–Kovalevsky theorem and Holmgren's uniqueness theorem. (See [35, 11] for further reading.)

Exercises

2.11 Classify and find the canonical forms for the following equations.

a. $u_{xx} - 2u_{xy} + 3u_{yy} + u_y = 0$

b. $u_{xx} - 6u_{xy} + 10u_{yy} + u_x - 3u_y = 0$

2.12 Apply the transformations $\xi = x + ct$ and $\eta = x - ct$ to the wave equation $u_{tt} = c^2 u_{xx}$ to obtain $u_{\xi\eta} = 0$. Find the general solution of the latter equation and use this to show that any solution of the wave equation has the form $u(x, t) = F(x + ct) + G(x - ct)$ for some twice-differentiable functions F and G. (This provides justification of the heuristic argument used in the discussion on page 27.)

2.13 Find regions in the xy-plane where the following equations are hyperbolic, elliptic, or parabolic.

a. $u_{xx} + 2xu_{xy} + u_{yy} + \cos(xy)u_x = 0$

b. $xu_{xx} + 2xyu_{xy} - yu_{yy} = 0$

c. $xu_{xx} + 2xyu_{xy} + yu_{yy} = 0$

2.14 Show that the characteristic curves for a first-order linear PDE are given by the integral curves.

2.4 Application to Vector Fields

Given a vector field $\mathbf{F} = (P, Q, R)$ that is C^1 on some region $D \subset \mathbb{R}^3$, the **divergence** of $\mathbf{F}$ is defined by

$$\text{div } \mathbf{F} = P_x + Q_y + R_z$$

and the **curl** of $\mathbf{F}$ by

$$\text{curl } \mathbf{F} = (R_y - Q_z, P_z - R_x, Q_x - P_y).$$

Introducing the vector differential operator $\boldsymbol{\nabla} = (\frac{\partial}{\partial x}, \frac{\partial}{\partial y}, \frac{\partial}{\partial z})$, convenient formal formulas for the divergence and curl are

$$\text{div } \mathbf{F} = \boldsymbol{\nabla} \cdot \mathbf{F}, \quad \text{curl } \mathbf{F} = \boldsymbol{\nabla} \times \mathbf{F}.$$

Recall that the divergence is a gauge of the rate of expansion of a vector field and the curl a gauge of its circulation. The following vector identities should be familiar. Let f be a scalar function and $\mathbf{F}$ and $\mathbf{G}$ vector fields, all assumed to be C^1 on a region.

$$\begin{aligned}
&\boldsymbol{\nabla} \cdot (f\mathbf{F}) = \boldsymbol{\nabla} f \cdot \mathbf{F} + f\boldsymbol{\nabla} \cdot \mathbf{F} \\
&\boldsymbol{\nabla} \cdot (\mathbf{F} \times \mathbf{G}) = \boldsymbol{\nabla} \times \mathbf{F} \cdot \mathbf{G} - \boldsymbol{\nabla} \times \mathbf{G} \cdot \mathbf{F} \\
&\boldsymbol{\nabla} \times \boldsymbol{\nabla} f = 0 \\
&\boldsymbol{\nabla} \times (f\mathbf{F}) = \boldsymbol{\nabla} f \times \mathbf{F} + f\boldsymbol{\nabla} \times \mathbf{F}
\end{aligned} \tag{2.4.1}$$

A vector field is said to be **divergence free** or **solenoidal** if div $\mathbf{F} = 0$ throughout D. Notice that $\boldsymbol{\nabla} \cdot \boldsymbol{\nabla} \times \mathbf{G} = 0$ for any vector field, i.e., the curl of any vector field is solenoidal. We will see below, as an application of a first-order linear PDE, that any solenoidal field $\mathbf{F}$ is the curl of a field $\mathbf{G}$; the vector field $\mathbf{G}$ is called a **vector potential** of $\mathbf{F}$.

Theorem 2.13 Let $\mathbf{F} = (P, Q, R)$ be C^1 and nonzero on a region D. If $\mathbf{F}$ is solenoidal on D, then for any point $P_0 \in D$, there is a neighborhood $D' \subset D$ containing P_0 and a C^1 vector field G on D' such that $\mathbf{F} = \text{curl } \mathbf{G}$ on D'.

Proof The proof of this result is constructive and gives a method for finding the vector potential of a solenoidal field. Let $\xi = \xi(x, y, z)$ and $\eta = \eta(x, y, z)$ be two first integrals of the vector field $\mathbf{F}$ that satisfy the gradient condition (2.1.5) in some neighborhood $D'' \subset D$ of the point P_0. The vector field $\mathbf{F}$ is parallel to $\boldsymbol{\nabla}\xi \times \boldsymbol{\nabla}\eta$, hence there is a function $\lambda = \lambda(x, y, z)$ such that

$$\mathbf{F} = \lambda(\boldsymbol{\nabla}\xi \times \boldsymbol{\nabla}\eta).$$

Solving for λ gives

$$\lambda = \frac{\mathbf{F} \cdot (\boldsymbol{\nabla}\xi \times \boldsymbol{\nabla}\eta)}{|\boldsymbol{\nabla}\xi \times \boldsymbol{\nabla}\eta|^2}.$$

This shows that $\lambda \in C^1(D'')$ since the first integrals are C^1. Using the vector identities in (2.4.1) gives

$$0 = \nabla \cdot \mathbf{F} = \nabla\lambda \cdot (\nabla\xi \times \nabla\eta). \tag{2.4.2}$$

In other words, $\nabla\lambda$ is orthogonal to $\nabla\xi \times \nabla\eta$ and hence orthogonal to the vector field $\mathbf{F}$. Thus λ is a solution of the first-order linear PDE

$$Pu_x + Qu_y + Ru_z = 0$$

and is therefore a first integral of the vector field $\mathbf{F}$. We can now apply Theorem 2.4 and conclude that there exists a neighborhood $D' \subset D''$ and a C^1 function f of two variables such that

$$\lambda(x, y, z) = f(\xi(x, y, z), \eta(x, y, z))$$

on D'. Now, we define a function g such that

$$f(\xi, \eta) = g_\xi(\xi, \eta).$$

It now follows that

$$\mathbf{F} = \lambda(\nabla\xi \times \nabla\eta) = g_\xi \nabla\xi \times \nabla\eta = \nabla g \times \nabla\eta, \tag{2.4.3}$$

the last step since $\nabla g = g_\xi \nabla\xi + g_\eta \nabla\eta$. Applying the fourth identity in (2.4.1) to $g\nabla\eta$ we obtain $\mathbf{F} = \nabla \times (g\nabla\eta)$, and the result follows by choosing $\mathbf{G} = g\nabla\eta$.

□

Example 2.20 Consider the field $\mathbf{F} = (x, y, -2z)$, which is solenoidal everywhere; we assume $z > 0$. To determine first integrals, we consider the system of equations

$$\frac{dx}{x} = \frac{dy}{y} = \frac{dz}{-2z}.$$

The first pair of equations has solution $y/x = c_1$ and the second, $y^2 z = c_2$. Setting $\xi = y/x$ and $\eta = y^2 z$, these are two first integrals of the field and

$$\nabla\xi \times \nabla\eta = \left(\frac{y^2}{x}, \frac{y^3}{x^2}, -\frac{2y^2 z}{x^2}\right).$$

Notice that $\mathbf{F} = \frac{x^2}{y^2}\nabla\xi \times \nabla\eta$, i.e., $\lambda = \frac{x^2}{y^2} = \xi^{-2}$. Following the proof, we take $g(\xi, \eta) = -\xi^{-1}$ and the desired vector field is

$$\mathbf{G} = -\frac{x}{y}\nabla\eta = (0, -2xz, -xy).$$

The fact that $\nabla \times \mathbf{G} = \mathbf{F}$ can easily be checked.

Remark 2.14 A vector field $\mathbf{F}$ is said to be conservative if $\boldsymbol{\nabla} \times \mathbf{F} = \mathbf{0}$. As a consequence of Stokes' theorem in multivariate calculus it is known that conservative vector fields are gradient fields, i.e., $\mathbf{F} \in C^1(D)$ is conservative if and only if $\mathbf{F} = \boldsymbol{\nabla} f$ for some scalar function f defined on D. The function f is called the scalar potential of the vector field $\mathbf{F}$. This is a companion statement to Theorem 2.13.

Exercises

2.15 Verify the vector identities (2.4.1).

2.16 Verify the vector computations in the proof of Theorem 2.13. In particular, verify formulas (2.4.2) and (2.4.3).

2.17 Determine the vector potentials for the following vector fields.

a. $(x, -y, 0)$

b. $(1, 1, 1)$

c. $\left(\dfrac{x}{x^2+y^2}, -\dfrac{y}{x^2+y^2}, 0\right)$ in $\mathbb{R}^3$ with the z-axis removed.

Chapter 3

The Classical Trinity Revisited

Beginning with an overview of the divergence theorem and associated Green's identities, we examine the classical trinity of partial differential equations in the context of physics and higher dimensions and expand on the qualitative study of these equations introduced in the first chapter. Along the way we will define the Laplace operator and explore its fundamental symmetry: invariance under translations and rotations. Such invariance lends itself to geometric insight into solution methodologies for equations involving the Laplace operator; in particular, the classical trinity.

3.1 The Divergence Theorem and Its Application

The Fundamental Theorem of Calculus reads: Let f be a function defined on the interval $[a, b]$ with continuous derivative, then

$$\int_a^b f'(x)dx = f(b) - f(a).$$

From a mathematical perspective, it is completely natural to ponder the analog of this basic formula in higher dimensions. More to the point, the demands of analyzing physical phenomena such as fluid flow, diffusion, and vibrations led to the development of analogs in higher dimension. One analog in three dimensions is contained in the following formula concerning a vector field $\mathbf{F}$ on a bounded set D:

$$\int_D \nabla \cdot \mathbf{F}\, dV = \int_{\partial D} \mathbf{F} \cdot \mathbf{n}_o\, dS. \qquad (3.1.1)$$

Here, dV is the volume element in $\mathbb{R}^3$, $\mathbf{n}_0$ is the outward unit normal to the boundary of the region ∂D, and dS is the surface element on ∂D. The integral on the right-hand side of this formula is the **flux integral** of the vector field out of the region and is a gauge of the vector field leaving the region. For example, if $\mathbf{F}$ is the velocity field of a fluid and $\rho = \rho(x, y, z)$ is the fluid's mass density, then the integral

$$\int_{\partial D} \rho \mathbf{F} \cdot \mathbf{n}_o \, dS$$

has units of mass/time and measures the mass flow of the fluid out of the region. The integrals in formula (3.1.1) are actually multiple integrals; we make the convention in this text of expressing all integrals using a single integral symbol unless we need to put them in explicit computational iterated form.

We need to specify conditions on the types of regions D and regularity on the vector field $\mathbf{F}$ for the validity of formula (3.1.1). For the former, the following definition will suffice for our purposes.[1]

Definition 3.1 A **domain** in $\mathbb{R}^3$ (or $\mathbb{R}^2$) is a bounded, open, connected set D that satisfies the following two conditions:

1. The boundary of D is a finite union of smooth surfaces (or curves), i.e., $\partial D = \cup_{k=1}^{m} S_k$, where each S_k is a surface (or curve) given as the level surface (or curve) of a function in class C^1 with nonvanishing gradient.

2. Any straight line parallel to one of the coordinate axes either intersects the boundary in a finite number of points or the intersection contains a line segment.

While these conditions may seem technical, their purpose is to avoid eccentric behavior of the boundary. The first condition guarantees that each boundary component S_k has a continuously varying nonzero normal vector. Indeed, if S_k is defined as the level surface $f_k(x, y, z) = 0$, then ∇f_k give the normal vector at each point of the surface. The second condition prevents wild oscillation of the boundary surface. A simple example of a domain in $\mathbb{R}^3$ is a rectangular box with sides parallel to the coordinate planes. Other examples include open balls, the regions between concentric spheres, the region enclosed by a cylinder, and the region between concentric cylinders.

With these preparations, we now give a precise statement of the divergence theorem, a.k.a. Gauss's theorem.

---

1. We have chosen this definition for a domain for its intuitive appeal. More general formulations of the nature of sets for which the divergence theorem holds can be found in the classic work of Kellog [19].

Theorem 3.2 (The Divergence Theorem) Let $D \subset \mathbb{R}^3$ be a domain and let $\mathbf{F}$ be a vector field in $C^1(\bar{D})$. Then (3.1.1) holds.

An outline of the proof of this result is as follows. First, in the case where D is a rectangular box with sides parallel to the coordinate planes, the result follows by the Fundamental Theorem of Calculus. The second step is to show that the conclusion holds for domains that consist of a finite union of such rectangular boxes. Such a domain has boundary made up of rectangles each of which is parallel to some coordinate plane. Finally, the general case follows by a limiting argument; a general domain can be approximated in a suitable sense by a domain considered in the second step.

In two dimensions, the surface integral in the divergence theorem is a path integral so orientation must be taken into account. Here the result is known as Green's theorem and takes two forms. Let $\mathbf{T}$ denote the unit tangent vector to the boundary, $\text{curl}_s\mathbf{F} = \boldsymbol{\nabla} \times \mathbf{F} \cdot \mathbf{e}_3$ be the scalar curl of the vector field $\mathbf{F}$, dA be the area element in the plane, and ds be the arc element on ∂D.

Theorem 3.3 (Green's Theorem) Let $D \subset \mathbb{R}^2$ be a domain whose boundary is a simple closed curve and let $\mathbf{F} \in C^1(\bar{D})$. Then

$$\text{(i)} \int_D \text{curl}_s\mathbf{F}\, dA = \oint_{\partial D} \mathbf{F} \cdot \mathbf{T}\, ds, \tag{3.1.2}$$

$$\text{(ii)} \int_D \boldsymbol{\nabla} \cdot \mathbf{F}\, dA = \oint_{\partial D} \mathbf{F} \cdot \mathbf{n}_o\, ds. \tag{3.1.3}$$

The integral on the right-hand side of (i) is called the circulation integral; it is a gauge of the vector field around the curve C. Formula (i), sometimes called Stokes' theorem in the plane, is stated here because of its easily shown equivalence to (ii). Indeed, given a vector field $\mathbf{F} = (P, Q)$ in component form, then (ii) follows from (i) by applying the latter to the vector field $(-Q, P)$. There is a version of Stokes' theorem in $\mathbb{R}^3$; however we will not need it in this text.

Throughout the sequel a particular differential operator will play a fundamental role. This is the **Laplace operator**, or just simply the **Laplacian**, $\nabla^2 = \boldsymbol{\nabla} \cdot \boldsymbol{\nabla}$, which is the divergence of the gradient. As applied to a function u it is given by

$$\nabla^2 u = \boldsymbol{\nabla} \cdot \boldsymbol{\nabla} u = \partial_{11}u + \partial_{22}u + \partial_{33}u. \tag{3.1.4}$$

The following identities involving the Laplace operator are a consequence of the divergence theorem; they are known as Green's identities. In one dimension, the analog of the Laplace operator is the second-derivative operator, and the analog of Green's identities can be obtained using integration by parts.

Theorem 3.4 **(Green's Identities)** Let $D \subset \mathbb{R}^3$ be a domain and let u and $v \in C^2(\bar{D})$. Then

$$\text{(i)} \int_{\partial D} u\nabla v \cdot \mathbf{n}_o \, dS = \int_D (u\nabla^2 v + \nabla u \cdot \nabla v) dV. \tag{3.1.5}$$

$$\text{(ii)} \int_{\partial D} (u\nabla v - v\nabla u) \cdot \mathbf{n}_o \, dS = \int_D (u\nabla^2 v - v\nabla^2 u) dV. \tag{3.1.6}$$

The first identity is obtained by applying the divergence theorem to the vector field $\mathbf{F} = u\nabla v$. The second is obtained by interchanging the roles of u and v in the first identity and then subtracting the result from the first.

The divergence theorem and Green's identities will be fundamental tools in the qualitative and quantitative study of partial differential equations. Alternatively and from a physical perspective, concepts and laws are often expressed via volume and surface integrals and the divergence theorem provides a tool for combining the two types of integrals. To obtain a partial differential equation representing the physical law, we need a mechanism to localize integrals. This is provided by the following mean value theorem.

Theorem 3.5 Let $D \subset \mathbb{R}^3$ be bounded, open, and connected and let $f \in C(\bar{D})$. Then there is a point $p^0 \in D$ such that

$$\int_D f \, dV = f(p^0)\text{vol}(D).$$

The Classical Trinity in Higher Dimensions

The classical trinity of PDE make their appearance in higher dimensions in a variety of physical contexts. The following paragraphs exemplify this.

Couloumb's Law and Gauss's Law Couloumb's law states that the force (of attraction or repulsion) of a pair of charges in space varies directly as the product of the charges and inversely as the square of the distance between them. The electric field is a force field in space defined simply as the force per unit charge. Thus by Coulomb's law, the electric field $\mathbf{E}(x)$ of a point charge q placed at position ξ is given by

$$\mathbf{E} = q\frac{\boldsymbol{x} - \boldsymbol{\xi}}{|x - \xi|^3}.$$

As an application of the divergence theorem, if D is a domain containing ξ, then the flux of the electric field $\mathbf{E}$ through the boundary of D is equal to $4\pi q$, an observation due to Gauss. Extending this, if we let $\sigma = \sigma(x)$ be the charge density in a domain D and $\mathbf{E} = \mathbf{E}(x)$ be the electric field, then Gauss's law states that the flux of the electric field is a multiple of the total charge. Explicitly this becomes

$$\int_{\partial D} \mathbf{E} \cdot \mathbf{n}_o \, dS = 4\pi \int_D \sigma \, dV.$$

Applying the divergence theorem yields a formula involving only volume integrals, known as Gauss's Law:

$$\int_D \boldsymbol{\nabla} \cdot \mathbf{E} \, dV = 4\pi \int_D \sigma \, dV.$$

We apply this formula where D is a ball of radius $r > 0$ centered at some point p and rewrite the formula as

$$\int_D [\boldsymbol{\nabla} \cdot \mathbf{E} - 4\pi\sigma] dV = 0.$$

By the mean value theorem, there is a point p^0 in the ball such that

$$[\boldsymbol{\nabla} \cdot \mathbf{E}(p_o) - 4\pi\sigma(p^0)]\frac{4}{3}\pi r^3 = 0.$$

Dividing out by the volume and letting $r \to 0$ yields the vector partial differential equation

$$\boldsymbol{\nabla} \cdot \mathbf{E} = 4\pi\sigma,$$

an equation representing the infinitesimal version of Gauss's law. It turns out that the electric field possesses a scalar potential, i.e., $\mathbf{E} = -\boldsymbol{\nabla} u$, for some scalar function u. In this case, the preceding vector PDE becomes Poisson's equation for the potential,

$$\nabla^2 u = -4\pi\sigma.$$

The Heat Equation Let us now demonstrate that the temperature $u = u(x,t)$, $x = (x_1, x_2, x_3) \in D$, $t > 0$ in a domain D obeys the heat equation

$$u_t = \alpha^2 \nabla^2 u \tag{3.1.7}$$

under suitable physical assumptions. Imagine a ball $B(x) \subset D$ centered at x. Let $E(x,t)$ be the heat energy density measured in energy per unit volume (e.g., calories/cm^3). Then $\int_{B(x)} E \, dV$ represents the energy content of $B(x)$ at time t, and the rate of change of heat energy with respect to time in $B(x)$ would be given by $\int_{B(x)} E_t \, dV$. The heat flux is a vector quantity denoted $\boldsymbol{\phi}$ and represents the rate of heat flow per unit area per unit time. *The notion of heat flux as a vector quantity rests on the empirical observation that heat energy flows from high temperatures to low temperatures.* Denoting heat source terms within D by $Q(x,t)$, we now write the conservation of heat energy as

$$\int_{B(x)} E_t\, dV = -\int_{\partial B(x)} \boldsymbol{\phi} \cdot n_o\, dS + \int_{B(x)} Q\, dV. \tag{3.1.8}$$

Applying the divergence theorem to the second integral and rearranging terms yields

$$\int_{B(x)} [E_t + \nabla \cdot \boldsymbol{\phi} - Q] dV = 0.$$

By the mean value theorem and continuity of the various functions (an assumption), we obtain an infinitesimal version of (3.1.8):

$$\frac{\partial E}{\partial t} = -\nabla \cdot \boldsymbol{\phi} + Q.$$

The energy density is related to temperature $u(x,t)$ via $E(x,t) = c\rho u(x,t)$, where c is the heat capacity of the object and $\rho = \rho(x,u)$ is the mass density. The former is measured (defined) as the heat energy required to raise one mass unit one degree in temperature. In general, $c = c(x,u)$. Depending on the material substance, there are ranges of temperature where one can assume c to be independent of temperature; this assumption we make for both c and ρ. Fourier's law (an experimental fact) takes the form $\boldsymbol{\phi} = -k\nabla u$, where $k = k(x) > 0$ is the thermal conductivity of the material. Conceptually the heat flux gives the direction of heat flow, which explains the presence of the minus sign. Substituting these expressions into the infinitesimal heat conservation formula and assuming thermal and material properties are constant relative to temperature, we arrive at the **heat equation with source term,**

$$u_t = \alpha^2 \nabla^2 u + \frac{Q}{c\rho},$$

where $\alpha^2 = \frac{k}{c\rho}$. In the absence of source terms this becomes (3.1.7).

Remark 3.6 (1) An equation analogous to the heat equation is the **diffusion equation**, which models the diffusion of a substance in a fluid (see Exercise 3.4).

(2) If the material properties depend on position, i.e. $k = k(x)$, etc., then the heat equation takes the form

$$c\rho\, u_t = \nabla \cdot (k\nabla u).$$

The Vibrating Membrane The two-dimensional version of a vibrating string is an elastic, homogeneous, flexible drum head, i.e., a membrane stretched over a frame that is assumed to lie in the $x_1 x_2$−plane. Let $u = u(x, t)$ be the vertical displacement under vibration (we assume no horizontal displacement). A key assumption is that $|\boldsymbol{\nabla} u|$ is small, corresponding to the small-slope assumption used in deriving the wave equation in one dimension (vibrating string). Suppose the frame is given by a simple closed curve enclosing a domain D, as in Figure 3.1. Let $x \in D$ and take $B(x)$ to be a circular disk contained in D. Then in the vertical direction, Newton's law gives

$$\int_{B(x)} \rho u_{tt}\, dA = \oint_{\partial B(x)} T \boldsymbol{\nabla} u \cdot \mathbf{n}_o\, ds, \tag{3.1.9}$$

where ρ is the mass density of the membrane and T is the magnitude of the tension, both assumed constant. The argument now proceeds as in the case of the heat equation. Green's theorem applies to the right-hand side, and application of the mean value theorem yields the equation

$$u_{tt} = c^2 \nabla^2 u, \quad x \in D, \quad t > 0.$$

This is the wave equation in two dimensions. Given the setup of this problem, u also satisfies the Dirichlet boundary condition: $u(\cdot, t)|_{\partial D} = 0$.

Stretched Membrane in Equilibrium Instead of considering vibrations in a membrane, let's deform the boundary vertically to some shape defined by a space curve and ask what shape the membrane takes. Here we again assume that $|\boldsymbol{\nabla} u|$ is small, so that the magnitude of the tension can be assumed to be constant. In this case, equation (3.1.9) becomes

$$\oint_{\partial B(x)} T \boldsymbol{\nabla} u \cdot \mathbf{n}_o\, ds = 0.$$

Here the argument yields that the displacement u satisfies Laplace's equation

$$\nabla^2 u = 0, \quad x \in D.$$

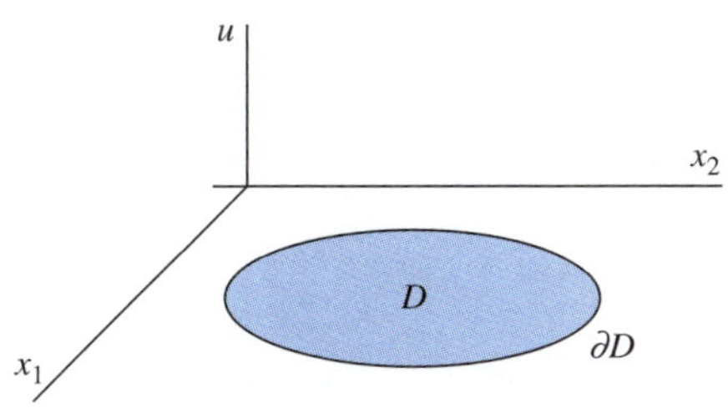

Figure 3.1 The Drum Head Frame

Further, there is a Dirichlet boundary condition $u|_{\partial D} = f$, where f is a function defined on ∂D representing the shape of the frame.

Sound Sound disturbances in the air as picked up by our ears are modeled by the equations of gas dynamics that form a system of nonlinear PDEs with velocity $\mathbf{V}$ and density ρ as the unknowns. Small sound disturbances are modeled well by linearization of these equations. The latter take the form

$$\frac{\partial \mathbf{V}}{\partial t} + \frac{c_0^2}{\rho_0}\boldsymbol{\nabla}\rho = 0$$
$$\frac{\partial \rho}{\partial t} + \rho_0 \boldsymbol{\nabla}\cdot\mathbf{V} = 0,$$

where ρ_0 and c_0 are the density and speed of sound in still air, respectively. Notice that this is a vector system of equations. In scalar form it represents four equations, one for each component of the velocity $\mathbf{V} = (v_1, v_2, v_3)$ and one for the density. We make the assumption that the curl of the velocity is zero, meaning that the sound waves do not circulate or form eddies. This implies that $\boldsymbol{\nabla}(\boldsymbol{\nabla}\cdot\mathbf{V}) = \nabla^2\mathbf{V}$ (see Exercise 3.5; note that $\nabla^2\mathbf{V}$ means componentwise application of the Laplace operator to the components of $\mathbf{V}$). Therefore, differentiating the first equation above with respect to t and substituting from the second equation gives

$$\frac{\partial^2 \mathbf{V}}{\partial t^2} = c_0^2\boldsymbol{\nabla}(\boldsymbol{\nabla}\cdot\mathbf{V}) = c_0^2\nabla^2\mathbf{V},$$

i.e., the velocity satisfies the wave equation. This is the vector form of the wave equation in three dimensions and means that each component of the velocity satisfies the same equation. In a similar fashion we can show that the density satisfies the wave equation

$$\frac{\partial^2 \rho}{\partial t^2} = c_0^2\nabla^2\rho. \tag{3.1.10}$$

Maxwell's Equations In the late 1800s J. C. Maxwell formulated a system of vector partial differential equations that encompass all known electromagnetic phenomena. The focus here is the connection between Maxwell's equations with the wave equation and Laplace's equation. We will consider the following forms of Maxwell's equations.

$\boldsymbol{\nabla}\cdot\mathbf{E} = 4\pi\sigma$	Coulomb's/Gauss's law
$c\boldsymbol{\nabla}\times\mathbf{E} + \frac{\partial \mathbf{B}}{\partial t} = 0$	Faraday's law
$c\boldsymbol{\nabla}\times\mathbf{B} - \frac{\partial \mathbf{E}}{\partial t} = 4\pi\mathbf{J}$	Ampére–Maxwell law
$\boldsymbol{\nabla}\cdot\mathbf{B} = 0$	Gauss's law

Here, $\mathbf{E}$ and $\mathbf{B}$ are the electric and magnetic fields, respectively; $\mathbf{J}$ is the current; and c is the speed of light. We have already discussed the first equation; it is a continuous version of the fact that a point charge produces an electric field inversely proportional to the distance from the charge (Coulomb's law). The fourth equation reflects the fact that there are no magnetic charges. Both equations are well supported by experiments.

The second equation states that a changing magnetic field produces an electric field—a fact extensively verified experimentally by Faraday. One implication of the third equation is that an electric current produces a magnetic field, a fact exploited in the design of electric motors. Originally, this law was expressed by Ampére in the form

$$c\nabla \times \mathbf{B} = 4\pi \mathbf{J},$$

valid when all quantities are independent of time. Indeed, since $\nabla \cdot \nabla \times \mathbf{B} = 0$ (the divergence of a curl is zero), this equation implies that $\nabla \cdot \mathbf{J} = 0$, which is not true in the time-dependent case. Maxwell introduced the appropriate modification of Ampére's law, completing the mathematical description of the electromagnetic fields.

Following are the implications/reductions that can be made to Maxwell's equations in special circumstances.

- **Conservation of Charge.** Applying the divergence operator to the Ampére–Maxwell equation we have $c\nabla \cdot \nabla \times \mathbf{B} - \nabla \cdot \frac{\partial \mathbf{E}}{\partial t} = 4\pi \nabla \cdot J$, which simplifies to $\frac{\partial}{\partial t}\nabla \cdot \mathbf{E} = -4\pi \nabla \cdot \mathbf{J}$. Applying Gauss's law we arrive at the **continuity equation**

  $$\frac{\partial \sigma}{\partial t} + \nabla \cdot \mathbf{J} = 0.$$

  Mathematically, this is a first-order partial differential equation that is a necessary condition for a solution of Maxwell's equations to exist. Physically, the continuity equation implies that charge cannot be created or destroyed. Indeed, integrating the continuity equation over a domain D and applying the divergence theorem, we obtain

  $$\frac{\partial}{\partial t}\int_D \sigma \, dV = -\int_{\partial D} \mathbf{J} \cdot \mathbf{n}_o \, dS.$$

  This equation implies that the rate of change of total charge within a given domain depends only on the presence of a current.

- **Absence of Charge: The Wave Equation.** The assumption here is that $\rho = 0$. Applying the curl operator to Faraday's law and a vector identity (see Exercise 3.7), we have

$$\begin{aligned} -\frac{\partial}{\partial t}\boldsymbol{\nabla} \times \mathbf{B} &= c\boldsymbol{\nabla} \times (\boldsymbol{\nabla} \times \mathbf{E}) \\ &= c\boldsymbol{\nabla}(\boldsymbol{\nabla} \cdot \mathbf{E}) - \nabla^2 \mathbf{E} \\ &= -\nabla^2 \mathbf{E}, \end{aligned}$$

where Gauss's law was used in the last step. Substituting from the Ampére–Maxwell law and rearranging gives

$$\frac{\partial^2 \mathbf{E}}{\partial t^2} = c^2 \nabla^2 \mathbf{E} - 4\pi \frac{\partial \mathbf{J}}{\partial t}.$$

This is a vector form of the wave equation.

- **Time Independence: Poisson's Equation.** Here it is assumed that charge and current are time independent. It follows that Maxwell's equations reduce to a decoupled system of PDEs given as

$$\begin{aligned} \boldsymbol{\nabla} \cdot \mathbf{E} &= 4\pi\sigma \\ \boldsymbol{\nabla} \times \mathbf{E} &= \mathbf{0} \\ c\boldsymbol{\nabla} \times \mathbf{B} &= 4\pi \mathbf{J} \\ \boldsymbol{\nabla} \cdot \mathbf{B} &= 0. \end{aligned}$$

The second and fourth equations imply that the electric and magnetic fields possess potentials (see Theorem 2.13 and the subsequent discussion), i.e., $\mathbf{E} = -\nabla u$ and $\mathbf{B} = \boldsymbol{\nabla} \times \boldsymbol{\Phi}$. As we have seen, the potential u satisfies Poisson's equation. It is left to the reader to show that

$$c\nabla^2 \boldsymbol{\Phi} = -4\pi \mathbf{J}, \tag{3.1.11}$$

a vector form of Poisson's equation.

Summary Table 3.1 gives a summary of the classical trinity with associated boundary and initial conditions. In this table the equations are written in homogeneous form. The boundary condition is written to encompass all cases; a, b, and g are functions possibly of $x \in \partial D$ and t. A Dirichlet condition arises when $a = 0$, Neumann condition when $b = 0$. The bulk of the rest of this text is focused on the study of qualitative and quantitative properties of these equations. Now would be a good time for the reader to review the basic questions of interest as outlined in Chapter 1.

Table 3.1 The Classical Trinity

	Heat/Diffusion	Wave	Laplace
PDE	$u_t = \alpha^2 \nabla^2 u$	$u_{tt} = c^2 \nabla^2 u$	$\nabla^2 u = 0$
Domain	$x \in D,\ t > 0$		$x \in D$
BC	$a\frac{du}{dn} + bu\big\rvert_{\partial D} = g$		
IC	$u(x,0) = f(x)$	$\begin{cases} u(x,0) = f(x) \\ u_t(x,0) = g(x) \end{cases}$	None

Classification in Higher Dimensions[2]

A general second-order linear partial differential operator can be written in the form

$$L = \sum_{i,j=1}^{d} a_{ij}\partial_{ij} + \sum_{i=1}^{d} b_i \partial_i + c, \tag{3.1.12}$$

where in general, the coefficients a_{ij}, b_i, and c are functions of the independent variables $x = (x_1, x_2, \ldots, x_d)$ and are assumed to be continuous functions on some domain in $D \subset \mathbb{R}^d$, called the domain of the operator. The principal symbol is given by

$$Q(x, \xi) = \sum_{i,j=1}^{d} a_{ij}\xi_i\xi_j$$

and has symmetric matrix $A = [a_{ij}]$. It is known from linear algebra that a symmetric matrix has d real eigenvalues (some may be repeated). The eigenvalues $\lambda_i = \lambda_i(x)$ are continuous functions on the domain of the operator. In light of the alternative formulation of classification for second-order PDEs in two variables, we make the following definition.

Definition 3.7 The operator L is said to be *elliptic, parabolic, hyperbolic,* or *ultra-hyperbolic* at a point x if, respectively, (i) all eigenvalues are nonzero and have the same sign, (ii) at least one eigenvalue is zero, (iii) all eigenvalues are nonzero and all but one have the same sign, (iv) all eigenvalues are nonzero and at least two of them are positive and at least two are negative. The operator is said to be elliptic, parabolic, hyperbolic, or ultra-hyperbolic in a domain if it is respectively, elliptic, parabolic, hyperbolic, or ultra-hyperbolic at every point in the domain. The partial differential equation $L(u) = f$ is said to be

2. This section requires knowledge of linear algebra.

elliptic, parabolic, hyperbolic, or ultra-hyperbolic at a point or in a domain if the differential operator L is of the same type at the point or domain.

Example 3.1 The Laplace operator ∇^2 is elliptic in $\mathbb{R}^3$. More generally, the operator

$$\partial_{11} + \partial_{22} + \cdots + \partial_{dd},$$

called the Laplace operator on $\mathbb{R}^d$, is elliptic. The wave operator

$$\partial_{11} - (\partial_{22} + \partial_{33} + \partial_{44})$$

is hyperbolic on $\mathbb{R}^4$. The eigenvalues of the coefficient matrix are $1, -1, -1, -1$. Note, x_1 stands for the time variable. The heat operator

$$\partial_1 - (\partial_{22} + \partial_{33} + \partial_{44})$$

is parabolic with corresponding eigenvalues $0, -1, -1, -1$. The simplest ultra-hyperbolic operator in dimension 4 has the form

$$\partial_{11} + \partial_{22} - \partial_{33} - \partial_{44}.$$

Here the eigenvalues of the coefficient matrix are $1, 1, -1, -1$.

Remark 3.8 In dimensions less than 4, the ultra-hyperbolic case is not possible. Furthermore, ultra-hyperbolic operators rarely appear in applications and therefore will not be further studied in this text.

Example 3.2 Consider the differential operator

$$2\partial_{11} + 2\partial_{22} - 15\partial_{33} + 8\partial_{12} - 12\partial_{23} - 12\partial_{13}.$$

The principal symbol is $Q(\xi) = 2\xi_1^2 + 2\xi_2^2 - 15\xi_3^2 + 8\xi_1\xi_2 - 12\xi_2\xi_3 - 12\xi_1\xi_3$ and the corresponding matrix is

$$A = \begin{bmatrix} 2 & 4 & -6 \\ 4 & 2 & -6 \\ -6 & -6 & -15 \end{bmatrix}.$$

The eigenvalues are computed from the equation $\det[A - \lambda I] = 0$. This cubic equation has roots -2, -18, and 9. Thus the differential operator is hyperbolic in $\mathbb{R}^3$.

Since the eigenvalues are continuous functions, if an operator is elliptic or hyperbolic at one point, then it will be of the same type in some open subset of the domain D. In the case of two dimensions, we saw that it is possible to introduce a change of variables to transform the equation and/or the operator into one of three canonical forms. However, in higher dimensions this is virtually impossible except in the case where the differential operator has constant coefficients where linear algebra is applicable. Indeed, a result from linear algebra states that there is a nonsingular matrix P such that $P^{-1}AP$ is a diagonal matrix whose diagonal entries are the eigenvalues λ_i of A. Introducing the change of variables $y = Px$ (x and y viewed as column vectors), the differential operator (3.1.12) goes over to one of the same form:

$$\sum_{i,j=1}^{d} A_{ij}\partial_{ij} + \sum_{i=1}^{d} B_i\partial_i + C,$$

where the differential operators ∂_i are now with respect to the y-variables and

$$A_{ij} = \begin{cases} 0, & \text{if } j \neq i \\ \lambda_i, & \text{if } j = i. \end{cases}$$

Hence, the second-order terms of the transformed operator have form

$$\sum_{i=1}^{d} \lambda_i \partial_{ii}.$$

Further, by rescaling the y-variables, we can further transform this to the case where the coefficients are $+1$, -1, or 0. The bottom line of this discussion is that in the case of dimension 3 or 4, any elliptic or hyperbolic partial differential operator can be transformed into the Laplace operator or wave operator, respectively, possibly with lower-order terms.

Exercises

3.1 Let $\mathbf{F} = (P, Q)$ be a C^1 vector field in a domain $D \subset \mathbb{R}^2$. Show that $\nabla \times \mathbf{F} = 0$ in D if and only if $\mathbf{F} = \nabla f$ for some scalar function on D. HINT: This is a consequence of Theorem 3.3(i). Vector fields whose curl is zero are called conservative and this result is typically found in any multivariate calculus text (e.g., [21]).

3.2 Write and prove the analogs of Green's identities in one dimension.

3.3 Write the analog of Green's identities in two dimensions.

3.4 Suppose a fluid (say, water) occupies a domain $D \subset \mathbb{R}^3$ and has velocity field $\mathbf{V} = \mathbf{V}(x,t)$. A substance (say, a dye) is suspended into the fluid and will be transported by the fluid as well as diffused within it; let $u = u(x,t)$ be the concentration of the substance (mass per unit volume). Let $\boldsymbol{\phi} = \boldsymbol{\phi}(x,t)$ be the concentration flux (mass per unit area per unit time, analogous to heat flux). Let $B(x)$ be a ball of radius $r > 0$ contained in D. Derive the conservation law

$$\int_{B(x)} u_t \, dV = -\int_{\partial B(x)} \boldsymbol{\phi} \cdot \mathbf{n}_o \, dS.$$

Fick's law for diffusion states that the concentration flux due to diffusion is proportional to the gradient of the concentration. Deduce that $\boldsymbol{\phi} = -k\boldsymbol{\nabla} u + u\mathbf{V}$. Apply the divergence theorem to the conservation law and substitute the flux formula to arrive at the **diffusion–transport equation**:

$$u_t = \boldsymbol{\nabla} \cdot (k\boldsymbol{\nabla} u) + \boldsymbol{\nabla} \cdot (u\mathbf{V}).$$

In the absence of diffusion ($k = 0$), this is the higher-dimensional transport equation. If the fluid is motionless, it is called the diffusion equation.

3.5 Verify that if $\boldsymbol{\nabla} \times \mathbf{V} = \mathbf{0}$, then $\boldsymbol{\nabla}(\boldsymbol{\nabla} \cdot \mathbf{V}) = \nabla^2 \mathbf{V}$.

3.6 Demonstrate formula (3.1.10).

3.7 Show that $\boldsymbol{\nabla} \times (\boldsymbol{\nabla} \times \mathbf{F}) = \boldsymbol{\nabla}(\boldsymbol{\nabla} \cdot \mathbf{F}) - \nabla^2 \mathbf{F}$.

3.8 Demonstrate formula (3.1.11).

3.9 (Requires linear algebra.) Show that the following equation is elliptic.

$$3\partial_{11}u - 2\partial_{12}u + 2\partial_{22}u - 2\partial_{23}u + 3\partial_{33}u + 12\partial_2 u - 8\partial_3 u = 0$$

Find a linear transformation to transform its principal part to $\partial'_{11}u + 3\partial'_{22} + 2\partial'_{33}$. Then find a scaling transformation that converts the principal part into the Laplacian.

3.2 The Amazing Laplace Operator

We have encountered the Laplace operator, also called the Laplacian, in the development of the classical trinity. Our purpose here is to point out fundamental invariance properties of this operator that are key to separation of variables in higher dimensions. We will also present formulas for the Laplacian in cylindrical and spherical coordinates.

Invariance Property

Let f be a function defined on a domain D. For a fixed $h \in \mathbb{R}^d$, we define the translated function f_h by

$$f_h(x) = f(x+h).$$

Note that h must be chosen so that $x+h$ remains in the domain of the function f. It is easy to see that

$$\nabla^2 f_h = \left(\nabla^2 f\right)_h. \tag{3.2.1}$$

In other words, we can compute the Laplacian of the shifted function f_h by first computing the Laplacian of f and then performing the shift. Another way to state this property comes from the fact that $\mathbb{R}^n$ is a group under addition, i.e. is closed under addition and each $h \in \mathbb{R}^n$ has an additive inverse, namely $-h$. Property (3.2.1) can be stated thus: *The Laplacian is invariant under the translation group.*

The same sort of property holds for the group of all rotations about the origin in $\mathbb{R}^d$, a group we now describe. A rotation on $\mathbb{R}^d$ is a linear transformation ($d \times d$ matrix) A such that for any $x, y \in \mathbb{R}^d$,

$$Ax \cdot Ay = x \cdot y.$$

Geometrically, rotations preserve length and the angle between vectors x and y. If A and B are rotations, then so is the composition or matrix product AB; for $d \geq 3$, however, this composition is not, in general, commutative: $AB \neq BA$. From properties of matrix algebra the above formula may be written

$$A^T Ax \cdot y = x \cdot y, \tag{3.2.2}$$

where A^T is the transpose of A. As a consequence, $A^T A = I = AA^T$, i.e., A^T is the inverse of A. Thus, the set of all rotations forms a group under composition of linear transformations, i.e., matrix multiplication. It also follows that $\det A = \pm 1$. We let $SO(d)$ be the set of all rotations whose determinant is $+1$; this forms a group under matrix multiplication. (Rotations with plus-one determinant preserve orientation.)

The elements of $SO(2)$ are easy to determine. They are simply 2×2 matrices of the form

$$A(\theta) = \begin{bmatrix} \cos\theta & -\sin\theta \\ \sin\theta & \cos\theta \end{bmatrix},$$

where $\theta \in \mathbb{R}$ gives the angle of rotation relative to the positive x_1-axis. Notice that $A(\theta)A(\theta') = A(\theta+\theta')$ and $A(\theta)^{-1} = A(-\theta)$. Furthermore, the action of $A(\theta)$ on the vector (x_1, x_2) is

$$\begin{bmatrix} \cos\theta & -\sin\theta \\ \sin\theta & \cos\theta \end{bmatrix} \begin{bmatrix} x_1 \\ x_2 \end{bmatrix} = \begin{bmatrix} x_1\cos\theta - x_2\sin\theta \\ x_1\sin\theta + x_2\cos\theta \end{bmatrix}.$$

With some effort, rotations in any number of dimensions can be expressed in terms of angles; however, we will not need the explicit formulas here. Rather, we write $A = [a_{ij}]$, and then by matrix multiplication the formula $AA^T = I$ in component form is

$$\sum_{k=1}^{d} a_{ik}a_{jk} = \delta_i^j,$$

where δ_i^j is the Kronecker delta: $\delta_i^j = 1$ if $j = i$ and is zero otherwise. Thus, rotations in d-dimensions are characterized as $d \times d$ matrices whose rows are unit vectors and, further, the rows are mutually perpendicular. (The same is true of the columns.)

Let $A \in SO(d)$ and for a function $f(x)$ define

$$f_A(x) = f(Ax).$$

Analogous to (3.2.1), the claim is

$$\nabla^2[f(Ax)] = \nabla^2 f(Ax). \tag{3.2.3}$$

To verify this invariance property, we set $x' = Ax$. We will now compute the Laplacian of f in terms of the primed variables. We write $\partial_i' = \frac{\partial}{\partial x_i'}$, then since

$$x_i' = \sum_{j=1}^{d} a_{ij}x_j,$$

the chain rule gives

$$\partial_i f(x') = \sum_{j=1}^{d} a_{ij}\partial_j' f(x').$$

Iterating this formula provides the second derivative:

$$\partial_{ii} f(x') = \sum_{j,k=1}^{d} a_{ij}a_{ki}\partial_{jk}' f(x').$$

Summing on i yields

$$\nabla^2 f(x') = \sum_{j,k=1}^{d}\left(\sum_{i=1}^{d} a_{ij}a_{ki}\right)\partial_{jk}' f(x') = \sum_{j=1}^{d} \partial_{jj}' f(x').$$

The right-hand side is the Laplacian of f in the primed variables, so this formula is the desired invariance property. We have carried out this computation using arbitrary dimension d; hence this works in both two and three dimensions (in fact any number).

The invariance under rotations also suggest that the Laplacian will have special form when expressed in terms of cylindrical or spherical coordinates. These facts are outlined in the following paragraphs.

Cylindrical Coordinates Cylindrical coordinates in $\mathbb{R}^3$ (polar coordinates in $\mathbb{R}^2$) are defined through the transformations

$$x_1 = r\cos\theta, \quad x_2 = r\sin\theta, \quad x_3 = x_3$$

or in the form

$$r^2 = x_1^2 + x_2^2, \quad \tan\theta = \frac{x_2}{x_1}, \quad x_3 = x_3.$$

The volume element is given by

$$dV = r\,dr\,d\theta\,dx_3.$$

Expressing the gradient in cylindrical form we have

$$\nabla = \left(\cos\theta\frac{\partial}{\partial r} - \frac{\sin\theta}{r}\frac{\partial}{\partial\theta}, \sin\theta\frac{\partial}{\partial r} - \frac{\cos\theta}{r}\frac{\partial}{\partial\theta}, \frac{\partial}{\partial x_3}\right). \tag{3.2.4}$$

Straightforward computation using the chain rule gives

$$\nabla^2 = \frac{\partial^2}{\partial r^2} + \frac{1}{r}\frac{\partial}{\partial r} + \frac{1}{r^2}\frac{\partial^2}{\partial\theta^2} + \frac{\partial^2}{\partial x_3^2}. \tag{3.2.5}$$

Spherical Coordinates Spherical coordinates are defined through the transformations

$$x_1 = \rho\cos\theta\sin\phi, \quad x_2 = \rho\sin\theta\sin\phi, \quad x_3 = \rho\cos\phi$$

or in the form

$$\rho^2 = x_1^2 + x_2^2 + x_3^2, \quad \tan\theta = \frac{x_2}{x_1}, \quad \cos\phi = \frac{x_3}{\rho}.$$

In this case the volume element has the form

$$dV = \rho^2\sin\phi\,d\rho\,d\phi\,d\theta.$$

Note the conversion equations between spherical and cylindrical coordinates:

$$r = \rho\sin\phi, \quad \theta = \theta, \quad x_3 = \rho\cos\phi$$

$$\rho^2 = r^2 + x_3^2, \quad \theta = \theta, \quad \tan\phi = \frac{r}{x_3}.$$

Using these and (3.2.4) or (3.2.5) we arrive at the desired expression of the Laplace operator in spherical coordinates:

$$\nabla^2 = \frac{\partial^2}{\partial \rho^2} + \frac{2}{\rho}\frac{\partial}{\partial \rho} + \frac{1}{\rho^2}\nabla^2_{\text{sp}},$$

where

$$\nabla^2_{\text{sp}} = \frac{\partial^2}{\partial \phi^2} + \frac{\cos\phi}{\sin\phi}\frac{\partial}{\partial \phi} + \frac{1}{\sin^2\phi}\frac{\partial^2}{\partial \theta^2}. \tag{3.2.6}$$

The two invariance properties may seem rather abstract. However, such properties underlie the method of separation of variables when using cylindrical and spherical coordinates, a basic tool in solving PDEs. Further, translating or rotating a coordinate system often will simplify computations involving the Laplacian or functions related to it. The invariance properties are the reason for the simplification. The following two examples illustrate the role of the invariance property.

Example 3.3 (Laplace's Equation in an Annulus) This example illustrates the utility of cylindrical and/or polar coordinates in domains with axial symmetry. Consider the annular domain $D = \{(x_1, x_2) \,|\, a^2 < x_1^2 + x_2^2 < b^2\}$, $0 < a < b$, and Laplace's equation in D with boundary conditions $u = A$ on the inner circle and $u = B$ on the outer boundary. Switching to polar coordinates, $u = u(r, \theta)$, the boundary conditions can be written $u(a, \theta) = A$ and $u(b, \theta) = B$. Given the rotation invariance of the boundary conditions and the rotation invariance of the Laplacian, common sense tells us that $u(r, \theta) = u(r)$ is independent of the angular variable θ. (In Chapter 7 we will see that this BVP has a unique solution, so if we find a radial solution, our commonsense approach is justified.) By (3.2.5), the BVP can be explicitly formulated as

$$\text{PDE: } u'' + \frac{1}{r}u' = 0, \quad a < r < b$$

$$\text{BC: } \begin{cases} u(a) = A \\ u(b) = B, \end{cases}$$

which is a BVP for an ODE. The general solution to the differential equation is $u(r) = c_1 + c_2 \log r$. Applying the boundary conditions yields two linear equations in two unknowns, specifically,

$$c_1 + c_2 \log a = A \quad \text{and} \quad c_1 + c_2 \log b = B.$$

Solving these provides the formula for the solution to the BVP:

$$u(r) = \frac{A \log b - B \log a + (B - A) \log r}{\log \frac{b}{a}}.$$

Remark 3.9 Two remarks regarding this example are worth mentioning:

1. The role of the invariance property of the Laplacian under rotations is clear; it provides the commonsense approach for transforming a boundary value problem for a PDE into one for an ODE.

2. This BVP can be interpreted as providing the steady-state temperature distribution in the annular region. Notice that the solution is a linear function of $\log r$. In one dimension, steady-state temperature distributions are linear functions of the spatial variable.

Example 3.4 In this example, we construct solutions to Laplace's equation in $\mathbb{R}^2$ by exploiting the invariance under rotations in a broader context. As before, we write $u(x) = u(r, \theta)$. Under a rotation $A = A(\theta')$, we have $u_A(r, \theta) = u(r, \theta + \theta')$. Now let's consider the one-dimensional vector space V_α of all functions defined on $\mathbb{R}^2$ of the form

$$u(r, \theta) = g(r)e^{\alpha\theta}.$$

For such a function, $u_A(r, \theta) = e^{\alpha\theta'}u(r, \theta)$. In other words, the vector space V_α is invariant under rotations. For such functions to be continuous, it is necessary that $u(r, -\pi) = u(r, \pi)$. This in turn forces $\alpha = in$, for some integer n. (Why?) The action of the Laplacian on V_{in} preserves V_{in}. Indeed, using (3.2.5) we have

$$\nabla^2 u(r, \theta) = \left(g'' + \frac{1}{r}g' - \frac{n^2}{r^2}g\right) e^{in\theta} \in V_{in}.$$

An alternative view of this formula is: The action of the Laplacian on V_{in} reduces to an ordinary differential operator. To find the functions in V_{in} that satisfy Laplace's equation, it suffices to solve the ODE

$$g'' + \frac{1}{r}g' - \frac{n^2}{r^2}g = 0.$$

This is an example of a Euler equation; the linearly independent solutions are $r^{|n|}$ and $r^{-|n|}$ if $n \neq 0$ and the constant function 1 and $\log r$ if $n = 0$. We now claim that the family of functions

$$u(r, \theta) = \begin{cases} r^{|n|}e^{in\theta}, & n \in \mathbb{Z} \\ \log r,\ r^{-|n|}e^{in\theta}, & n \in \mathbb{Z},\ n \neq 0 \end{cases}$$

are strong solutions to Laplace's equation, the first row in the entire plane and the second in the plane with origin removed. What needs to be checked is that these functions are in fact C^2. The point here is that functions expressed in polar coordinates (cylindrical and spherical also in three dimensions) may be smooth in the (r, θ) variables, yet may not be when written as functions of the Cartesian variables x_1 and x_2. Indeed, the function $v(r, \theta) = r$ is a smooth function of the polar variables, however, as a function of $x = (x_1, x_2)$, $v(x) = \sqrt{x_1^2 + x_2^2}$, which is not differentiable at the origin. For the functions $u(r, \theta)$ constructed above, using Euler's identity we obtain for the first row

$$u(r, \theta) = \begin{cases} (x_1 + ix_2)^n, & n \geq 0 \\ (x_1 - ix_2)^{|n|}, & n < 0, \end{cases}$$

which are polynomials and hence continuously differentiable. For the second row we have

$$u(r, \theta) = \begin{cases} \log \sqrt{x_1^2 + x_2^2}, & \\ \frac{(x_1 + ix_2)^n}{(x_1^2 + x_2^2)^n}, & n \geq 1 \\ \frac{(x_1 - ix_2)^{|n|}}{(x_1^2 + x_2^2)^{|n|}}, & n \leq -1, \end{cases}$$

which are all continuously differentiable away from the origin.

Remark 3.10 The solutions found in Example 3.14 are written as complex-valued functions. In some applications real-valued functions are more useful. A set of real-valued solutions to Laplace's equation can be obtained from these by using Euler's formula ($e^{in\theta} = \cos n\theta + i \sin n\theta$) and the fact that linear combinations of solutions are again solutions (the superposition principle). Applying this, the resulting real-valued solutions to Laplace's equation take the form

$$u_n(r, \theta) = \begin{cases} r^n \cos n\theta,\ r^n \sin n\theta, & n = 0, 1, \ldots \\ \log r,\ r^{-n} \cos n\theta,\ r^{-n} \sin n\theta, & n = 1, 2, \ldots. \end{cases} \tag{3.2.7}$$

Remark 3.11 The solutions to Laplace's equation found in Example 3.14 are called **separated solutions**, i.e., solutions of the form of a product of functions of the independent variables r and θ: $u(r, \theta) = R(r)\Theta(\theta)$. They were constructed based on the invariance of the Laplacian under rotations. This geometric view of separated solutions of PDEs underlies the method of separation of variables, a method discussed further in Chapter 5 and used throughout the rest of the text.

Exercises

3.10 Use the form of the gradient in cylindrical coordinates (3.2.4) to show (3.2.5).

3.11 Use the conversion equations between cylindrical and spherical coordinates to verify (3.2.6).

3.12 In Example 3.3, find the solution to the boundary value problem for Laplace's equation where the boundary condition on the inner circle is replaced with $u_r(a) = A$. Interpret physically.

3.13 For a function of form $g(r)e^{\alpha\theta}$ to be continuous in the plane, explain why it is necessary that $\alpha = n \in \mathbb{Z}$.

3.14 Determine all solutions of Laplace's equation in the plane of the form $u(r, \theta) = R(r)\Theta(\theta)$ using the following steps.

a. Substitute u of this form into Laplace's equation and rewrite to obtain

$$\frac{r^2 R'' + rR'}{r^2} = -\frac{\Theta''}{\Theta},$$

where the primes on the left indicate differentiation with respect to r, and those on the right with respect to θ.

b. Argue that the above ratios must be constant, call it μ. This provides a pair of ODEs:

$$r^2 R'' + rR' - \mu r^2 R = 0 \quad \text{and} \quad \Theta'' + \mu\Theta = 0.$$

Note that the variable $\theta \in [-\pi, \pi]$ and $r \geq 0$.

c. Explain why $\Theta(\theta)$ must satisfy the conditions $\Theta(-\pi) = \Theta(\pi)$ and $\Theta'(-\pi) = \Theta'(\pi)$.

d. Determine all solutions of the equation for Θ. You need to take cases $\mu < 0$, $\mu = 0$, and $\mu > 0$.

e. Show that the equation for Θ has nontrivial solutions that also satisfy the conditions in part (c) if and only if $\mu = m^2$, where $m \in \mathbb{Z}$. For each integer m, these solutions can be written as $\Theta_m(\theta) = e^{im\theta}$ (or any multiple thereof).

f. For the values of μ determined in part (e), solve the equation for R.

g. The final answer to this problem is: For each $m \in \mathbb{Z}$, separated solutions of Laplace's equation have form $u_m(r,\theta) = r^{|m|}e^{im\theta}$.

3.15 Determine all separated solutions of Laplace's equation in the plane that are bounded outside the unit circle.

3.16 Determine the ODE that the function $g(r)$ must satisfy for the function $g(r)e^{in\theta} \in V_{in}$ to be a solution of the PDE

$$\nabla^2 u + \lambda u = 0$$

in the plane. Here λ is a constant.

3.17 Determine the steady-state temperature distribution for the domain between two concentric spheres of radii $0 < a < b$ where the temperature on the inner spherical boundary is A and the temperature on the outer spherical boundary is B.

3.18 Which of the following functions of (r, θ) defines a smooth function in the plane? Explain.

a. $r^4 \cos 2\theta$

b. $r^2 \cos 2\theta$

c. r^n, $n \geq 0$

d. $r^n \cos m\theta$, where $n, m = 1, 2, \ldots$

3.19 The operator ∇^2_{sp} acts on functions $f(\omega)$ defined on the unit sphere S^2 in $\mathbb{R}^3$. It is easy to see that this operator is invariant under rotations, i.e., $\nabla_{\text{sp}} f(A\omega) = (\nabla_{\text{sp}} f)(A\omega)$. Parametrize functions on the sphere in terms of the angles (ϕ, θ); consider the vector space V_n of all functions of the form $g(\phi)e^{in\theta}$. Then V_n is invariant under all rotations that leave the x_3-axis fixed; this is a subgroup of $SO(3)$ and can be identified with $SO(2)$. Determine a differential equation that g must satisfy such that a function in V_n will satisfy the equation $\nabla_{\text{sp}} u + \lambda u = 0$, where λ is a constant.

3.3 Qualitative Methods for Diffusion and Waves

The focus here is to introduce qualitative methods for the heat and wave equation. Uniqueness of solutions for a variety of boundary value problems are an immediate consequence.

One of the main themes of this section is the extension of the energy method introduced in Chapter 1. The extension is carried out as an application of

Green's first identity. Generally speaking, the energy method consists of multiplying the equation by an appropriate quantity, usually u or u_t, integrating over the domain, applying Green's identity, and interpreting certain resulting quantities in terms of physical energy. In effect, we will be relating the total energy of the solution to the boundary data.

Maximum Principle and Energy for the Heat Equation

The Maximum Principle In loose terms, the maximum principle[3] states that solutions of the heat equation achieve their maximum (and minimum) values on the space–time boundary. A heuristic argument for this was given on page 21 in Chapter 1. The precise statement of the principle is as follows.

Theorem 3.12 (Maximum Principle for the Heat Equation) Let $D \subset \mathbb{R}^d$ be a bounded domain. Let $u = u(x,t)$ be a C^2 solution of the homogeneous heat equation $u_t = \alpha^2 \nabla^2 u$ in $D \times (0,T)$ that is continuous on $\overline{D} \times [0,T]$ for some $T > 0$. Then the maximum of u on $\overline{D} \times [0,T]$ is achieved in the cylinder

$$\Gamma = \{(x,t) \,:\, x \in \partial D,\, 0 \le t \le T\} \cup \{(x,0) \,:\, x \in \overline{D}\}.$$

We may attempt to prove this result as follows. Suppose the maximum of $u(x,t)$ occurs in the set $D \times (0,T)$. Then from calculus we know that all first partials of u will be zero at this point and further, all second partials $\partial_{ii} u \le 0$ at this point. If, in fact, strict inequality holds for all i, then clearly $u_t \ne \alpha^2 \nabla^2 u$ at this point. This contradiction would then force the maximum to occur on the boundary set Γ. The argument breaks down if $\partial_{ii} u = 0$ at this point for some i and careful modification must be employed.

Proof (of the Maximum Principle). Let $R = \max_{x \in \overline{D}} |x|$ (the maximum of the distance from points in $\overline{D}$ to the origin) and let

$$M = \max_{(x,t) \in \Gamma} u(x,t).$$

The latter exists as u is continuous. Let $\varepsilon > 0$ and introduce the auxiliary function $v(x,t) = u(x,t) + \varepsilon |x|^2$. Observe that $|v(x,t)| \le M + \varepsilon R^2$ on Γ. On the set $D \times (0,T)$, the function v satisfies the inequality

$$v_t - \alpha^2 \nabla^2 v = u_t - \alpha^2 (\nabla^2 u + 2\varepsilon) = -2\alpha^2 \varepsilon < 0. \tag{3.3.1}$$

If $v(x,t)$ has a maximum at $(x^0, t^0) \in D \times (0,T)$, then $v_t(x^0,t^0) = 0$ and $\nabla^2 v(x^0,t^0) \le 0$ (all second partial derivatives of v at this point would be

3. Maximum principles have a long history, not only for the heat equation, but also parabolic and elliptic partial differential equations (we will see a version for Laplace's equation in Chapter 7), and various classes of ordinary differential equations; a classic reference is [26].

less than or equal 0 due to concavity). However, this contradicts the above inequality. If $v(x,t)$ has a maximum at (x^0,T) where $x^0 \in D$, then $\partial_i v(x^0,T) = 0$, $i = 1,\dots d$, and $\nabla^2 v(x^0,T) \le 0$. Furthermore,

$$v_t(x^0,T) = \lim_{h\to 0+} \frac{v(x^0,T) - v(x^0,T-h)}{h} \ge 0.$$

This again contradicts the inequality (3.3.1). Since v has a maximum on $\overline{D} \times [0,T]$, it now follows that it must occur in the set Γ. In other words, $v(x,t) \le M + \varepsilon R^2$ throughout $\overline{D} \times [0,T]$. Going back to the relation between u and v, we conclude that

$$u(x,t) = v(x,t) - \varepsilon|x|^2 \le M + \varepsilon(R^2 - |x|^2).$$

The desired conclusion follows letting $\varepsilon \to 0$. □

Remark 3.13 A few remarks regarding the method of auxiliary functions and the result are in order.

1. The choice of the auxiliary function used in the preceding argument is not unique. For example, the argument can be modified when using the function $v(x,t) = u(x,t) - \varepsilon t$, where $\varepsilon > 0$.
2. Theorem 3.12 is actually known as the weak maximum principle. There is a strong maximum principle with the same hypothesis and conclusion that the maximum of u is achieved only in the set Γ or the solution u must be constant. The proof of the strong version also uses the method of auxiliary functions and is considerably harder than the proof given here (see [22]).

Corollary 3.14 (Minimum Principle) Let D, u, and Γ be as in Theorem 3.12. Then the minimum of u is achieved in the cylinder Γ.

This result is immediate by applying the maximum principle to the function $-u$.

Taken together, the maximum and minimum principles easily imply a uniqueness result for the Dirichlet IBVP for the heat equation as well as a stability result relative to the initial data.

Proposition 3.15 (Uniqueness Theorem) Let D be a domain, let $f(x)$ and $F(x)$ be functions defined on D, and let $g(t)$ be a function defined on $(0,\infty)$. Then there is at most one solution $u = u(x,t)$ of the nonhomogeneous IBVP

$$\begin{aligned} \text{PDE: } & u_t = \alpha^2\nabla^2 u + F(x), \quad x \in D, \quad t > 0 \\ \text{BC: } & u(x,t) = g(t), \quad x \in \partial D \\ \text{IC: } & u(x,0) = f(x) \end{aligned} \tag{3.3.2}$$

that is continuous on $\overline{D} \times [0,\infty)$.

Proof Suppose $u_1(x,t)$ and $u_2(x,t)$ are solutions to this IBVP. Then the function $u = u_1 - u_2$ is a solution to the homogeneous heat equation that satisfies a homogeneous Dirichlet boundary condition and initial condition $u(x,0) = 0$. For any $T > 0$, the maximum and minimum of $u(x,t)$ is achieved on the cylinder Γ. Since the maximum and minimum value is zero, it follows that $u(x,t) = 0$ on $D \times (0,\infty)$, i.e., $u_1 = u_2$. □

Proposition 3.16 (Stability Theorem) Let D be a domain, let f_1 and f_2 be continuous functions on $\overline{D}$, and let u_1 and u_2 be solutions of IBVP (3.3.2) corresponding to initial data $f = f_1(x)$ and $f = f_2(x)$, respectively. Assume that both u_1 and u_2 are continuous on $\overline{D} \times [0,\infty)$. Then

$$|u_1(x,t) - u_2(x,t)| \le \max_{y\in D} |f_1(y) - f_2(y)|.$$

Proof Let $v(x,t) = u_1(x,t) - u_2(x,t)$. Then v satisfies the homogeneous heat equation with homogeneous Dirichlet boundary condition, and initial condition $v(x,0) = f_1(x) - f_2(x)$. Further, v satisfies the hypothesis of the maximum principle for all $T > 0$. By the maximum and minimum principles,

$$\min\{0, \min_{y\in D}[f_1(y) - f_2(y)]\} \le v(x,t) \le \max\{0, \max_{y\in D}[f_1(y) - f_2(y)]\}.$$

For any function g, $\min g(y) \ge -\max(-|g(y)|)$, it follows that

$$-\max_{y\in D} |f_1(y) - f_2(y)| \le v(x,t) \le \max_{y\in D} |f_1(y) - f_2(y)|$$

for all $t > 0$. Substituting for $v(x,t)$ we can write this as

$$|u_1(x,t) - u_2(x,t)| \le \max_{y\in D} |f_1(y) - f_2(y)|,$$

an inequality that holds for all $x \in D$ and $t > 0$. □

A third application of the maximum principle is a variation for the Cauchy problem for the heat equation on $\mathbb{R}^d$:

$$\begin{aligned} \text{PDE: } & u_t = \alpha^2 \nabla^2 u, \quad x \in \mathbb{R}^d, \quad t > 0 \\ \text{IC: } & u(x,0) = f(x), \quad x \in \mathbb{R}^d. \end{aligned} \tag{3.3.3}$$

Proposition 3.17 Let $u = u(x,t)$ be a solution to IVP (3.3.3) that is continuous on $\mathbb{R}^d \times [0,T]$ for some $T > 0$ and satisfies $u(x,t) \le M$ there. If $f(x) \le M_0$ for all $x \in \mathbb{R}^d$, then $u(x,t) \le M_0$ on $\mathbb{R}^d \times [0,T]$.

Proof There is nothing to prove if $M \le M_0$, so assume $M > M_0$. To demonstrate this, consider the auxiliary function $v(x,t) = u(x,t) + \varepsilon(2\alpha^2 t + |x|^2)$, where $\varepsilon > 0$. It is easy to see that v is a solution of the heat equation. Let $R > 0$ (to be chosen shortly) and let $Q_{R,T} = B_R(O) \times (0, T)$. Note that $Q_{R,T}$ is the interior of a cylinder. Part of the boundary of this cylinder is denoted

$$\Gamma_{R,T} = \{(x,t) \,:\, |x| = R,\, 0 \le t \le T\} \cup \{(x,0) \,:\, |x| < R\}.$$

By the maximum principle, we have

$$\max_{Q_{R,T}} v(x,t) \le \max_{\Gamma_{R,T}} v(x,t) = \max_{\Gamma_{R,T}} [u(x,t) - \varepsilon(2\alpha^2 t - |x|^2)].$$

Looking on the boundary, the latter is less than or equal to the larger of the two numbers $M - \varepsilon R^2$ and M_0. At this point we choose $R = R(\varepsilon)$ so that $M - \varepsilon R^2 = M_0$. It follows that $\max_{Q_{R,T}} v(x,t) \le M_0$. Notice that as $\varepsilon \to 0$, $R(\varepsilon) \to \infty$. Since $\lim_{\varepsilon \to 0} v(x,t) = u(x,t)$, we arrive at the conclusion that $u(x,t) \le M_0$, as desired. □

A minimum principle can also be formulated for the Cauchy problem on $\mathbb{R}^d$. Combined, they imply uniqueness of solutions; details are left to the interested reader.

Energy Estimates for the Heat Equation An energy identity for the heat equation in higher dimensions is obtained by multiplying the equation by u, integrating over the spatial domain D, and applying Green's first identity. The result of this computation is as follows:

$$\begin{aligned}\int_D u u_t \, dV &= \alpha^2 \int_D u \nabla^2 u \, dV \\ &= \alpha^2 \int_{\partial D} u \boldsymbol{\nabla} u \cdot \mathbf{n}_o \, dS - \alpha^2 \int_D |\boldsymbol{\nabla} u|^2 \, dV.\end{aligned}$$

Now, the term on the left can be written $\frac{d}{dt}\left[\frac{1}{2}\int_D u^2 \, dV\right]$. As in Chapter 1, the term in brackets is denoted $E(t)$ and serves as a gauge on the energy. (Recall that the heat energy per unit volume is given by the expression $c\rho u$, where c is the heat capacity and ρ is the material density. Hence the quantity $c\rho E^{1/2}$ would carry the units of energy. However, since c and ρ are constants, we simply use E as the gauge on energy.) The identity can now be written as

$$\frac{dE}{dt} = \alpha^2 \int_{\partial D} u \boldsymbol{\nabla} u \cdot \mathbf{n}_o \, dS - \alpha^2 \int_D |\boldsymbol{\nabla} u|^2 \, dV. \tag{3.3.4}$$

In the case of a homogeneous Dirichlet or Neumann boundary condition, the boundary term vanishes, leading to the conclusion that $\frac{dE}{dt} \le 0$. The same conclusion can be reached in the case of a Robin boundary condition:

$$\left[\frac{du}{dn} + au\right|_{\partial D} = 0, \tag{3.3.5}$$

in the case where $a = a(x)$ is a nonnegative function on the boundary of D. Thus, under any of the three types of homogeneous boundary conditions, $E(t)$ is a nonincreasing function. In particular, we can conclude that $E(t) \le E(0) = \frac{1}{2}\int_D f^2\,dV$, where f is the initial data for the problem. Our observations are summarized as follows.

Proposition 3.18 Let $u = u(x,t)$ be a solution to the homogeneous heat equation on $D\times(0,\infty)$ that is continuous on $\overline{D}\times[0,\infty)$. Assume that u satisfies a homogeneous Dirichlet, Neumann, Robin boundary condition, or mixture of the three (for the Robin condition we assume $a \ge 0$ as above). Then the energy gauge $E(t)$ is a nonincreasing function and satisfies $E(t) \le E(0)$. Moreover, the IBVP (3.3.2), where the boundary condition can be any of the four types, has at most one solution.

Uniqueness of solutions follows from the method we have used in previous discussions and is left to the reader. The regularity assumptions placed on u is this result were made so as to apply Green's identity; they can be relaxed using more advanced techniques.

Refined Energy Estimates In the case of a homogeneous Dirichlet boundary condition, the zero function is the steady-state solution of the heat equation (independent of initial condition), and hence $E(t) \to 0$ as $t \to \infty$. It is natural to ponder how fast $E(t)$ tends to zero in this case. The following beautiful inequality provides an approach to this question. For simplicity, we assume that the domain D is convex, i.e., if x and y are in D, then the line segment between them is wholly contained in D. Throughout the following discussion we assume that $d = 2$ or 3; the one-dimensional case is derived in the exercises.

Lemma 3.19 (Poincaré Inequality: Version I) Let D be a bounded convex domain and let $\delta = \max_{x,y\in D}|x-y|$ be its diameter. Let $F(x)$ be a C^1 function on $\overline{D}$ with $F(x) = 0$ on ∂D. Then

$$\int_D F^2\,dV \le \delta^2 \int_D |\nabla F|^2\,dV. \tag{3.3.6}$$

Proof Let $x_0 \in \partial D$ be fixed and take $x \in D$. Let γ be the straight line segment from x_0 to x, i.e. $\gamma(t) = (1-t)x_0 + t\,x$. Note that the length of γ is $|x - x_0| \le \delta$. For convenience we define $F(x)$ on all of $\mathbb{R}^d$ by setting $F(x) = 0$ for x in the complement of $\overline{D}$; note that F is now C^1 on $\mathbb{R}^d$. Denoting the arc element for a path by ds (which is dt for the line segment), it follows that

$$F(x) = \int_\gamma \nabla F\cdot\mathbf{T}\,ds,$$

where $\mathbf{T}$ is the unit tangent vector for the path. Taking the absolute value of both sides we have $|F(x)| \le \int_\gamma |\nabla F|\,ds$. Applying the Cauchy–Schwarz inequality[4] we obtain

$$|F(x)| \leq \left[\int_\gamma |\nabla F|^2\, ds\right]^{1/2} \left[\int_\gamma ds\right]^{1/2} \leq \delta^{1/2}\left[\int_\gamma |\nabla F|^2\, ds\right]^{1/2}.$$

Squaring both sides and integrating over $\mathbb{R}^d$, we get

$$\int_D F^2(x)dV \leq \delta \int_{\mathbb{R}^d}\int_\gamma |\nabla F|^2\, ds\, dV.$$

Now, $\int_\gamma |\nabla F|^2\, ds = \int_0^1 |\nabla F((1-t)x_0 + t\,x)|^2 |x-x_0|\, dt$. Substituting this on the right-hand side of the inequality and interchanging orders of integration yields

$$\begin{aligned}\int_D F^2(x)dV &\leq \delta \int_0^1 \int_{\mathbb{R}^d} |\nabla F((1-t)x_0 + t\,x)|^2\, dV(x)|x-x_0|dt \\ &\leq \delta^2 \int_0^1 \int_{\mathbb{R}^d} |\nabla F(x)|^2 t^d\, dV(x)dt \leq \delta^2 \int_D |\nabla F(x)|^2\, dV(x),\end{aligned}$$

where a change of variable was made going from the first line to the second. □

Returning to the heat equation, assume that $u = u(x,t)$ is a solution of the homogeneous heat equation and satisfies a homogeneous Dirichlet boundary condition, then (3.3.4) can be written

$$\frac{dE}{dt} \leq -\alpha^2 \int_D |\nabla u|^2\, dV. \tag{3.3.7}$$

Apply the Poincaré inequality to the right-hand side to obtain

$$\frac{dE}{dt} \leq -\frac{\alpha^2}{\delta^2}\int_D u^2(x,t)dV(x) = -\frac{2\alpha^2}{\delta^2}E.$$

This is a first-order differential inequality and can be handled just like first-order linear ordinary differential equations. Indeed, writing the inequality in the form $E' + \frac{2\alpha^2}{\delta^2}E \leq 0$ and multiplying both sides by $e^{\frac{2\alpha^2}{\delta^2}t}$ we obtain $\left[e^{\frac{2\alpha^2}{\delta^2}t}E\right]' \leq 0$. Upon integration we get $E(t) \leq E_0 e^{-\frac{2\alpha^2}{\delta^2}t}$, where E_0 is a constant. It should be clear that E_0 is given by the initial value of the energy $E(0)$. The following result sumarizes these observations.

4. The Cauchy–Schwarz inequality states that

$$\int |f(x)g(x)|dx \leq \left(\int f^2(x)dx\right)^{1/2}\left(\int g^2(x)dx\right)^{1/2},$$

where the integrals are over an interval, a path, or a domain (in higher dimensions). This inequality is more fully discussed in Chapter 5.

Proposition 3.20 Let D be a convex bounded domain and let $u = u(x,t)$ be a solution of the heat equation that is C^1 on $\overline{D} \times [0,\infty)$. Then the energy gauge satisfies the estimate $E(t) \leq E(0)\exp\left(-\frac{2\alpha^2}{\delta^2}t\right)$.

Example 3.5 (Rectangular Domain) If $D = [0,L] \times [0,M]$ is a rectangle in two dimensions, then $\delta = \sqrt{L^2 + M^2}$ and the estimate in Proposition 3.20 becomes

$$E(t) \leq E(0)\exp\left(-\frac{2\alpha^2}{L^2 + M^2}t\right).$$

In one spatial dimension, the domain would be an interval $[0,L]$ and $\delta = L$. In this case the estimate reads $E(t) \leq E(0)\exp\left(-\frac{2\alpha^2}{L^2}t\right)$. We will see later (Chapter 6) that in this case, the estimate is sharp.

In the case of a homogeneous Neumann boundary condition, a nonzero steady-state solution exists and the preceding ideas need modification. The idea here is carried in the following variation of the Poincaré inequality.

Lemma 3.21 (Poincaré Inequality: Version II) Let D be a bounded convex domain and let δ be its diameter as before. Let $F(x)$ be a C^1 function on $\overline{D}$ and set

$$F_0 = \frac{1}{\mathrm{Vol}(D)} \int_D F \, dV.$$

(F_0 is the average value of F on D.) Then

$$\int_D [F(x) - F_0]^2 \, dV \leq \delta^2 \int_D |\nabla F|^2 \, dV. \tag{3.3.8}$$

The proof of this is left to the interested reader; it is strikingly similar to the previous case. (See Exercise 3.23 for the analog in one dimension.) The following result provides the analog of Proposition 3.20.

Proposition 3.22 Let D be a bounded convex domain and let $u = u(x,t)$ be a solution of the heat equation that is C^1 on $\overline{D} \times [0,\infty)$ and satisfies a homogeneous Neumann boundary condition. Let u_0 (a constant) be the steady-state solution. Then there is a constant C such that

$$\int_D [u(x,t) - u_0]^2 \, dV \leq C \exp\left(-\frac{2\alpha^2}{\delta^2}t\right).$$

The proof of this is similar to the previous case with one modification, which we now outline. First, it is clear that $u_0 = \frac{1}{\text{vol}(D)} \int_D u(x,t)dV$. Since u_0 satisfies the heat equation, we modify the energy method applied above by writing

$$(u - u_0)u_t = \alpha^2(u - u_0)\nabla^2 u.$$

Integrating over the domain, applying Green's first identity and the Neumann boundary condition gives

$$\frac{d}{dt}\int_D [u(x,t) - u_0]^2\, dV = -2\alpha^2 \int_D |\boldsymbol{\nabla} u|^2\, dV.$$

This is a variation of the identity (3.3.4) appropriate to handle the (possibly) nonzero steady state.

Energy and Causality for the Wave Equation

In analogy with the heat equation, the energy method applied to the wave equation begins by multiplying the equation by u_t, integrating over the bounded domain D of spatial variables, and applying Green's identity. We will work in two dimensions for this computation and use the physics of the vibrating membrane as a guide for interpretation; the same computation can be carried out in three dimensions.

$$\begin{aligned}\int_D u_t u_{tt}\, dV &= c^2 \int_D u_t \nabla^2 u\, dA \\ &= c^2 \oint_{\partial D} u_t \boldsymbol{\nabla} u \cdot \mathbf{n}_o\, ds - \int_D \boldsymbol{\nabla} u_t \cdot \boldsymbol{\nabla} u\, dA.\end{aligned}$$

The term on the left-hand side is equal to $\frac{d}{dt}\frac{1}{2}\int_D u_t^2\, dA$, whereas the second term on the right-hand side is $\frac{d}{dt}\frac{1}{2}\int_D |\boldsymbol{\nabla} u|^2\, dA$. Combining these terms together on the right-hand side gives one form of the energy identity for the wave equation:

$$\frac{d}{dt}\left[\frac{1}{2}\int_D (u_t^2 + c^2|\boldsymbol{\nabla} u|^2)dA\right] = c^2 \oint_{\partial D} u_t \boldsymbol{\nabla} u \cdot \mathbf{n}_o\, ds. \tag{3.3.9}$$

The quantity in brackets on the left is denoted $E(t)$. It is the sum of two terms, $\text{KE}(t) = \frac{1}{2}\int_D u_t^2\, dA$, which is proportional to the total kinetic energy of the string, and $\text{PE}(t) = \frac{c^2}{2}\int_D |\boldsymbol{\nabla} u|^2\, dA$, which is proportional to the potential energy. (Recall that $c^2 = T/\rho$, where ρ is the density and T is the tension, both assumed constant. The density is in fact the proportionality constant.) If the boundary conditions are of homogeneous Dirichlet, Neumann, or a mixture of the two, then the right-hand side of (3.3.9) is zero. In this situation, it follows that $\frac{dE}{dt} = 0$ and hence the energy $E(t)$ is a constant function. In physical terms, conservation of energy is valid. The above observations are summarized as follows.

Proposition 3.23 (Conservation of Energy) Let $u = u(x,t)$ be a (strong) solution of the wave equation on $D \times (0,\infty)$ that is C^1 on $\overline{D} \times [0,\infty)$ and satisfies homogeneous Dirichlet, Neumann, or mixed boundary conditions. Then the total energy $E(t)$ is a constant function, in fact $E(t) = E(0)$. Furthermore, an IBVP for the wave equation can have at most one solution satisfying the above hypothesis.

In the case of a Robin boundary condition, a conservation law also holds but takes a different form due to the exchange of energy at the boundary; see Exercise 1.16.

The Causality Principle Recall d'Alembert's formula

$$u(x,t) = \frac{1}{2}[f(x+ct) + f(x-ct)] + \frac{1}{2c}\int_{x-ct}^{x+ct} g(s)ds,$$

which gives the solution to the wave equation on the real line with initial data $u(x,0) = f(x)$ and $u_t(x,0) = g(x)$. One property of the solution readily apparent from this formula is the following.

> *The value of the solution at the point (x_0, t_0) in space-time only depends on the values of the initial data in the interval $(x_0 - ct_0, x_0 + ct_0)$.*

This is known as the **causality principle** for the wave equation. Geometrically it is depicted in Figure 3.2 where the characteristic lines $x \pm ct = x_0 \pm ct_0$ through the point (x_0, t_0) are depicted. The lines define a (two-dimensional) characteristic cone through this point. The bottom part of the cone (which is what is depicted in the figure) is called the backward characteristic cone, as

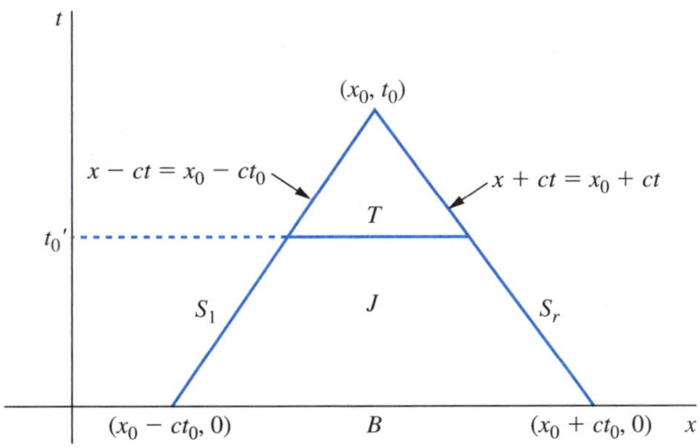

Figure 3.2 Causality Principle

the time values are less than t_0. An alternative statement of this principle reformulates in terms of the forward characteristic cone through the point $(x_0, 0)$ as follows.

> *Initial data at the point* $(x_0, 0)$ *can influence the solution only in the cone* $(x - x_0)^2 = c^2t^2$, $t > 0$.

The region inside this forward cone is the **domain of influence** of this point. From either perspective it follows that the wave moves with speed no more than c.

What is interesting is that the causality principle can be derived directly from the wave equation via energy methods and is applicable to other hyperbolic equations. Here is the argument for the case of the wave equation.

To demonstrate, we let u be a C^2 solution to the wave equation. Recall the energy identity (1.3.8), rewritten here in divergence form:

$$\begin{aligned} 0 &= \frac{\partial}{\partial t}\left(\frac{1}{2}u_t^2 + \frac{c^2}{2}u_x^2\right) + \frac{\partial}{\partial x}\left(-c^2 u_x u_t\right) \\ &= \left\langle \frac{\partial}{\partial x}, \frac{\partial}{\partial t} \right\rangle \cdot \left\langle -c^2 u_x u_t, \frac{1}{2}u_t^2 + \frac{c^2}{2}u_x^2 \right\rangle. \end{aligned}$$

The second line recognizes the structure of the energy identity as the divergence in space–time of a vector field. Let's fix $t_0^{'} < t_0$ and consider the trapezoidal region J depicted in Figure 3.2. Integrating the formula over the trapezoidal region J and applying Green's theorem yields

$$\begin{aligned} 0 &= \int_J \operatorname{div}\left\langle -c^2 u_x u_t, \frac{1}{2}u_t^2 + \frac{c^2}{2}u_x^2 \right\rangle dA \\ &= \oint_{\partial J} \left\langle -c^2 u_x u_t, \frac{1}{2}u_t^2 + \frac{c^2}{2}u_x^2 \right\rangle \cdot \mathbf{n}_o \, ds \\ &= I_B + I_{S_l} + I_{S_r} + I_T, \end{aligned} \tag{3.3.10}$$

where in the last line the boundary integral is broken into four integrals, one for each side of the trapezoid as marked in the figure, $\partial J = B \cup S_l \cup S_r \cup T$. Keep in mind that each of the four integrals is a path integral with direction indicated in the figure. The integrals are computed separately as follows.

- On B, $\mathbf{n}_o = \langle 0, -1 \rangle$ and hence,

$$I_B = -\int_B \left(\frac{1}{2}u_t^2 + \frac{c^2}{2}u_x^2\right) ds.$$

  Similarly, on T, $\mathbf{n}_o = \langle 0, 1 \rangle$ and $I_t = \int_T \left(\frac{1}{2}u_t^2 + \frac{c^2}{2}u_x^2\right) ds$.

- On I_{S_r}, $\mathbf{n}_o = \frac{\langle 1,c\rangle}{\sqrt{1+c^2}}$. In this case an application of completing the square is useful as follows:

$$\begin{aligned} I_{S_l} &= \frac{c}{\sqrt{1+c^2}} \int_{S_r} \left(\frac{1}{2}u_t^2 + \frac{c^2}{2}u_x^2 - c\,u_t u_x\right) ds \\ &= \frac{c}{2\sqrt{1+c^2}} \int_{S_r} (u_t - c\,u_x)^2 ds. \end{aligned}$$

A similar expression holds for I_{S_l}. In particular, I_{S_r} and I_{S_l} are both ≥ 0.

Returning to (3.3.10), we have

$$\int_B \left(\frac{1}{2}u_t^2 + \frac{c^2}{2}u_x^2\right) ds = I_{S_l} + I_{S_r} + \int_T \left(\frac{1}{2}u_t^2 + \frac{c^2}{2}u_x^2\right) ds,$$

and, in particular,

$$\int_T \left(\frac{1}{2}u_t^2 + \frac{c^2}{2}u_x^2\right) ds \leq \int_B \left(\frac{1}{2}u_t^2 + \frac{c^2}{2}u_x^2\right) ds$$

by the second bullet item. Now if the initial data are zero along B, then the integral on the right is zero, and hence

$$\int_T \left(\frac{1}{2}u_t^2 + \frac{c^2}{2}u_x^2\right) ds = 0.$$

Further, this is true for every $t_0' < t_0$. It follows that $u_t(x,t) = u_x(x,t) = 0$ inside the cone in Figure 3.2 and by continuity, $u(x,t) = 0$. This concludes the proof of the causality principle.

An interesting application of the causality principle is in the case where the initial data f and g vanish outside a fixed closed interval $[-R, R]$. In this case, the characteristic lines through the points $(\pm R, 0)$ divide the space–time view of the solution into regions, as shown in Figure 3.3. In the two triangular regions T2 and T3, $u(x,t) = 0$. Physically, the wave has yet to arrive and the characteristic lines $x - ct = R$ and $x + ct = -R$ are the leading edges of the wave. This also holds in T1 if $g = 0$ or more generally, $\int_{-R}^{R} g(s)ds = 0$. Under these conditions, the lines $x - ct = -R$ and $x + ct = R$ are the trailing edges of the wave.

Remark 3.24 You should wonder if a causality principle holds for the wave equation in two or three dimensions. It states the following: On $\mathbb{R}^d \times (0,\infty)$, the solution $u(x,t)$ of the wave equation $u_{tt} = c^2\nabla^2 u$ at a point (x', t') in space–time depends only on the values of the initial data in the sphere $|x - x'| = ct'$. The proof is based on the same ideas as given previously for the one–dimensional wave equation and is an application of the divergence theorem. Consult [29] for the details.

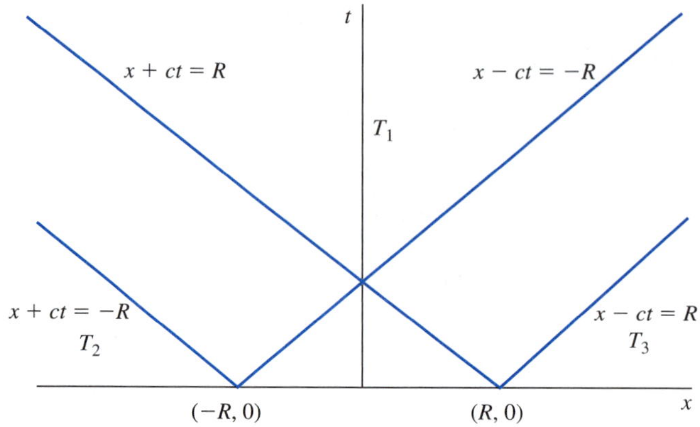

Figure 3.3 Causality Illustration

Exercises

3.20 Let $u = u(x,t)$ be a solution of the heat equation on $D \times (0, \infty)$ and satisfy the homogeneous Robin boundary condition (3.3.5) where $a(x) \geq 0$ on the boundary. Show that $\frac{dE}{dt} \leq 0$ in this case.

3.21 Let D be a bounded domain and assume that $k = k(x)$ is a C^1 function on $\overline{D}$.

a. Verify the identity $\nabla \cdot [ku\nabla u] = k|\nabla u|^2 + u\nabla \cdot [k\nabla u]$.

b. Let $u(x,t)$ be a solution of the equation $u_t = \nabla \cdot (k\nabla u)$ on $D \times (0, \infty)$ that is C^1 on $\overline{D} \times [0, \infty)$. With $E(t)$ defined as above, show that

$$\frac{dE}{dt} = \int_{\partial D} ku\nabla u \cdot \mathbf{n}_o \, ds - \int_D k|\nabla u|^2 \, dV.$$

c. Use part (a) to conclude that $\frac{dE}{dt} \leq 0$ if u satisfies a homogeneous Dirichlet, Neumann, or Robin boundary condition.

3.22 Version I of the Poincaré inequality in one dimension takes the following form. Let $F \in C^1[a,b]$ with $F(a) = F(b) = 0$. Then

$$\int_a^b F^2(x)dx \leq (b-a)^2 \int_a^b [F'(x)]^2 \, dx.$$

Demonstrate this. HINT: $F(x) = \int_a^x F'(s)ds$ by the Fundamental Theorem of Calculus. Apply the Cauchy–Schwarz inequality.

3.23 Version II of the Poincaré inequality in one dimension takes the following form. Let $F \in C^1[a,b]$ and set $F_0 = \frac{1}{b-a}\int_a^b F(x)dx$. Then

$$\int_a^b [F(x) - F_0]^2 \, dx \leq (b-a)^2 \int_a^b [F'(x)]^2 \, dx.$$

Demonstrate this result. HINT: Here take $c \in [a,b]$ such that $F(c) = F_0$ (this is justified by the intermediate value theorem). The Fundamental Theorem of Calculus then can be written $F(x) - F(c) = \int_c^x F'(s)ds$ and the argument proceeds as in the previous exercise.

3.24 Write down the analogs of Propositions 3.20 and 3.22 in the case of one spatial dimension. For the domain, take $D = (0, L)$.

3.25 Let D be a bounded domain and consider the equation $u_{tt} = \nabla \cdot (k\nabla u)$ on $D \times (0, \infty)$ where $k = k(x)$ is a C^1 function on $\overline{D}$.

- **a.** Formulate the analog of (3.3.9) for this equation. HINT: An identity similar to that in Exercise 3.21 part (a) will be useful here.
- **b.** Formulate a conservation law for this equation assuming that $u(x,t)$ satisfies a homogeneous Dirichlet boundary condition.

3.26 Consider the wave equation $u_{tt} = u_{xx}$. The energy density is defined as $e(x,t) = \frac{1}{2}(u_t^2 + u_x^2)$ and the momentum density as $p(x,t) = u_t u_x$. Show that $e_t = p_x$ and $e_x = p_t$. Further, show that e and p satisfy the wave equation.

3.27 Consider the damped wave equation $u_{tt} = c^2\nabla^2 u + ru_t$ for $x \in D$ and where $r > 0$. Show that $E(t)$ as defined by (3.3.9) is nonincreasing.

3.28 Consider the Klein–Gordon equation $u_{tt} = c^2 u_{xx} - m^2 u$ for $0 < x < L$ with homogeneous Dirichlet boundary conditions. Show that

$$E(t) = \frac{1}{2}\int_0^L \left[u_t^2(s,t) + c^2 u_s^2(s,t) + m^2 u^2(s,t)\right] ds$$

is conserved.

3.29 Determine the causality principle for the Klein–Gordon equation in one dimension. Deduce that the speed of propagation is at most c.

3.30 Reformulate and rework Exercise 3.28 in higher dimensions.

Chapter 4

An Introduction to Distributions

As we have already seen, the solution methodology for some partial differential equations leads to functions defining the solution that are not differentiable. Moreover, this can arise in physically realizable situations; in fact, this is typical for wave-type phenomena. In this chapter we present a brief guide to the theory of distributions. Historically, the concept was born in the 1920s with the Dirac delta function introduced formally by the physicist Paul Dirac in his work on quantum mechanics. The notion of a generalized function was furthered by Sobolev in the 1930s, and in the 1940s Laurent Schwartz made rigorous the mathematical theory of distributions. In a sense, the theory of distributions is the completion of differential calculus providing a generalized notion of the derivative quite suitable for the study of partial differential equations. We will apply this theory to develop fully the notion of weak solutions of partial differential equations discussed heuristically in Chapter 1. We will also apply the theory to the fundamental solution of partial differential operators, an idea that aids in mathematical and physical interpretation of solution formulas.

4.1 The Dirac Delta and Distributions

Motivated by the heuristic discussion of Section 1.3, we devote this section to introducing a generalized notion of a function, often called a generalized function or distribution. We begin by reformulating this heuristic discussion as follows. Suppose x represents the position of a point in space and $f(x)$ represents a measured quantity—say, temperature at the point x. Temperature is measured using a thermometer placed at the point x. In contrast to the point x, the bulb of a thermometer has a nonzero size, so the act of measuring the temperature at the point x is, in effect, measuring an average temperature near this point. Thus, the thermometer can be represented as an integral,

$$\int f(x)\phi(x)dV,$$

where $\phi(x)$ depends on the properties of the thermometer. For this to be an average, it is natural to assume $\phi(x) \geq 0$ for all x and $\int \phi(x)dV = 1$. We also want $\phi(x)$ to be concentrated near a given point, namely the point where we are attempting to measure the temperature. This means that $\phi(x)$ is basically zero sufficiently far away from this point. Temperature measurement can now be thought of as a **functional**, a function defined on the class of functions ϕ:

$$\phi \to \int f(x)\phi(x)dV.$$

Henceforth, we will write this assignment as $\phi \to (f, \phi)$. A key property is that this functional is linear,

$$(f, \alpha\phi_1 + \beta\phi_2) = \alpha(f, \phi_1) + \beta(f, \phi_2),$$

where α and β are scalars. Notice that we have replaced our usual notion of a function $f(x)$ with this new assignment $\phi \to (f, \phi)$ defined on functions ϕ. This assignment generalizes the notion of a function defined at points of space.

The functions ϕ are usually called **test functions,** a notion that we need to make more precise. The underlying point set will be $\mathbb{R}^d$ or a domain $D \subset \mathbb{R}^d$. A test function is an element $\phi \in C^\infty(D)$, which vanishes outside a closed, bounded subset of D. The closed bounded subset outside of which ϕ vanishes is called the support of ϕ. So a test function is infinitely differentiable and vanishes near the boundary of the domain. The class of test functions is denoted $\mathcal{D}(D)$.

Example 4.1 Here is an example on $\mathbb{R}$. Let

$$\psi(x) = \begin{cases} e^{-1/x^2}, & x > 0 \\ 0, & x \leq 0. \end{cases}$$

The claim is that $\psi \in C^\infty(\mathbb{R})$ and, further, that $\psi^{(n)}(0) = 0$ for all n. To see this, direct computation will show that

$$\frac{d^n}{dx^n} e^{-1/x^2} = \frac{p(x)}{x^{3n}} e^{-1/x^2},$$

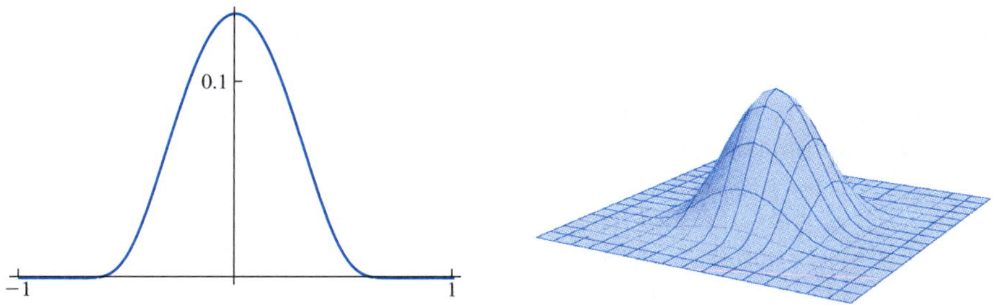

(a) One dimension (b) Two dimensions

Figure 4.1 Examples of Test Functions

where $p(x)$ is a polynomial. Since this tends to zero as $x \to 0^+$, the claim follows. Now we define

$$\phi(x) = \psi(1+x)\psi(1-x).$$

From our claim, $\psi(1+x)$ is smooth and vanishes for $x \leq -1$ and $\psi(1-x)$ is smooth and vanishes for $x \geq 1$. Hence the function $\phi \in C^\infty(\mathbb{R})$ and $\phi(x) = 0$ for $|x| \geq 1$, i.e. the support of ϕ is the interval $[-1, 1]$. A plot of ϕ is given in Figure 4.1(a).

Having constructed one test function, it is easy to generate many more through natural operations.

- **Translations.** $\phi(x+h)$ for any fixed $h \in \mathbb{R}$
- **Scaling.** $\phi(ax)$, where $a > 0$
- **Linear Combinations.** $\alpha_1\phi_1(x) + \alpha_2\phi_2(x)$

From test functions on $\mathbb{R}$, test functions on $\mathbb{R}^d$ can be constructed by taking products. Indeed, if $\phi_k \in \mathcal{D}(\mathbb{R})$, then the function $\phi(x) = \phi_1(x_1)\phi_2(x_2)\ldots\phi_d(x_d)$ is a test function on $\mathbb{R}^d$ (see Figure 4.1(b)). Again, the natural operations applied in $\mathbb{R}^d$ lead to many more examples.

Definition 4.1 A **distribution** is a continuous linear functional on $\mathcal{D}(D)$. In other words, a distribution u is a map from $\mathcal{D}(D) \to \mathbb{R}$, denoted $\phi \to (u, \phi)$, which is linear,

$$(u, \alpha_1\phi_1 + \alpha_2\phi_2) = \alpha_1(u, \phi_1) + \alpha_2(u, \phi_2),$$

where $\alpha_i \in \mathbb{R}$ and $\phi_i \in \mathcal{D}(D)$, and continuous in the sense that if $\{\phi_k\} \subset \mathcal{D}(D)$ is a sequence of test functions all of which vanish outside a fixed closed bounded set and ϕ_k and all of its derivatives converge uniformly, then $(u, \phi_k) \to (u, \phi)$, where $\phi = \lim_{k\to\infty} \phi_k$. The space of distributions is denoted $\mathcal{D}'(D)$.

Undoubtedly this definition seems technically obtuse. However, the condition that all the functions ϕ_k vanish outside a fixed closed bounded set merely localizes their convergence to a given region, and the condition that ϕ_k and all of its derivatives converge uniformly ensures that the limit of ϕ_k and all of its derivatives are continuous, i.e., that their limit function is a test function.

The two principal distributions used throughout this text are distributions defined through locally integrable functions and the Dirac delta functional. Given a function u defined on D, we define

$$(u, \phi) = \int_D u(x)\phi(x)dV.$$

The only requirement we need to place on u is that it is locally integrable on D; i.e., it is integrable on any closed bounded subset of the domain D. It is then easy to see that this mapping is continuous; it is obviously linear. The Dirac delta is defined on $\mathbb{R}^d$ as the evaluation functional at the origin, i.e.,

$$(\delta, \phi) = \phi(O).$$

This is trivially linear and continuous. More generally, for a fixed ξ, δ_ξ evaluates at $x = \xi$, i.e., $(\delta_\xi, \phi) = \phi(\xi)$.

Example 4.2 Take $D = \mathbb{R}^3$ and let $u(x) = |x|^{-1}$. Then $u(x)$ is locally integrable on $\mathbb{R}^3$. To see this, it suffices to show that $\int_{B_1(O)} \frac{dV}{|x|}$ is finite. Switching to spherical coordinates, (ρ, θ, ϕ), we have $dV = \rho^2 \sin\phi \, d\rho \, d\phi \, d\theta$. Hence, this integral becomes $\int_0^{2\pi} \int_0^{\pi} \int_0^1 \rho \sin\phi \, d\rho \, d\phi \, d\theta = 2\pi$. Thus, u defines the distribution

$$(u, \phi) = \int_{\mathbb{R}^3} \frac{\phi(x)}{|x|} \, dV.$$

As with functions, it is important to define convergence of a sequence of distributions. Given $\{u_n\} \subset \mathcal{D}'(D)$, we say that u_n converges in distribution to u if

$$(u_n, \phi) \to (u, \phi)$$

for every test function $\phi \in \mathcal{D}(D)$.

Example 4.3 On the real line, let

$$u_n(x) = \begin{cases} \frac{n}{2}, & -\frac{1}{n} \le x \le \frac{1}{n} \\ 0, & |x| > \frac{1}{n}. \end{cases}$$

Notice that $\int_{\mathbb{R}} u_n(x)dx = 1$ for all n. Then $u_n \in \mathcal{D}'(\mathbb{R})$ is given by

$$(u_n, \phi) = \frac{n}{2}\int_{-1/n}^{1/n} \phi(x)dx.$$

We claim that $u_n \to \delta$ in distribution. To show this, consider the estimate,

$$|(u_n, \phi) - (\delta, \phi)| = \left|\frac{n}{2}\int_{-1/n}^{1/n} [\phi(x) - \phi(0)]dx\right| \le \frac{n}{2}\int_{-1/n}^{1/n} |\phi(x) - \phi(0)|dx.$$

By continuity of functions, given $\varepsilon > 0$, there is a $\eta > 0$ such that if $|x| < \eta$, then $|\phi(x) - \phi(0)| < \varepsilon$. Now choose a natural number n_0 such that $1/n_0 < \eta$. From the above estimate, if $n \ge n_0$, then

$$|(u_n, \phi) - (\delta, \phi)| < \frac{n}{2}\varepsilon \int_{-1/n}^{1/n} dx = \varepsilon.$$

This is precisely the desired convergence.

An important result in the theory of distributions is that any distribution can be approximated by a test function. This result enables one to carry natural linear operations on test functions over to distributions. The precise result is as follows.

Theorem 4.2 Let $u \in \mathcal{D}'(D)$. Then there exists a sequence of test functions $\{u_n\}$ such that $u_n \to u$ in distribution.

For a proof of this result see [30]; here is an important example.

Example 4.4 Let ϕ be the test function constructed in Example 4.1 normalized so that $\int_{\mathbb{R}} \phi(x)dx = 1$. Let $u_n(x) = n\phi(nx)$. Then $u_n \in \mathcal{D}(\mathbb{R})$ and $u_n \to \delta$ in distribution. To demonstrate the latter, let ψ be any test function. Then using $\int_{\mathbb{R}} u_n(x)dx = 1$ and the fact that $u_n(x) = 0$ for $|x| > 1/n$, we have

$$(u_n, \psi) - (\delta, \psi) = \int_{\mathbb{R}} u_n(x)\psi(x)dx - \psi(0) = \int_{-1/n}^{1/n} u_n(x)[\psi(x) - \psi(0)]dx.$$

Given $\varepsilon > 0$, let n_0 be an integer such that if $n \geq n_0$ and $|x| < 1/n$, $|\psi(x) - \psi(0)| < \varepsilon$, justified by the continuity of ψ. Then for $n \geq n_0$, we have

$$|(u_n, \psi) - (\delta, \psi)| < \varepsilon \int_{-1/n}^{1/n} u_n(x)dx = \varepsilon \int_{-1}^{1} \phi(x)dx = \varepsilon.$$

Hence, the convergence is proved.

Three important natural applications extending linear operations on test functions to distributions are the following.

1. **Translation.** Given a test function $\phi(x)$, the translation operator is defined by $\tau_h\phi(x) = \phi(x+h)$, where $h \in \mathbb{R}^d$ is fixed. Suppose u is a distribution and let $\{\phi_n\} \subset \mathcal{D}$ such that $\phi_n \to u$ in distribution. Then

$$(\tau_h\phi_n, \phi) = \int \phi_n(x+h)\phi(x)dV = \int \phi_n(x)\phi(x-h)dV = (\phi_n, \tau_{-h}\phi).$$

We can now take the limit as $n \to \infty$ and define the action of τ_h on u via

$$(\tau_h u, \phi) = (u, \tau_{-h}\phi).$$

This defines a distribution as $\tau_{-h}\phi \in \mathcal{D}$. Usually the translation of a distribution is denoted using the standard form, i.e., $u(x+h)$. For example, on the real line $\delta(x-1)$ is the evaluation functional at 1: $(\delta(x-1), \phi(x)) = (\delta(x), \phi(x+1)) = \phi(1)$. More generally, $\delta(x-\xi) = \delta_\xi(x)$.

2. **Derivatives.** Let $u \in \mathcal{D}'$ and let $\{\phi_n\} \subset \mathcal{D}$ such that $\phi_n \to u$ in distribution. Then using integration by parts, we have

$$(\partial_i\phi_n, \phi) = \int (\partial_i\phi_n)(x)\phi(x)dV = -\int \phi_n(x)\partial_i\phi(x)dV.$$

By taking the limit as $n \to \infty$ we effectively define the derivative of u via the formula

$$(\partial_i u, \phi) = -(u, \partial_i\phi).$$

3. **Convolution.** For our purposes here the domain is $\mathbb{R}^d$. Given two functions ϕ and u, their convolution is defined as

$$(\phi * u)(x) = \int \phi(x-y)u(y)dV(y).$$

This is defined for any function for which the integral makes sense and is certainly defined when both ϕ and u are test functions. Convolutions have a variety of applications, in particular in approximating a nonsmooth function with smooth functions. (Revisit the discussion on page 22 regarding the solution of the Cauchy problem for the heat equation on $\mathbb{R}$. The solution formula is a convolution.) The definition of convolution here can be extended to the case where one of the functions, say u, is a distribution. To see this, rewrite the formula as

$$(\phi * u)(x) = \int u(y)(\tau_{-x}\tilde{\phi})(y)dy = (u, \tau_{-x}\tilde{\phi}),$$

where $\tilde{\phi}(y) = \phi(-y)$. Since $\tau_{-x}\tilde{\phi}$ is a test function for any x, this formula is valid for $u \in \mathcal{D}'$. The formula defines $\phi * u$ as a function that is in fact C^∞, and hence $\phi * u$ is a distribution. A useful example is to take $u = \delta$, then

$$(\phi * \delta)(x) = (\delta, \tau_{-x}\tilde{\phi}) = \phi(x-y)|_{y=0} = \phi(x).$$

Notice the incredible and powerful consequence of item 2: Every distribution has a derivative, and in fact can be differentiated infinitely often! Also notice that the definition of the derivative of a distribution is completely consistent with that of test functions; this is a consequence of the integration by parts formula.

Example 4.5 The derivative of the delta function on $\mathbb{R}$ is given by $(\delta', \phi) = -(\delta, \phi') = -\phi'(0)$. In other words, δ' evaluates $-\phi'$ at zero. In higher dimensions a similar formula holds for derivatives: $(\partial_i \delta, \phi) = -(\delta, \partial_i \phi) = -\partial_i \phi(O)$.

Example 4.6 Consider the Heaviside function defined on the real line by

$$H(x) = \begin{cases} 0, & x < 0 \\ 1, & x \geq 0. \end{cases}$$

As a function, this has derivative zero everywhere except at $x = 0$, where the derivative is undefined. However, as a distribution,

$$(H', \phi) = -(H, \phi') = -\int_0^\infty \phi'(x)dx = \phi(0).$$

We conclude that $H' = \delta$ as distributions. More generally, if $u(x) = v(x) + AH(x - \xi)$ where A and ξ are constants and $v(x)$ is a smooth function, then $u' = v'(x) + A\delta(x - \xi)$. This shows the effect of differentiating a function with a single jump discontinuity.

Example 4.7 (The Constant Distribution) Let $u(x) = K$, a constant. Clearly u is locally integrable on the real line and $(u, \phi) = K \int_{\mathbb{R}} \phi(x)dx$. The derivative is given by $(u', \phi) = K \int_{\mathbb{R}} \phi'(x)dx = 0$, i.e., u' is, not surprisingly, the zero distribution. The converse is also true. Suppose $u \in \mathcal{D}'$ ($\mathcal{D}$ and $\mathcal{D}'$ are on $\mathbb{R}$ in this example) with $u' = 0$. Let $\phi_o \in \mathcal{D}$ such that $\int_{\mathbb{R}} \phi_0(x)dx = 1$. Let $\phi \in \mathcal{D}$, set $c = \int_{\mathbb{R}} \phi(x)dx$, and define

$$\psi(x) = \int_{-\infty}^{x} [\phi(s) - c\phi_0(s)]ds.$$

Then $\psi \in \mathcal{D}$. (Explain why.) Now $0 = (u', \psi) = -(u, \psi') = -(u, \phi - c\phi_0)$. Hence, $(u, \phi) = c(u, \phi_0)$ and the result follows with $K = (u, \phi_0)$.

Example 4.8 In general, the product of two distributions is not defined. For example, there is no way to make sense out of $\delta(x)\delta(x + h)$. However, multiplication of a distribution by a smooth bounded function does make sense and, in fact, is very useful. Let $u \in \mathcal{D}'$ and let $m(x)$ be a smooth bounded function. Then

$$(mu, \phi) = (u, m\phi)$$

defines the product mu as a distribution. The product rule holds: $(mu)' = m'u + mu'$ as is shown in the the following computation.

$$\begin{aligned}((mu)', \phi) &= -(mu, \phi') = -(u, m\phi') \\ &= -(u, (m\phi)') + (u, m'\phi) = (mu', \phi) + (m'u, \phi).\end{aligned}$$

A useful special case of this formula is the product $m(x)\delta(x-h)$. This is given by

$$\begin{aligned}(m(x)\delta(x - h), \phi(x)) &= (\delta(x - h), m(x)\phi(x)) \\ &= m(h)\phi(h) = (m(h)\delta(x - h), \phi(x)).\end{aligned}$$

We close this introduction to distributions with the following result of particular importance in distributions defined by infinite series.

Proposition 4.3 Let $\{u_k\} \subset \mathcal{D}'(D)$ with $u_k \to u$ in distribution. Then $\partial_i u_k \to \partial_i u$ in distribution.

The proof is straightforward from the definitions and left as an exercise. It follows from this that $L(u_k) \to L(u)$ for any constant coefficient differential operator L, in particular the cases $L = \partial_{ii}$, ∂_{ij}, or any higher-order partial derivative.

Exercises

4.1 Let $u \in \mathcal{D}'(\mathbb{R}^n)$. Show that $\partial_{jk}u = \partial_{kj}u$ for any j and k.

4.2 Let $A = A_\theta \in SO(2)$ and $u \in \mathcal{D}'(\mathcal{R}^2)$. Define the composition of u with a rotation by the formula $(u \circ A_\theta, \phi) = (u, \phi \circ A_{-\theta})$. Show that this definition is consistent with the definition for functions. If $(u, \phi) = \int_{\mathbb{R}} \phi(x_1, x_2)dx_2$, find $u \circ A_{\pi/2}$.

4.3 Suppose $u(x)$ is a distribution on $\mathbb{R}$ such that $u'(x) = \delta(x)$. Show that $u(x) = H(x) + A$, for some constant A.

4.4 We say that $u \in \mathcal{D}'(\mathbb{R}^2)$ is radial if $u \circ A_\theta = u$ for all θ.

a. Show that δ is radial.

b. Define u_{rad} by $(u_{\text{rad}}, \phi) = \frac{1}{2\pi}\int_0^{2\pi}(u \circ A_\theta, \phi)d\theta$. Show that u is radial if and only if $u_{\text{rad}} = u$.

4.5 On $\mathbb{R}^2$, define $(u, \phi) = \int_{\mathbb{R}}\int_0^\infty \phi(x_1, x_2)dx_1\, dx_2$.

a. Find $\partial_1 u$ and $\partial_2 u$.

b. Find $\partial_{12}u$.

4.6 Prove: If $u \in \mathcal{D}'(\mathbb{R})$ with $\frac{d^k u}{dx^k} = 0$, then u is a polynomial of degree $< k$.

4.7 Demonstrate Proposition 4.3.

4.2 Weak Solutions of Differential Equations

Given a linear partial differential operator L, we can now interpret its action on distributions and seek distributional solutions of the partial differential equation $L(u) = f$. Such solutions are called distributional solutions or, more commonly, weak solutions.

Example 4.9 On the real line consider the differential equation $u' = H$, where H is the Heaviside function. Thinking in terms of functions, we are led by integration to a proposed solution

$$u(x) = \begin{cases} c_1, & x < 0 \\ x + c_2, & x > 0, \end{cases}$$

where c_1 and c_2 are constants. However, this function is not differentiable at $x = 0$ for any choice of the constants c_i. Nonetheless, computing the derivative as a distribution we have

$$(u', \phi) = -(u, \phi') = -c_1 \int_{-\infty}^{0} \phi'(x)dx - \frac{1}{n} \int_{0}^{\infty} (x + c_2)\phi'(x)dx$$
$$= (c_2 - c_1)\phi(0) + \int_{0}^{\infty} \phi(x)dx,$$

where integration by parts has been used. It follows that $u' = (c_2 - c_1)\delta(x) + H(x)$. Thus, choosing $c_2 = c_1$, the locally integrable function $u(x)$ is then a weak solution of $u' = H$.

Example 4.10 Consider the equation $u' = \delta$ on the real line. It has already been shown that $H' = \delta$, hence $u = H$ is a weak solution of the differential equation $u' = \delta$. From previous considerations, any solution of this differential equation has the form $u(x) = H(x) + c$, for some constant c.

Weak Solutions of the Transport Equation

In Chapter 1 and in more detail in Chapter 2, it was shown that the Cauchy problem for the transport equation,

$$\text{PDE: } u_t - cu_x = 0, \quad x \in \mathbb{R}, \quad t > 0$$
$$\text{IC: } u(x, 0) = f(x),$$

has solution $u(x, t) = f(x + ct)$ provided f is continuously differentiable. Relaxing this latter condition, we claim that $u(x, t) = f(x + ct)$ is a weak solution to the transport equation provided $f \in C(\mathbb{R})$. To see this, we let $\phi = \phi(x, t)$ be a test function on $\mathbb{R} \times (0, \infty)$ and extend ϕ to a test function on $\mathbb{R}^2$ by setting $\phi(x, t) = 0$ for $t < 0$. Then

$$(u_t - cu_x, \phi) = (u, -\phi_t + c\phi_x) = -\int_{\mathbb{R}} \int_{\mathbb{R}} f(x + ct)[\phi_t(x, t) - c\phi_x(x, t)]dx\, dt.$$

Hence to prove the claim, it must be shown that the above double integral is zero. To effect this, we introduce a change of variables. Let $\xi = x + ct$ and choose $\eta = t$. The determinant of the Jacobian matrix of this transformation is one. Now $\phi_t = \phi_\xi \xi_t + \phi_\eta \eta_t = c\phi_\xi + \phi_\eta$ and $\phi_x = \phi_\xi \xi_x + \phi_\eta \eta_x = \phi_\xi$ and hence $\phi_t - c\phi_x = \phi_\eta$. Hence, the preceding double integral can be rewritten as

$$-\int_{\mathbb{R}} \int_{\mathbb{R}} f(\xi)\phi_\eta(\xi, \eta)d\eta\, d\xi = -\int_{\mathbb{R}} f(\xi) \left[\int_{\mathbb{R}} \phi_\eta(\xi, \eta)d\eta \right] d\xi.$$

Now the inner integral is zero by the Fundamental Theorem of Calculus and the fact that ϕ is zero outside a closed bounded set. This proves the claim.

Example 4.11 In Example 1.2 of Chapter 1, it was shown that for the transport equation $u_t - u_x = 0$ with initial data

$$u(x,0) = g(x) = \begin{cases} \cos x, & |x| \leq \pi/2 \\ 0, & \text{otherwise}, \end{cases}$$

the formula $u(x,t) = g(x+ct)$ does not define a strong solution to the equation. From the preceding discussion, it is a weak solution. Furthermore, the singularities of the initial data—that is, the points $x = \pm\pi/2$ where $g'(x)$ does not exist in the classical sense—propagate in the weak solution along the characteristic lines $x + t = \pm\pi/2$. This behavior of propagation of singularities along characteristics is typical of wave-type equations and the associated physical phenomena.

Weak Solutions of the Wave Equation

Here we reconsider the discussion in Chapter 1 leading to d'Alembert's solution to the Cauchy problem for the wave equation. Specifically, it was shown that the Cauchy initial value problem

$$\text{PDE: } u_{tt} = c^2 u_{xx}, \quad x \in \mathbb{R}, \quad t > 0$$

$$\text{IC: } \begin{cases} u(x,0) = f(x) \\ u_t(x,0) = 0 \end{cases}$$

has a strong solution given by

$$u(x,t) = \frac{1}{2}[f(x+ct) + f(x-ct)], \tag{4.2.1}$$

provided that f is a C^2 function. However, if the initial data are given by the function

$$f(x) = \begin{cases} 1 - |x|, & |x| < 1 \\ 0, & |x| \geq 1, \end{cases} \tag{4.2.2}$$

the formula (4.2.1) does not provide a strong solution. It is a weak solution, as the following computations show. It is left to the reader (Exercise 4.10) to show that as a distribution, $f'(x) = H(x+1) - H(x-1) + 2H(x)$ and $f''(x) = \delta(x+1) - \delta(x-1) + 2\delta(x)$. Now consider the distribution defined on space–time by $u(x,t) = f(x+ct)$, (this is a locally integrable function, hence a distribution). By the chain rule,

$$\begin{aligned} u_{xx}(x,t) &= \delta(x+ct+1) - \delta(x+ct-1) + 2\delta(x+ct) \\ u_t(x,t) &= cH(x+ct+1) - cH(x+ct-1) + 2cH(x+ct) \\ u_{tt}(x,t) &= c^2\delta(x+ct+1) - c^2\delta(x+ct-1) + 2c^2\delta(x+ct) \end{aligned}$$

It follows by comparing the formulas that u provides a weak solution of the wave equation $u_{tt} = c^2 u_{xx}$. In a similar fashion, $f(x-ct)$ also is a weak solution of the wave equation. Hence (4.2.1) defines a weak solution of the wave equation that satisfies the initial conditions $u(x,0) = f(x)$ and $u_t(x,0) = 0$. In other words, by broadening our view of the derivative concept, we can give meaning to solutions of the wave equation for initial data that are not differentiable in the usual sense. As in the case of the transport equation, the singularities of the initial data propagate along the characteristic lines $x \pm ct = 0, \pm 1$.

Weak solutions of the wave equation can be described in the same manner as the transport equation. If $u(x,t)$ is a distribution on space–time that satisfies the wave equation, then for any test function $\phi = \phi(x,t)$,

$$0 = (u_{tt} - c^2 u_{xx}, \phi) = (u, \phi_{tt} - c^2 u_{xx}).$$

Thus, a distribution $u(x,t)$ is a weak solution of the wave equation if and only if for every test function ϕ, $(u, \phi_{tt} - c^2\phi_{xx}) = 0$. If the distribution u is given by a continuous function, this latter condition is equivalent to

$$\int_0^\infty \int_{\mathbb{R}} u(x,t)[\phi_{tt}(x,t) - c^2\phi_{xx}(x,t)]dx\,dt = 0. \tag{4.2.3}$$

In this form, as we did for the transport equation, we can work with integral formulas instead of the differentiation formulas in verifying that $u(x,t) = f(x \pm ct)$ are weak solution of the wave equation. The specific result of interest for the wave equation is the following.

Proposition 4.4 Let $F(x)$ be a continuous function on $\mathbb{R}$. Then the functions $u(x,t) = F(x \pm ct)$ are weak solutions of the wave equation.

The proof is achieved by substituting into (4.2.3) and carrying out a similar change-of-variables argument that was used for the transport equation. Details are left to the reader (Exercise 4.11).

Exercises

4.8 Find the weak solution of the differential equation $u' + \lambda u = H$ where λ is a constant.

4.9 The function $u(x, y) = g(y^2 - 2x)$ is a solution of the first-order PDE $yu_x + u_y = 0$ for any function g that is continuously differentiable. This follows by the methods of Chapter 2 or can be verified by the chain rule. If g is merely assumed to be continuous, show that $u(x, t)$ defines a weak solution of this equation.

4.10 For the distribution $f(x)$ defined in (4.2.2), find formulas for $f'(x)$ and $f''(x)$.

4.11 Let $F \in C(\mathbb{R})$ and define $u(x, t) = F(x + ct)$. Show that u is a weak solution of the wave equation by demonstrating (4.2.3). A similar argument applies to the function $F(x - ct)$. Use these facts to show that if $f(x)$ and $g(x)$ are continuous functions on $\mathbb{R}$, then d'Alembert's formula (1.3.15) gives a weak solution to the wave equation.

4.3 Fundamental Solutions for Differential Operators

Given a linear differential operator L acting on functions defined on $\mathbb{R}^d$, a fundamental solution for the operator is a distribution K that satisfies the equation $L(u) = \delta$. The case where L has constant coefficients is particularly nice since $u(x) = (\phi * K)(x)$ then provides a smooth solution to the equation $L(u) = \phi$, at least when ϕ is a test function.

Example 4.12 On the real line, consider $L = \frac{d^2}{dx^2}$. We have already seen that $H'(x) = \delta(x)$. Hence, $K(x) = xH(x)$ is a fundamental solution for L. To verify this, the product rule gives $(xH(x))' = x\delta(x) + H(x)$. Further, by Example 4.8, $x\delta(x) = 0$. Differentiating again verifies the conclusion. It follows that the function given by

$$u(x) = (\phi * K)(x) = \int_0^\infty y\phi(x - y)dy \tag{4.3.1}$$

is a solution to the equation $u'' = \phi$, provided ϕ is a test function. This can be checked by differentiating the formula for u twice and integrating the result by parts. On the other hand, a fundamental set for the equation $u'' = 0$ is $\{1, x\}$. Applying the method of variation of parameters to the equation $u'' = \phi$ leads to a solution formula equivalent to formula (4.3.1). (Keep in mind, any solution of the equation $u'' = \phi$ will have the form of $u(x) = c_1 + c_2x + u_p(x)$ where c_1 and c_2 are constants and $u_p(x)$ is a particular solution to the equation; see Exercise 4.12.)

Example 4.13 (Fundamental Solutions of the Laplacian) The purpose here is to show that the function

$$K(x) = K_d(x) = \begin{cases} -\frac{1}{2\pi}\log|x|, & d = 2 \\ -\frac{1}{4\pi}\frac{1}{|x|}, & d = 3, \end{cases}$$

is a fundamental solution to the Laplacian on $\mathbb{R}^d$, i.e., $\nabla^2 K = \delta$. The details for $d = 2$ are carried out; $d = 3$ is left for the exercises. Let $\phi(x)$ be a test function on $\mathbb{R}^2$. We need to compute $(\nabla^2 K, \phi)$. Let $R > 0$ be large enough so that $\phi(x) = 0$ for $|x| \geq R$. Choose $\varepsilon > 0$ with $\varepsilon < R$ and let $D_{\varepsilon,R} = \{\varepsilon < |x| < R\}$. Using Green's second identity

$$\int_{D_{\varepsilon,R}} (\phi(x)\nabla^2 K(x) - K(x)\nabla^2\phi(x))\,dA$$
$$= \oint_{\partial D_{\varepsilon,R}} \left(\phi(x)\frac{dK}{dn}(x) - K(x)\frac{d\phi}{dn}(x)\right) ds.$$

The boundary of $D_{\varepsilon,R}$ consists of two circles $S_\varepsilon(O)$ and $S_R(O)$. Because of our choice of $R > 0$, the boundary integrals on the right vanish on the larger sphere. Further, using the polar coordinate formula for the Laplacian, it is easy to see that $\nabla^2 \log|x| = 0$ for $x \neq O$. Hence, the preceding formula collapses to

$$-\int_{D_{\varepsilon,R}} K(x)\nabla^2\phi(x)dA(x) = -\oint_{S_\varepsilon(O)} \left(\phi(x)\frac{dK}{dn}(x) - K(x)\frac{d\phi}{dn}(x)\right) ds. \tag{4.3.2}$$

On $S_\varepsilon(O)$, $\frac{dK}{dn} = \frac{1}{2\pi\varepsilon}$ and $\left|\frac{d\phi}{dn}\right| \leq M$, for some constant M. It follows that as $\varepsilon \to 0$

$$\int_{S_\varepsilon(O)} \phi(x)\frac{dK}{dn}(x)ds = \frac{1}{2\pi\varepsilon}\int_{S_\varepsilon(O)} \phi(x)ds \to \phi(O),$$

by continuity and

$$\left|\int_{S_\varepsilon(O)} K(x)\frac{d\phi}{dn}(x)ds\right| \leq M(\log\varepsilon)2\pi\varepsilon \to 0.$$

Returning to (4.3.2) and taking the limit we obtain

$$\int_{\mathbb{R}^2} K(x)\nabla^2\phi(x)dA = \phi(O) = \langle\delta,\phi\rangle\,.$$

The result follows since $(\nabla^2 K, \phi) = (K, \nabla^2\phi) = \int_{\mathbb{R}^2} K(x)\nabla^2\phi(x)dV$.

Physical Interpretation The fundamental solution of the Laplacian given in Example 4.13 has nice physical interpretation in the case $d = 3$ as follows.

1. **Potential Due to a Charge Distribution.** Mathematically, a unit point charge placed at the origin can be represented as a delta functional. Thus, from Gauss's law, the potential due to this point charge is a solution of $\nabla^2 u = -4\pi\delta$. From Example 4.2, a solution to this problem would be $u(x) = \frac{1}{|x|}$. Any solution to the equation would have form $\frac{1}{|x|} + v(x)$, where $v(x)$ is a solution to $\nabla^2 v = 0$ in $\mathbb{R}^3$. Physically, the potential must decay as $|x| \to \infty$, which forces v to be a bounded function. In Chapter 7 it will be seen that this forces v to be a constant, and hence zero. Thus, the potential created by a unit charge at the origin is given by $-4\pi K(x) = \frac{1}{|x|}$. (This is equivalent to Couloumb's law.) Now if $\sigma(x)$ is the density of a charge distribution in space, the resultant potential (weak solution of Poisson's equation) is given by

$$u(x) = -4\pi(\sigma * K)(x) = \int_{\mathbb{R}^3} \frac{\sigma(y)}{|x-y|}\, dV(y),$$

which is the convolution of σ and $\frac{1}{|x|}$. This formula would be valid if σ is a test function and can be extended to more general charge distributions that are behaved well enough so that the integral makes sense.

2. **Steady-State Heat Flow.** Consider heat flow in free space governed by the equation $u_t = \nabla^2 u + \phi(x)$, where ϕ represents a time-independent heat source. In steady state, the temperature would be governed by $\nabla^2 u = -\phi$. Mathematically, we can imagine a unit of heat energy at the origin as being represented by δ. (Recall total energy is given by $E = \int c\rho u\, dx$, where c is the specific heat and ρ is the density. In this example we assume $c\rho = 1$ for simplicity.) The solution of the equation $\nabla^2 u = -\delta$ would be interpreted as the resulting steady-state temperature distribution, which from the preceding considerations is given by $u(x) = \frac{1}{4\pi|x|} + v(x)$, where $\nabla^2 v = 0$. In Chapter 7 it will be seen that solutions of Laplace's equation in all of space must be unbounded or constant. Both cases can be ruled out on physical grounds. Therefore, the steady-state temperature distribution being sought is $u(x) = \frac{1}{4\pi|x|}$. Furthermore, the steady state temperature distribution resulting from a time-independent heat source ϕ should be given by the convolution

$$u(x) = \frac{1}{4\pi} \int_{\mathbb{R}^3} \frac{\phi(y)}{|x-y|}\, dV(y).$$

Kernel Functions

In the case of initial value problems, a variation on the idea of a fundamental solution is useful. We illustrate here for the heat and wave equations on $\mathbb{R}$. In

Chapter 1 we showed that

$$u(x,t) = \int_{\mathbb{R}} f(y)\mathcal{S}(x-y,t)dy, \tag{4.3.3}$$

where $\mathcal{S}(x,t) = \frac{1}{2\alpha\sqrt{\pi t}}e^{-\frac{x^2}{4\alpha t}}$, gives the solution to the heat equation on $\mathbb{R}$ with initial condition $u(x,0) = f(x)$. The initial condition is satisfied in the sense that

$$\lim_{t\to 0^+} u(x,t) = f(x).$$

The **heat kernel** for this IVP is the distributional solution of the heat equation on $\mathbb{R}$ that satisfies the initial condition $u(x,0) = \delta(x)$. From the preceding discussion, the heat kernel is given by $\mathcal{S}(x,t)$. Indeed, $\mathcal{S}(x,t)$ is a strong solution to the heat equation and

$$\lim_{t\to 0^+} \langle \mathcal{S}(x,t), \phi(x)\rangle = \phi(0) = \langle \delta, \phi\rangle$$

for any test function ϕ. Thus, the function $\mathcal{S}(x,t)$ can be interpreted as the heat distribution corresponding to a unit source of heat placed at the origin initially. This gives further intuitive appeal to (4.3.3) as a superposition of initial heat sources $f(y)$.

In the case of the wave equation, the **wave kernel** is defined as the weak solution of the initial value problem

$$\begin{aligned} &\text{PDE: } u_{tt} = c^2 u_{xx}, \quad x\in\mathbb{R}, \quad t>0 \\ &\text{IC: } \begin{cases} u(x,0) = 0 \\ u_t(x,0) = \delta(x). \end{cases} \end{aligned} \tag{4.3.4}$$

The solution, which can easily be read from d'Alembert's formula (1.3.15), is given by

$$\mathcal{W}(x,t) = \frac{1}{2c}\left[H(x-ct) - H(x+ct)\right].$$

Replacing the second initial condition by $u_t(x,0) = \delta(x-\xi)$; then the kernel of interest is given by

$$\mathcal{W}(x,\xi,t) = \frac{1}{2c}\left[H(x-\xi-ct) - H(x-\xi+ct)\right].$$

An alternative form of d'Alembert's formula is now clear; namely,

$$u(x,t) = \int_{\mathbb{R}} g(\xi)\mathcal{W}(x,\xi,t)d\xi$$

gives the weak solution of the IBVP (4.3.4) where the second initial condition is replaced by $u_t(x,0) = g(x)$, $g \in C(\mathbb{R})$. In this form, d'Alembert's formula is a superposition of initial velocity sources specified by the function g. Discussion of a modified wave kernel where the initial conditions are replaced by $u(x,0) = \delta(x)$ and $u_t(x,0) = 0$ is left to the exercises.

Kernel Functions and the Method of Reflection

Let us now determine the kernel functions for the heat and wave equations on the half-line with a homogeneous Dirichlet condition at the origin. The construction is based on the previous discussion on the real line and introduces the method of reflection.

For the heat equation, we are interested in the solution to the following IBVP, where $\xi > 0$ is fixed.

$$\begin{aligned} \text{PDE: } & u_t = \alpha^2 u_{xx}, \quad x > 0, \quad t > 0 \\ \text{BC: } & u(0,t) = 0 \\ \text{IC: } & u(x,0) = \delta(x-\xi) \end{aligned} \tag{4.3.5}$$

The method is based on the fact that if the initial data for the heat equation on the real line is an odd function, then the solution is odd in the x-variable. Since an odd continuous function is automatically zero at the origin, we extend the initial data to an odd distribution and consider the following IVP:

$$\begin{aligned} \text{PDE: } & u_t = \alpha^2 u_{xx}, \quad x \in \mathbb{R}, \quad t > 0 \\ \text{IC: } & u(x,0) = \delta(x-\xi) - \delta(x+\xi). \end{aligned}$$

The solution to this problem is given by $\mathcal{S}_d(x,\xi,t) = \mathcal{S}(x-\xi,t) - \mathcal{S}(x+\xi,t)$. Notice that $\mathcal{S}_d(0,\xi,t) = 0$ and that it is symmetric in the x- and ξ-variables, i.e., $\mathcal{S}_d(\xi,x,t) = \mathcal{S}_d(x,\xi,t)$. Restricting to the half-line, $x > 0$, $\mathcal{S}_d(x,\xi,t)$ gives the solution to the initial boundary value problem of interest (4.3.5). To check the initial condition, observe that $(\delta(x+\xi),\phi(x)) = 0$ for test functions defined on $(0,\infty)$.

The function $\mathcal{S}_d(x,\xi,t)$ is the heat kernel on the half-line with homogeneous Dirichlet condition at the origin. Indeed, if f is a continuous bounded function on $(0,\infty)$, the function

$$u(x,t) = \int_0^\infty f(\xi)\mathcal{S}_d(x,\xi,t)d\xi$$

is a solution to the heat equation on $(0,\infty)$ and satisfies the boundary condition $u(0,t) = 0$ and the initial condition $u(x,0) = f(x)$. This essentially follows from Theorem 1.2. To see that the initial condition is satisfied, we introduce the odd extension of the function f via

$$f^o(x) = \begin{cases} f(x), & x > 0 \\ -f(-x), & x < 0. \end{cases}$$

Then the function u can be expressed by a change of variable as

$$u(x,t) = \int_{\mathbb{R}} f^0(\xi)\mathcal{S}(x-\xi,t)d\xi.$$

The same approach can be applied to the wave equation on the half-line:

$$\begin{aligned} \text{PDE: } & u_{tt} = c^2 u_{xx}, \quad x > 0, \quad t > 0 \\ \text{BC: } & u(0,x) = 0 \\ \text{IC: } & \begin{cases} u(x,0) = 0 \\ u_t(x,0) = \delta(x-\xi). \end{cases} \end{aligned} \tag{4.3.6}$$

The extension to a problem on the line is done precisely as in the case of the heat equation. We seek the weak solution of the wave equation that satisfies the initial conditions

$$\text{IC: } \begin{cases} u(x,0) = 0 \\ u_t(x,0) = \delta(x-\xi) - \delta(x+\xi). \end{cases}$$

This problem has a weak solution given by the formula

$$\mathcal{W}_d(x,\xi,t) = \mathcal{W}(x-\xi,t) - \mathcal{W}(x+\xi,t),$$

where $\mathcal{W}(x,t)$ is the wave kernel on $\mathbb{R}$ determined earlier. The characteristic line $x = ct$ through the origin divides space–time into two regions; this is depicted in the left side of Figure 4.2. If $x > ct$, then the second term does not contribute for any value of ξ, whereas if $x < ct$, the second term contributes for $0 < \xi < ct - x$. It follows that the wave kernel on the half-line has the form

$$\mathcal{W}_d(x,\xi,t) = \begin{cases} \frac{1}{2c}, & \text{if } x > ct \text{ and } x - ct < \xi < x + ct \\ \frac{1}{2c}, & \text{if } x < ct \text{ and } ct - x < \xi < x + ct \\ 0, & \text{otherwise.} \end{cases}$$

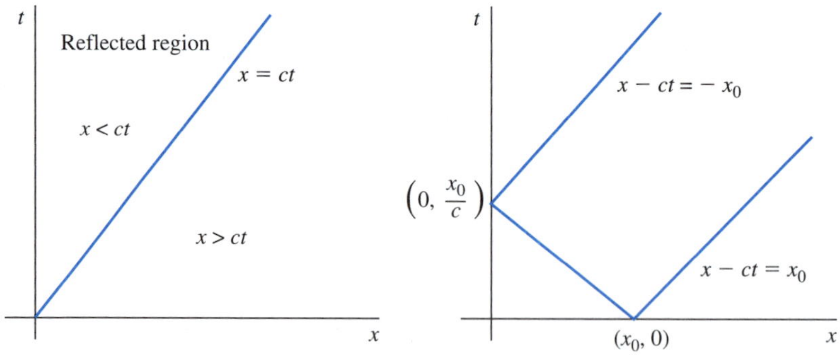

Figure 4.2 Characteristics and Reflection

Thus, replacing the second initial condition in (4.3.6) by $u_t(x,0) = g(x)$ where $g \in C[0,\infty)$, the weak solution of the resulting IBVP is given by the formula

$$u(x,t) = \begin{cases} \frac{1}{2c}\int_{x-ct}^{x+ct} g(s)ds, & x > ct \\ \frac{1}{2c}\int_{ct-x}^{x+ct} g(s)ds, & x < ct. \end{cases} \tag{4.3.7}$$

From a physical perspective, the second term is the result of a reflected wave from the boundary $x = 0$. The domain of influence in space–time of a point $(x_0, 0)$ can be read from this formula; it is the diagonal strip as depicted in the right side of Figure 4.2. As we did for the heat equation, the solution formula can also be expressed in terms of the odd extension of the function g:

$$u(x,t) = \frac{1}{2c}\int_{x-ct}^{x+ct} g^o(s)ds.$$

Remark 4.5 The preceding ideas can be adapted to the case where the boundary condition is replaced by the Neumann condition $u_x(0,t) = 0$ for both the heat and wave equations. The method also applies to the wave equation for the modified wave kernel. These considerations are left as exercises.

Green's Functions for ODEs

Consider the following BVP on the interval $[a,b]$, where p and q are continuous functions on $[a,b]$ and $p(x) > 0$ for all $x \in [a,b]$.

$$\begin{aligned} &\text{ODE: } (p(x)u')' + q(x)u = f(x), \quad x \in [a,b] \\ &\text{BC: } \begin{cases} a_{11}u'(a) + a_{12}u(a) = 0 \\ a_{21}u'(b) + a_{22}u(b) = 0 \end{cases} \end{aligned} \tag{4.3.8}$$

If we let $L = \frac{d}{dx}(p\frac{d}{dx}) + q$, the equation can be written in operator form as $L(u) = f$. The idea is to use variation of parameters and find a solution formula in the form

$$u(x) = \int_a^b G(x,\xi)f(\xi)d\xi. \tag{4.3.9}$$

We will then interpret the kernel $G(x,\xi)$ in terms of distributions and give an alternative derivation using distributional formulas. The following simple example serves to point out a subtlety.

Example 4.14 Consider the problem $u'' = f(x)$ on the interval $[0,1]$ with boundary conditions $u'(0) = u'(1) = 0$. By integration, the general solution to this ODE is

$$u(x) = \int_0^x \int_0^s f(z)dz\, ds + Ax + B.$$

Since $u'(x) = \int_0^x f(z)dz + A$, the first boundary condition is satisfied by choosing $A = 0$. But then, the second boundary condition can be satisfied if and only if $\int_0^1 f(z)dz = 0$. It follows that the problem has a solution if and only if this last condition holds and, further, the solution is not unique as B can be arbitrary.

To circumvent the difficulty portrayed in this example, assume for (4.3.8) that no solution of the corresponding homogeneous equation, $L(u) = 0$, satisfies both of the boundary conditions. With this assumption in place, we can solve the problem by the method of variation of parameters. Let $u_a(x)$ and $u_b(x)$ be solutions of the corresponding homogeneous equation where u_a satisfies the first boundary condition and u_b the second. Notice that any multiple of u_a is also a solution of $L(u) = 0$ and satisfies the boundary condition at $x = a$. Furthermore, the only solution of $L(u) = 0$ that satisfies the first boundary condition is a multiple of u_a. The reason for this is that a function u satisfies the boundary condition at $x = a$ if and only if the vector $(u(a), u'(a))$ is a multiple of the vector $(-a_{12}, a_{11})$ and a solution of $L(u) = 0$ is uniquely determined by the values of $u(a)$ and $u'(a)$. A similar analysis holds for u_b, and it follows that $\{u_a(x), u_b(x)\}$ are linearly independent on $[a, b]$. In particular, the Wronskian $W(x) = u_a(x)u_b'(x) - u_a'(x)u_b(x)$ is never zero in the interval.

The method of variation of parameters proceeds to seek a solution of the equation $L(u) = f$ of the form

$$u(x) = C_a(x)u_a(x) + C_b(x)u_b(x) \tag{4.3.10}$$

under the constraint that $C_a'u_a + C_b'u_b = 0$. Substituting this form for u into the equation and simplifying yields

$$p(C_a'u_a' + C_b'u_b') = f.$$

This equation along with the constraint is a system of linear equations that is easily solved to give

$$C_a'(x) = -\frac{f(x)u_b(x)}{p(x)W(x)}, \quad C_b'(x) = \frac{f(x)u_a(x)}{p(x)W(x)}. \tag{4.3.11}$$

By computation (Exercise 4.20), $pW' = -p'W$ and hence $(pW)' = 0$. It follows that $p(x)W(x) = \mu_0^{-1}$ is a constant. This constant can be determined by evaluating the Wronskian at one point. Now from these formulas, the functions

C_a and C_b can be determined up to integration constants; imposing the conditions $C_a(b) = 0$ and $C_b(a) = 0$ determines the constants and these functions uniquely. Applying this procedure, we obtain

$$C_a(x) = \mu_0 \int_x^b f(\xi) u_b(\xi) d\xi, \quad C_b(x) = \mu_0 \int_a^x f(\xi) u_a(\xi) d\xi,$$

and a solution to the equation $L(u) = f$ is given by

$$u(x) = u_a(x)\mu_0 \int_x^b f(\xi) u_b(\xi) d\xi + u_b(x)\mu_0 \int_a^x f(\xi) u_a(\xi) d\xi.$$

Now, we introduce the function

$$G(x,\xi) = \begin{cases} \mu_0 u_b(\xi) u_a(x), & a \le x \le \xi \\ \mu_0 u_a(\xi) u_b(x), & \xi \le x \le b, \end{cases} \tag{4.3.12}$$

and the solution can be written more concisely as

$$u(x) = \int_a^b G(x,\xi) f(\xi) d\xi.$$

The function $G(x,\xi)$ has the following properties:

1. $x \to G(x,\xi)$ satisfies the homogeneous equation $L(u) = 0$ except at $x = \xi$.
2. $x \to G(x,\xi)$ satisfies the boundary conditions.
3. $G(x,\xi)$ is continuous for $a \le x, \xi \le b$.
4. The partial derivative $\frac{\partial G}{\partial x}$ is continuous except when $x = \xi$, and at this point there is a jump discontinuity given by
$$\frac{\partial G}{\partial x}(\xi + 0, \xi) - \frac{\partial G}{\partial x}(\xi - 0, \xi) = \frac{1}{p(\xi)}.$$
5. The function $G(x,\xi)$ is symmetric, i.e., $G(x,\xi) = G(\xi, x)$.

From properties 1 and 4, we can write $p(x)\frac{\partial G}{\partial x} = v(x) + H(x - \xi)$, where $v(x)$ is twice continuously differentiable on the interval $a \le x \le b$ and further $v' + qG = 0$ for $x \in (a,b)$. Hence, the jump discontinuity in the partial derivative of G implies that

$$\frac{d}{dx}\left(p \frac{\partial G}{\partial x}\right) + qG = \delta(x - \xi), \tag{4.3.13}$$

a formula akin to those earlier in this section. The function $G(x,\xi)$ is called the Green's function for the BVP (4.3.8). The delta function viewpoint lends interpretation of the solution formula (4.3.9) as a superposition of sources via the delta functional with coefficients coming from the function f. The Green's function is characterized by properties 1 through 4; alternatively, properties 1 and 4 can be replaced by (4.3.13).

Before giving an example, we outline how to derive the formula for the Green's function using distributional notions and formulas. Essentially we are going to use variation of parameters interpreting the resulting derivatives as distributions. The starting point is the same: formula (4.3.10) where now we want $L(u) = \delta(x-\xi)$. Assuming the previously used constraint equation, the analog of (4.3.11) becomes

$$C_a'(x) = -\mu_0 u_b(x)\delta(x-\xi), \quad C_b'(x) = \mu_0 u_a(x)\delta(x-\xi).$$

Now, as distributions we have $u_b(x)\delta(x-\xi) = u_b(\xi)\delta(x-\xi)$ and $u_a(x)\delta(x-\xi) = u_a(\xi)\delta(x-\xi)$ (see Example 4.8). It follows that

$$C_a(x) = -\mu_0 u_b(\xi)[H(x-\xi)+A], \quad \text{and} \quad C_b(x) = \mu_0 u_a(\xi)[H(x-\xi)+B],$$

for some constants A and B. The condition $C_a(b) = 0$ implies $0 = -\mu_0 u_b(\xi)[1+A]$, and hence $A = -1$. Similar reasoning produces $B = 0$. Thus $C_a(x)$ and $C_b(x)$ are completely determined and the desired Green's function $u(x) = G(x, \xi)$ is

$$G(x,\xi) = \mu_0 u_b(\xi)u_a(x)[1-H(x-\xi)] + \mu_0 u_a(\xi)u_b(x)H(x-\xi).$$

This is an alternative way to write our original formula (4.3.12).

Example 4.15 We are going to find the Green's function for the equation $u'' + u = f$ with boundary conditions $u(0) = 0$ and $u'(0) = 0$. Any solution of the corresponding homogeneous equation has the form $A\cos x + B\sin x$. Choosing $A = 0$ and $B = 1$, we can take $u_0(x) = \sin x$. To find u_π, we need to choose the constants so that

$$-A\sin\pi + B\cos\pi = 0.$$

It follows that $B = 0$ and A can be arbitrary, so we set $u_\pi(x) = \cos x$. The Wronskian of these functions computes as $W(x) = -1$. Thus $\mu_0 = -1$ and the Green's function for this problem is then given by

$$G(x,\xi) = \begin{cases} -\cos\xi\sin x, & 0 \le x \le \xi \\ -\sin\xi\cos x, & \xi \le x \le \pi. \end{cases}$$

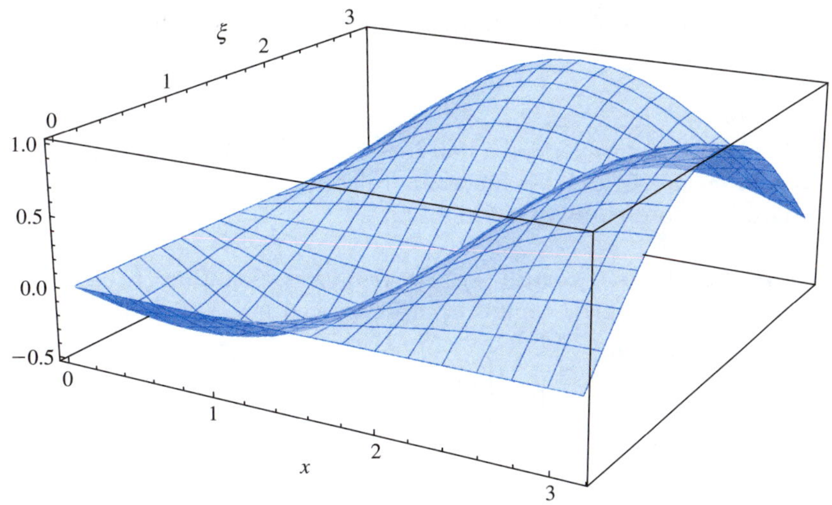

Figure 4.3 Green's Function from Example 4.15

The solution of the BVP will be

$$u(x) = -\cos x \int_0^x f(\xi)\sin\xi\, d\xi - \sin x \int_x^\pi f(\xi)\cos\xi\, d\xi.$$

We can easily check that $u'' = -u + f$ for any function $f \in C[a,b]$. Figure 4.3 provides a surface plot of the negative of this Green's function, which illustrates properties 2–5.

Exercises

4.12 a. Verify that (4.3.1) gives the solutions of $u'' = \phi$ when ϕ is a test function.

b. Obtain formula (4.3.1) by applying variation of parameters to the differential equation $u'' = \phi$.

4.13 Show that the $K(x) = -\frac{1}{4\pi|x|}$ is a fundamental solution of the Laplacian on $\mathbb{R}^3$.

4.14 Another physical interpretation of result of Example 4.13 comes from the gravitational field of a unit mass placed at the origin given by $\mathbf{F} = -G\frac{x}{|x|^3}$. The gravitational potential is given by $\mathbf{F} = -\boldsymbol{\nabla} u$. Show that $u(x) = -\frac{G}{|x|}$ and conclude that u satisfies $\nabla^2 u = 4\pi G\delta$. If $\rho(x)$ is a continuous distribution of mass over a region D of space, what formula gives the resulting gravitational potential?

4.15 Let $f(x) = 1$ for $1 < x < 3$ and $f(x) = 0$, otherwise. Determine the solution to the heat flow problem on a half-line with boundary condition $u(0,t) = 0$ explicitly in terms of the error function. Take $\alpha = 1$ and plot the solution and initial data for representative values of t.

4.16 A **modified wave kernel** is the distributional solution of the wave equation on $\mathbb{R}$ where the initial conditions are replaced by

$$\text{IC:}\ \begin{cases} u(x,0) = \delta(x) \\ u_t(x,0) = 0. \end{cases}$$

Show that the modified wave kernel is given by $\frac{\partial \mathcal{W}}{\partial t}$, where $\mathcal{W}$ is the wave kernel.

4.17 Let $g(x) = 1$ for $2 < x < 3$ and $g(x) = 0$, otherwise. Determine explicitly $u(x,t)$ as given by the formula (4.3.7). Take $c = 1$ and plot the solution in space–time.

4.18 Determine the heat kernel for the half-line where the boundary condition is given by $u_x(0,t) = 0$.

4.19 Determine the modified wave kernel on a half-line with boundary condition $u(0,t) = 0$.

4.20 For the differential equation in (4.3.8), verify that $pW' = -p'W$, where W is the Wronskian of the solutions u_a and u_b.

4.21 For the differential equation $u'' = f$ on $[0,\pi]$, find the Green's function associated with the following sets of boundary conditions.

a. $u(0) = u(\pi) = 0$

b. $u(0) = u'(\pi) + hu(\pi) = 0\ (h > 0)$

For $h = 2$, provide a surface plot of the Green's function.

4.22 Find the Green's function for the ODE $u'' - \mu^2 u = f$ for each of the following sets of boundary conditions.

a. $u(0) = u(\pi) = 0$

b. $u(0) = u_x(\pi) = 0$

4.23 Find the Green's function for the following BVP on the interval $[1, 2]$: $x^2u'' + 4xu' + 2u = f$, $u(1) = u'(2) = 0$.

4.24 Find the Green's function for the following BVP on the interval $[0, \infty)$: $u'' + \mu^2 u = f$, $u(0) = 0$, $\lim_{x\to\infty} u(x) = 0$. (Although this is on an unbounded interval, work the problem in fashion similar to those above.)

Chapter 5

Fourier Expansions

Expanding on the geometrical and symmetry ideas presented in Chapter 3, we begin this chapter by introducing the method of separation of variables, which in effect reduces the problem of finding solutions of a partial differential equation and associated boundary conditions to solving a set of ordinary differential equations, at least one of which carries boundary conditions. This naturally leads to the study of Fourier series and, more generally, eigenfunction expansions.

5.1 Separation of Variables

In Example 3.4, we constructed solutions of Laplace's equation in the plane using the invariance property of the Laplacian under rotations. Specifically, this invariance lead to the conclusion that if $u(r,\theta)$ is a solution of Laplace's equation, then so is $u(r,\theta+\theta')$ for any θ'. A function of the form $g(r)e^{in\theta}$ is smooth in the θ variable and generates a one-dimensional vector space of functions that is preserved under rotations. This motivated the search for solutions of Laplace's equation of this form, i.e., determining the function g. The result of this process was that Laplace's equation has solutions of the form

$$u(r,\theta) = \begin{cases} 1, & r^{|n|}e^{in\theta} \\ \log r, & r^{-|n|}e^{in\theta} \end{cases}, \; n \in \mathbb{Z}.$$

The functions 1 and $\log r$ come from the case $n = 0$. The first row shows solutions of Laplace's equation in the entire plane; the second row shows solutions in the plane minus the origin.

The vein of thought applied to Laplace's equation can be applied to other equations as well. Like the case of ordinary differential equations, linear homogeneous partial differential equations with constant coefficients possess exponential solutions. The geometrical reason is that the equation is invariant under translations in the independent variables, i.e., solutions remain so under translation. For example, if $u(x,t)$ is a solution of the heat equation $u_t = \alpha^2 u_{xx}$, then so is $u(x+h, t+\tau)$ for any constants h and τ. We say that translations are a symmetry of the heat equation. In general, a symmetry of a PDE is a transformation in the independent variables that preserves solutions.

In seeking solutions of the heat equation, we are going to look in vector spaces of functions that already possess the translation symmetry. Such a vector space would be generated by a differentiable function $g(x,t)$ such that

$$g(x+h, t+\tau) = C(h,\tau)g(x,t),$$

for some function $C = C(h,\tau)$. Setting $\tau = 0$ and subtracting $g(x,t)$ from both sides yields

$$g(x+h,t) - g(x,t) = [C(h,0) - 1]g(x,t).$$

By dividing through by h and letting $h \to 0$, we obtain the following partial differential equation for g:

$$\frac{\partial g}{\partial x} = \mu g,$$

where $\mu = \frac{\partial C}{\partial h}(0,0)$ is a constant. Similarly, we have $\frac{\partial g}{\partial t} = \nu g$, for some constant ν. This is a simple system of partial differential equations that can be solved by integration; the result is

$$g(x,t) = C\exp(\mu x + \nu t),$$

where C is now a constant. The vector space generated by $\exp(\mu x + \nu t)$ is invariant under translation, so the idea continues by seeking solutions of the heat equation of this form. Direct substitution of g into the heat equation gives the algebraic relation between μ and ν: $\nu = \alpha^2\mu^2$. Therefore, for any scalar μ, the heat equation has solutions of the form

$$u(x,t) = \exp((\alpha\mu)^2 t + \mu x). \qquad (5.1.1)$$

For each μ, this defines a one-dimensional vector space of solutions of the homogeneous heat equation as scalar multiples of this function are again solutions.

A key observation regarding our geometric approach to finding solutions in these examples is that it leads to solutions in the form of a product of functions of the independent variables: $u(r,\theta) = R(r)\Theta(\theta)$ in the case of Laplace's equation and $u(x,t) = T(t)X(x)$ in the case of the heat equation. Such solutions are called **separated solutions**. Finding separated solutions of a linear homogeneous partial differential equation is a powerful technique for solving IBVPs and BVPs. Generally, the separated solutions sought also satisfy natural (homogeneous or periodic) boundary conditions. In the following examples, we revisit Laplace's equation and the heat equation using this method, generally called the **method of separation of variables**.

Example 5.1 Consider the following BVP for Laplace's equation on the unit disk written in polar form. Here the unknown function is $u = u(r, \theta)$.

$$\begin{aligned} &\text{PDE: } u_{rr} + \tfrac{1}{r}u_r + \tfrac{1}{r^2}u_{\theta\theta} = 0, \quad 0 \le r < 1, \quad -\pi \le \theta \le \pi \\ &\text{BC: } u(1, \theta) = f(\theta) \end{aligned} \tag{5.1.2}$$

In this example we will propose a solution to this problem using the method of separation of variables; the focus here is on the *method*, leaving rigorous justification for later. Separated solutions have the form $u(r, \theta) = R(r)\Theta(\theta)$. Substituting this form into the PDE and rearranging leads to

$$\left(r^2 R'' + rR'\right)\Theta + R\Theta'' = 0.$$

An abuse of notation occurs in this formula: The prime on R means differentiation with respect to r and on Θ with respect to θ. Dividing by the product $R\Theta$ and rearranging, we obtain

$$\frac{r^2 R'' + rR'}{R} = -\frac{\Theta''}{\Theta}.$$

Here, the left-hand side is a function of r and the right-hand side a function of θ. As these are independent variables, these ratios must be equal to a constant μ, called the **separation constant**. This provides a pair of ordinary differential equations for the functions $R(r)$ and $\Theta(\theta)$:

$$\begin{aligned} r^2 R'' + rR' - \mu R = 0, &\quad 0 \le r < 1 \\ \Theta'' + \mu\Theta = 0, &\quad -\pi \le \theta \le \pi. \end{aligned}$$

The first equation is an Euler equation and has solutions of the form r^α; the second has constant coefficients and has exponential solutions. In both cases, the form of their solution depends on the separation constant μ and the geometry of the domain will provide information on this constant. Ultimately $u(r, \theta) = R(r)\Theta(\theta)$ must be a solution to Laplace's equation in the disk, hence the function Θ must be continuous and have continuous first derivative. This provides matching conditions, or hidden boundary conditions, at the end-points of the θ interval, i.e.,

$$\Theta(-\pi) = \Theta(\pi)$$
$$\Theta'(-\pi) = \Theta'(\pi).$$

These are called **periodic boundary conditions**. The next step in the process is to determine solutions of the equation for Θ that satisfy these conditions. It should be noted that $\Theta(\theta) = 0$ is always a solution of the ODE and the Θ-boundary conditions. Hence we are interested in determining the values of the separation constant μ that yield nontrivial solutions of the Θ-BVP. There are three cases to consider: (i) $\mu = \lambda^2 < 0$; (ii) $\mu = 0$; (iii) $\mu = \lambda^2 > 0$.
Case (i). Here the solutions for Θ take the form

$$\Theta(\theta) = c_1 \cosh \lambda\theta + c_2 \sinh \lambda\theta.$$

Imposing the boundary conditions produces a linear system of two equations for the coefficients:

$$c_1 \cosh \lambda\pi - c_2 \sinh \lambda\pi = c_1 \cosh \lambda\pi + c_2 \sinh \lambda\pi$$
$$-\lambda c_1 \sinh \lambda\pi + \lambda c_2 \cosh \lambda\pi = \lambda c_1 \sinh \lambda\pi + \lambda c_2 \cosh \lambda\pi.$$

The first equation leads to $c_2 \sinh \lambda\pi = 0$. However, $\sinh \lambda\pi = 0$ only when $\lambda = 0$, which is excluded in this case. Therefore $c_2 = 0$. The second equation now becomes $\lambda c_2 \sinh \lambda\pi = 0$, which by similar reasoning yields $c_2 = 0$. We conclude that only trivial solutions are possible in the case $\mu < 0$.
Case (ii). In this case $\Theta(\theta) = c_1\theta + c_2$. The system of equations takes form

$$-c_1\pi + c_2 = c_1\pi + c_2$$
$$c_2 = c_2.$$

It is immediate that $c_1 = 0$ and c_2 can be an arbitrary constant. Therefore, corresponding to $\mu = 0$, the BVP has solution $\Theta(\theta) = 1$ (or any multiple thereof).
Case (iii). For this case, $\Theta(\theta) = c_1 e^{i\lambda\theta} + c_2 e^{-i\lambda\theta}$ and the system of equations becomes

$$c_1 e^{-i\lambda\pi} + c_2 e^{i\lambda\pi} = c_1 e^{i\lambda\pi} + c_2 e^{-i\lambda\pi}$$
$$i\lambda c_1 e^{-i\lambda\pi} - i\lambda c_2 e^{i\lambda\pi} = i\lambda c_1 e^{i\lambda\pi} - i\lambda c_2 e^{-i\lambda\pi}.$$

Dividing the second equation by $i\lambda$ and adding it to the first reduces the system to $c_1 e^{-i\lambda\pi} = c_1 e^{i\lambda\pi}$ or, after applying Euler's identity,

$$c_1 \sin \lambda\pi = 0.$$

Hence, c_1 can be arbitrary if $\lambda = n$, an integer. Likewise, with this choice of λ, c_2 can be arbitrary.

Summarizing the three cases, the BVP for Θ has nontrivial solutions if and only if $\mu = n^2$ for some $n \in \mathbb{Z}$ and the corresponding solutions are multiples of

$$\Theta_n(\theta) = e^{in\theta}.$$

Notice that all possibilities are included in this formula as n can be any integer, positive, zero, or negative. Having determined the values of the separation constant, the third step in the process is to return to the equation for R, solve it for these values of μ, and write the separated solutions of the PDE. In this case, we need to solve the Euler equation

$$r^2R'' + rR' - n^2R = 0$$

on the interval $0 \leq r < 1$. The solutions to this equation, presented in the last chapter, are specifically

$$R_n(r) = \begin{cases} 1,\ \log r, & \text{if } n = 0 \\ r^{|n|},\ r^{-|n|}, & \text{if } n \neq 0. \end{cases}$$

As the solution must exist at $r = 0$, the solutions $\log r$ and $r^{-|n|}$ must be discarded. Hence, the separated solutions of Laplace's equation for this domain are

$$u_n(r, \theta) = r^{|n|}e^{in\theta},\ n \in \mathbb{Z}.$$

The final step in the process is to build a solution of Laplace's equation that satisfies its boundary condition $u(1, \theta) = f(\theta)$. This is where the fact that the PDE is linear and homogeneous comes into play. The idea is to use the superposition principle and seek a solution to the BVP (5.1.2) of the form

$$u(r, \theta) = \sum_n c_n r^{|n|}e^{in\theta}.$$

In general, this sum will be an infinite series with the sum being over all integers n. Hence, we will have to consider questions of convergence and whether the series can be differentiated term by term so as to justify that the series does define a solution to Laplace's equation. These technical questions are set aside for later. Heuristically, if the coefficients $\{c_n\}$ form a bounded sequence, since $r < 1$ the factor $r^{|n|}$ makes the general term of the series tend to zero rapidly.

The same applies to the series of term-by-term derivatives in either the r or θ variables. Hence, our expectation is that the infinite series does indeed provide a solution to Laplace's equation. Thus, the final question is: Can the coefficients $\{c_n\}$ be determined so as to satisfy the boundary condition for u, i.e.,

$$f(\theta) = \sum_n c_n e^{in\theta}? \tag{5.1.3}$$

For functions of particular form the answer to this question is easy. For example, if $f(\theta) = \sin 2\theta - \sin 5\theta$, then by applying Euler's identity to both sine terms, we obtain the coefficients

$$c_{\pm 2} = \pm\frac{1}{2i}, \quad c_{\pm 5} = \mp\frac{1}{2i}, \quad \text{and} \quad c_n = 0 \quad \text{for } n \neq \pm 2, \pm 5.$$

Indeed, the function

$$u(r,\theta) = r^2 \sin 2\theta - r^5 \sin 5\theta$$

solves the BVP for Laplace's equation in this case. Similarly, a solution to the problem can be easily found if the boundary data $f(\theta)$ is a finite linear combination of the complex exponentials $e^{in\theta}$. The question of representing general functions f as in (5.1.3) is more difficult and questions of convergence of the series play a role; however, the idea can be gleaned from the following simple integration formula:

$$\int_{-\pi}^{\pi} e^{in\theta} e^{-im\theta}\, d\theta = \begin{cases} 0, & m \neq n \\ 2\pi, & m = n. \end{cases} \tag{5.1.4}$$

Formally, multiplying both sides of (5.1.3) by $e^{-im\theta}$ and integrating produces

$$\int_{-\pi}^{\pi} f(\theta) e^{-im\theta}\, d\theta = \sum_n c_n \int_{-\pi}^{\pi} e^{in\theta} e^{-im\theta}\, d\theta = 2\pi c_m,$$

where we have casually assumed that interchanging the sum and integral is permitted. This is a stunning formula that, if justified, provides a simple algorithm for the desired representation. We will explore this algorithm further in the next section.

Remark 5.1 A few comments are in order relative to the above example.

1. In the treatment of the Θ-BVP, the possibility of a complex separation constant μ was not considered. The reason for this will be seen in the general theory presented in Section 5.4; see also Exercise 5.1.

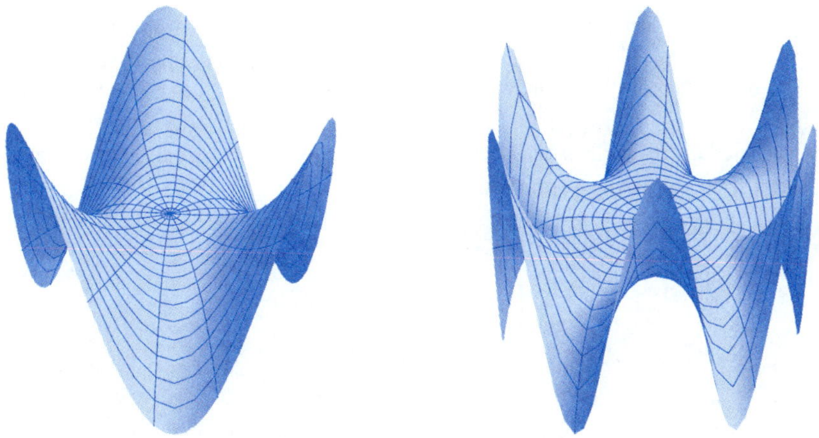

Figure 5.1 Separated Solutions of Laplace's Equation on a Disk

2. In the analysis of the Θ-BVP for $\mu = -\lambda^2$, the exponentials $\{e^{\lambda\theta}, e^{-\lambda\theta}\}$ could have been used instead of the hyperbolic cosine and sine functions. Often, the use of the latter simplifies computation; we used the the fact that hyperbolic cosine and sine are even and odd, respectively, in our computations.

3. Likewise, in the analysis of the Θ-BVP for $\mu = \lambda^2 > 0$, the complex exponentials could have been replaced with $\{\cos\lambda\theta, \sin\lambda\theta\}$. Using the latter, the separated solutions would take the form

$$u_n(r,\theta) = \begin{cases} 1, & n = 0 \\ r^n \cos n\theta,\ r^n \sin n\theta, & n \neq 0. \end{cases}$$

The choice of complex exponentials versus real-valued sine and cosine functions is a matter of choice, although complex arithmetic often facilitates computation. In either form, the functions $u_n(r,\theta)$ can be thought of as building blocks for more complicated solutions of Laplace's equation, specifically, those that satisfy prescribed boundary conditions. A plot of the building blocks $r^n \cos n\theta$ for $n = 3$ and 6 is given in Figure 5.1.

Example 5.2 This example parallels the last in the context of the heat equation. Consider the IBVP

$$\begin{aligned} \text{PDE: } & u_t = \alpha^2 u_{xx}, \quad 0 < x < \pi, \quad t > 0 \\ \text{BC: } & u(0,t) = u(\pi,t) = 0 \\ \text{IC: } & u(x,0) = f(x). \end{aligned} \tag{5.1.5}$$

The first step is to produce separated solutions of the heat equation that satisfy the homogeneous Dirichlet boundary conditions. These have the form $u(x,t) = T(t)X(x)$. Forcing the separated solution to satisfy the boundary conditions implies $T(t)X(0) = 0$ and $T(t)X(\pi) = 0$. Avoiding the trivial solution, this implies that $X(0) = X(\pi) = 0$. As in the preceding example we are led to the following equations:

$$T' + \alpha^2 \mu T = 0 \quad \text{and} \quad \begin{cases} \text{ODE:}\, X'' + \mu X = 0, & 0 < x < \pi \\ \text{BC:}\, X(0) = X(\pi) = 0, \end{cases}$$

where μ is the separation constant. The boundary value problem for X has solution $X(x) = 0$ for any value of μ. The focus is to determine values of μ for which this problem has nontrivial solutions. There are three cases to be explored: $\mu = 0$, $\mu = -\lambda^2 < 0$, and $\mu = \lambda^2 > 0$. The first case gives that $X(x) = c_1 x + c_2$, a linear function. Invoking the boundary conditions easily gives that $c_1 = c_2 = 0$, i.e., only trivial solutions arise in this case. For the second case the solutions of the ODE can be written in the form

$$X(x) = c_1 \cosh \lambda x + c_2 \sinh \lambda x,$$

and it is easy to see that the X-boundary conditions lead to the trivial solution for X. In the third case, solutions take the form

$$X(x) = c_1 \cos \lambda x + c_2 \sin \lambda x. \tag{5.1.6}$$

The first X-boundary condition implies $c_1 = 0$. From this and the second boundary condition we get the equation $c_2 \sin \lambda\pi = 0$. For meaningful solutions we must choose λ to satisfy $\sin \lambda\pi = 0$. Hence, there are an infinite number of solutions: $\lambda = n$, $n = 1, 2, 3, \ldots$. The corresponding solutions of the X-BVP are $X_n(x) = \sin nx$. Notice that any multiple of a solution is again a solution (the ODE is linear and homogeneous), explaining why we restricted n to the positive integers. With $\mu = n^2$, solving the equations for T yields separated solution of the heat equation satisfying the boundary conditions of the form

$$u_n(x,t) = e^{-(\alpha n)^2 t} \sin nx, \; n = 1, 2, \ldots .$$

Note that for each fixed n, this actually defines a one-dimensional vector space of solutions as any multiple of this function satisfies the heat equation and boundary conditions. We can build more solutions of the heat equation that satisfy the boundary conditions using superposition, i.e.,

$$u(x,t) = \sum_n a_n e^{-(\alpha n)^2 t} \sin nx, \tag{5.1.7}$$

and this sum is typically an infinite series. The question at this point is: Can the coefficients $\{a_n\}$ be chosen so as to satisfy the initial condition? If the initial data $f(x)$ has a simple form, say

$$f(x) = \sin x - 2\sin 3x,$$

then the answer is self-evident. Indeed in this case the unique solution to the IBVP would be

$$u(x,t) = e^{-\alpha^2 t}\sin x - 2e^{-9\alpha^2 t}\sin 3x.$$

In the general case, the problem is one of choosing the coefficients that provide the representation

$$f(x) = \sum_n a_n \sin nx. \qquad (5.1.8)$$

As in the previous example, the following formula from calculus provides a hint to the answer:

$$\int_0^\pi \sin nx\,\sin mx\,dx = \begin{cases} 0, \text{ if } n = m \\ \frac{\pi}{2}, \text{ if } n \neq m. \end{cases} \qquad (5.1.9)$$

Multiplying both sides of (5.1.8) by $\sin mx$ and integrating gives

$$\int_0^\pi f(x)\sin mx\,dx = \sum_n a_n \int_0^\pi \sin nx \sin mx\,dx = \frac{\pi}{2}a_m, m = 1, 2, \dots .$$

Conceivably, we have found a series solution to the IBVP of (5.1.5) with coefficients computed from this formula. As an explicit example, let $f(x) = 1$. Computing the coefficients we have

$$a_n = \frac{2}{\pi}\int_0^\pi \sin nx\,dx = \frac{2}{\pi}\frac{(1-(-1)^n)}{n}.$$

Notice that $a_n = 0$ if n is odd, thus the series solution can be written

$$u(x,t) = \frac{4}{\pi}\sum_{n=0}^{\infty}\frac{1}{2n+1}e^{-(\alpha(2n+1))^2 t}\sin(2n+1)x.$$

Because of the rapid decay of the exponential factor, the series can be differentiated term by term in both variables and hence does define a solution of the heat equation. The question remains regarding the recovery of the initial condition, i.e., does

$$1 = \frac{4}{\pi}\sum_{n=0}^{\infty}\frac{1}{2n+1}\sin(2n+1)x?$$

We can explore this question graphically by truncating the series. In Figure 5.2, the series for $u(x,t)$ is plotted for various values of t using the first twelve nonzero terms. Intuitively, we expect the solution to decay to zero as t gets large due to the fact that the ends are held at constant temperature zero. Thus, even the sixth partial sum of the series solutions gives a decent representation of the solution at $t = 0.05$. Figure 5.3 provides plots of representative partial sums of the above sine series. Notice the oscillatory behavior of the partial sums about the graph of $f(x) = 1$ and that the partial sums have peaks near $x = 0$ and π.

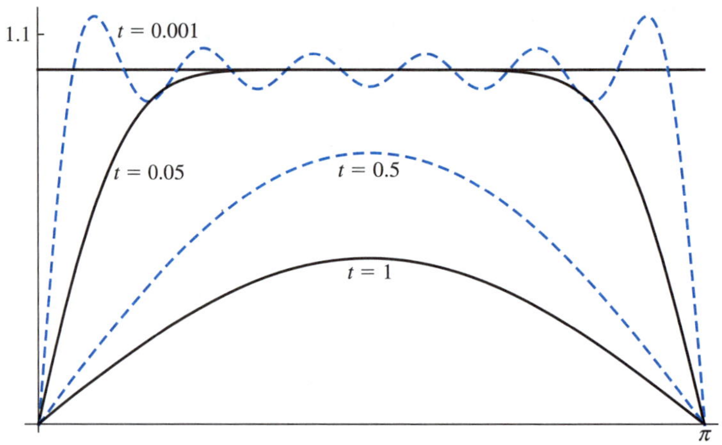

Figure 5.2 Solution of Heat Equation, IC: $u(x,0) = 1$

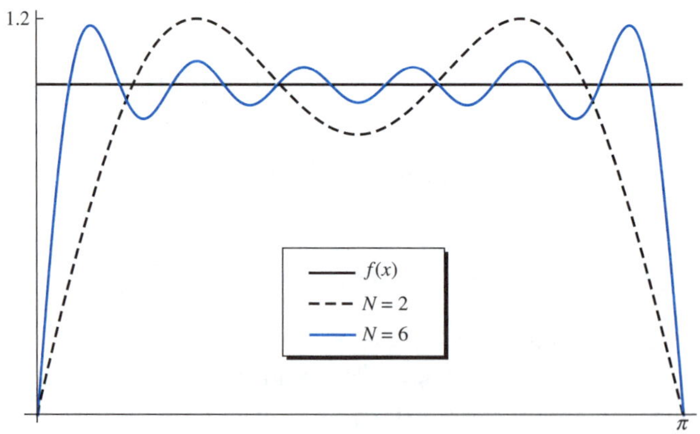

Figure 5.3 Partial Sums of Sine Series

The latter is an example of Gibb's phenomena (illustrated further in Section 5.3); this effect is also present in the previous figure for the smallest value of t.

Example 5.3 In Example 5.2, suppose the boundary conditions are replaced by

$$u(0,t) = 0 \quad \text{and} \quad u_x(\pi,t) + u(\pi,t) = 0.$$

In this case, the X-BVP becomes

$$\text{ODE: } X'' + \mu X = 0, \quad 0 < x < \pi$$

$$\text{BC: } \begin{cases} X(0) = 0 \\ X'(\pi) + X(\pi) = 0. \end{cases}$$

We leave it for the reader to check that the cases $\mu \leq 0$ generate only trivial solutions. If $\mu = \lambda^2 > 0$, then referring to (5.1.6), $c_1 = 0$ and λ must be chosen to satisfy the equation $\tan \pi\lambda = -\lambda$. This transcendental equation has an infinite number of solutions $\lambda_n \nearrow \infty$, as is indicated in Figure 5.4. For the corresponding solutions we choose $X_n(x) = \sin \lambda_n x$; the separated solutions of the heat equation satisfying the boundary conditions are

$$u_n(x,t) = e^{-(\alpha\lambda_n)^2 t} \sin \lambda_n x.$$

At this point, superposition can be applied following the ideas of the preceding two examples. The analogous calculus formula takes the form

$$\int_0^\pi \sin \lambda_n x \sin \lambda_m x \, dx = \begin{cases} 0, & n \neq m \\ \Lambda_m, & n = m, \end{cases} \tag{5.1.10}$$

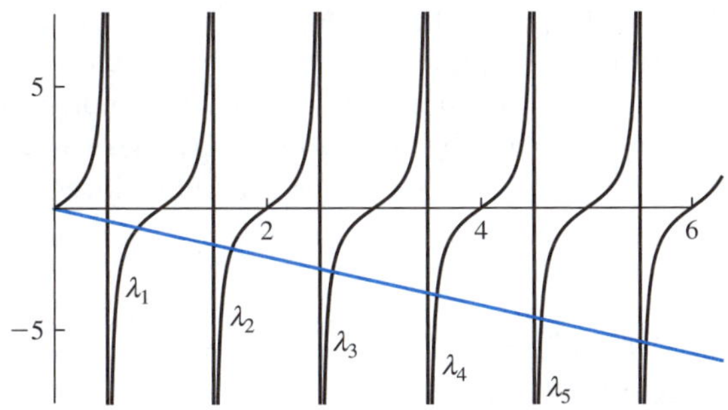

Figure 5.4 Solutions of $\tan \pi\lambda = -\lambda$

where $\Lambda_m \neq 0$ and can be computed. The reader is invited to obtain this formula via direct computation in Exercise 5.7.

Remark 5.2

1. The use of the interval $[0, \pi]$ in the previous two examples may seem to be artificial. However, the heat equation $u_t = \alpha^2 u_{xx}$ on the interval $[0, L]$ is transformed into the equation $u_t = \beta^2 u_{ss}$ where $0 < s < \pi$ under the transformation $s = \frac{\pi x}{L}$. Here $\beta = \left(\frac{\alpha \pi}{L}\right)$.

2. The calculus formulas (5.1.9) and (5.1.10) are called orthogonality conditions. On the vector space of continuous functions on $[0, \pi]$, we have an inner product

$$\langle f, g \rangle = \int_0^\pi f(x)\, g(x)\, dx.$$

An **inner product** *on a space of functions is what the dot product is for vectors in* $\mathbb{R}^d$. Two functions are said to be **orthogonal** if $\langle f, g \rangle = 0$. In this terminology, elements of the set of functions $\{\sin nx\}$ are orthogonal in the vector space of continuous functions on $[0, \pi]$. The method of separation of variables generically generates orthogonality conditions connected with the structure of the boundary value problem; we will formalize this in the fourth section.

3. The method of separation of variables works in general as long as the equations and boundary conditions are homogeneous and the equation is in canonical form. See Exercise 5.9 for an example of an equation in noncanonical form where it breaks down. For equations not in canonical form, often the idea of using symmetries of the equation will lead to solutions.

Historical Comment The question of representing functions using a series of sines or cosines arose in the work of Fourier in his extensive and pioneering study of heat phenomena in solids during the years 1803–1807. Interestingly, the definite integral was still decades away and Fourier's method for finding the coefficients in a series such as (5.1.8) was essentially algebraic. Bernhard Riemann in the 1850s considered the problem of representing a function by a trigonometric series and along the way conceived what we now refer to as the Riemann integral.

Exercises

5.1 Consider the BVP $X'' + \mu X = 0$ on the interval $(0, \pi)$ with boundary conditions $X(0) = X(\pi) = 0$. If $\mu = \lambda + i\eta$ with $\eta \neq 0$, show that the problem has only the trivial solution $X = 0$.

5.2 Consider the IBVP (5.1.5) with boundary conditions $u_x(0,t) = u_x(\pi,t) = 0$. Determine all separated solutions that satisfy the boundary conditions. What do you conjecture for the solution of the IBVP?

5.3 Consider IBVP (5.1.5) with boundary conditions $u(0,t) = u_x(\pi,t) = 0$. Determine all separated solutions that satisfy the boundary conditions. What do you conjecture for the solution of the IBVP?

5.4 Determine the separated solutions of the wave equation $u_{tt} = c^2 u_{xx}$ that satisfy the following sets of boundary conditions.

a. $u(0,t) = 0,\ u(\pi,t) = 0$

b. $u(0,t) = 0,\ u_x(\pi,t) = 0$

5.5 Consider IBVP (5.1.5) on the interval $[0, L]$. For each of the following boundary conditions, determine all separated solutions that satisfy the following boundary conditions and conjecture a form for the series solution of the IBVP.

a. $u(0,t) = u(L,t) = 0$

b. $u_x(0,t) = u_x(L,t) = 0$

c. $u(0,t) = u_x(0,t) = 0$

5.6 In Example 5.3, show that only the trivial solution occurs when $\mu \leq 0$.

5.7 Suppose $\{\lambda_n\}$ are chosen as in Example 5.3. Show by direct computation that if $n \neq m$, then

$$\int_0^\pi \sin \lambda_n x \, \sin \lambda_m x \, dx = 0.$$

HINT: Apply the product formula for the sine function, integrate, and use the defining relation for λ.

5.8 Consider Laplace's equation $u_{xx} + u_{yy} = 0$ on the square $\{0 < x < \pi,\ 0 < y < \pi\}$. Determine all separated solutions that satisfy the following boundary conditions.

a. $u(0,y) = 0$ and $u_x(\pi,y) = 0$

b. $u(x,0) = 0$ and $u(x,\pi) = 0$

5.9 Can separation of variables be applied to $u_{xx} + 2u_{xy} + u_{yy} = 0$? Put this equation in canonical form and answer the same question.

5.2 Inner Products and Orthogonality

Let $\rho \in C[a,b]$ be strictly positive. Given two functions f and $g \in C[a,b]$, consider the expression

$$\langle f, g \rangle_{\rho} = \int_a^b f(x)\overline{g(x)}\,\rho(x)dx, \tag{5.2.1}$$

called the **inner product** of f and g. If all functions considered are real valued, we can delete the conjugate over g. The function ρ is called the weight function. If there is no confusion regarding the weight, we will delete the subscript and simply write $\langle f, g \rangle$ for the inner product.

The inner product forms a map $C[a,b] \times C[a,b] \to \mathbb{C}$ with the following properties.

1. **Positive definiteness.** $\langle f, f \rangle \geq 0$ and $\langle f, f \rangle = 0$ if and only if $f = 0$.
2. **Conjugate symmetry.** $\langle g, f \rangle = \overline{\langle f, g \rangle}$ (symmetry in the case of real-valued functions).
3. **Linearity.** If α and β are scalars, then $\langle \alpha f + \beta g, h \rangle = \alpha \langle f, h \rangle + \beta \langle g, h \rangle$.

The proof of these is left as exercises. In mathematical language, an inner product is called a positive definite conjugate symmetric bilinear form on the space $C[a,b]$. (The phrase "conjugate symmetric" is replaced with "symmetric" if all functions are real valued.)

Remark 5.3 The notion of an inner product can be introduced on a general vector space V over the real or complex field as a map from $V \times V \to \mathbb{R}$ or $\mathbb{C}$ that satisfies the above three properties where f, g, and h are in V. *All of the results of this section can be formulated and proved in this abstract setting.* We have chosen to formulate the results in the concrete setting of vector spaces of functions defined on an interval because our applications are of this type.

A fourth fundamental property of the inner product, which follows from the first three, is the **Cauchy–Schwarz inequality**.

Proposition 5.4 Let f and g be in $C[a,b]$. Then $|\langle f, g \rangle| \leq \|f\|_2 \|g\|_2$, where $\|f\|_2 = \left(\int_a^b |f(x)|^2 \rho(x)dx\right)^{1/2}$ is called the norm of f.

Proof We may assume $g \neq 0$, since otherwise both sides of the inequality are zero. Also, neither $|\langle f, g \rangle|$ nor the product $\|f\|_2\|g\|_2$ is affected if we multiply f by a scalar of modulus 1. Therefore we can replace f by cf, where c is a complex number with $|c| = 1$ so that $\langle f, g \rangle$ is real valued. This reduces the issue to considering the case where $\langle f, g\rangle$ is a real number. Applying the properties of the inner product, for any real number λ,

$$0 \leq \|f + \lambda g\|_2^2 = \|f\|_2^2 + 2\lambda \langle f, g\rangle + \lambda^2 \|g\|_2^2.$$

The right-hand side is a quadratic function of λ and achieves its minimum at $\lambda = -(f, g)/\|g\|^2$. Substituting this value of λ into the inequality we obtain

$$0 \leq \|f\|_2^2 - 2\frac{\langle f, g\rangle^2}{\|g\|_2^2} + \frac{\langle f, g\rangle^2}{\|g\|_2^2} = \|f\|_2^2 - \frac{\langle f, g\rangle^2}{\|g\|_2^2}.$$

Rearranging this inequality gives the result. □

An inner product is a direct analog of the dot product on $\mathbb{R}^3$ and as such enables us to introduce geometric properties on the vector space $C[a, b]$. The first such is that of length or *norm* of f defined in the proposition. The norm of a function provides a gauge on the size of the function and has the following properties, where we write $\|\cdot\|$ in place of $\|\cdot\|_2$ as the concept of a norm arises in more general contexts.

N1 Positive definiteness. $\|f\| \geq 0$ and $\|f\| = 0$ if and only if $f = 0$.

N2 Homogeneity. For any scalar α, $\|\alpha f\| = |\alpha|\|f\|$.

N3 Triangle Inequality. $\|f + g\| \leq \|f\| + \|g\|$.

The first two properties are a direct consequence of the first and third properties of the inner product. The triangle inequality is an easy application of the Cauchy–Schwartz inequality and the third property of inner products and is left to the reader.

Remark 5.5 On the space $C[a, b]$ another norm can be defined by

$$\|f\|_\infty = \max_{a \leq x \leq b} |f(x)|.$$

It is very easy to demonstrate that $\|\cdot\|_\infty$ satisfies the three properties N1–N3. In many respects, this norm is more natural to consider on the space $C[a, b]$. For example, if $\{f_n(x)\} \subset C[a, b]$, then $f_n \to f$ uniformly on $[a, b]$ if and only if $\lim_{n\to\infty} \|f_n - f\|_\infty = 0$. There is a simple comparison between the infinity norm and the square norm given by the inequality

$$\|f\|_2 \leq (b - a)^{1/2} \|f\|_\infty.$$

The inner product can be extended to other vector spaces, in particular the vector space of piecewise continuous functions on $[a, b]$. It is clear that the integral defining an inner product exists if f and g are piecewise continuous. However, property 1 of the inner product and hence N1 of the norm are in question. This is because if $f \in \mathrm{PC}[a, b]$ and $\|f\|_2 = 0$, the most we can conclude is that $f = 0$ except at a finite number of points since the integral actually ignores the values of a function at those points. If we think of such a set of points as outliers in an experimental process, the natural thing to do is to disregard them. Mathematically, we say that two piecewise continuous functions are equivalent if they differ on a finite set. With this understanding, then if $f \in \mathrm{PC}[a, b]$ and $\|f\|_2 = 0$, then f is equivalent to the zero function, and hence may be regarded as the zero function.

Another such extension of the inner product is to the class of all square integrable functions with respect to the weight function $\rho(x)$ on the interval $[a, b]$. This class is denoted $L^2([a, b], \rho(x)dx)$ or simply as $L^2[a, b]$ if there is no confusion regarding the weight function.[1] In the case $\rho = 1$, this space includes such functions as $f(x) = (x - a)^{-1/4}$. As we saw in Chapter 1, square integrable functions enter into energy computations for heat flow and vibrations, so this is a natural space to consider. For future reference, note the inclusions: $C[a, b] \subset \mathrm{PC}[a, b] \subset L^2[a, b]$.

The second geometrical notion to be introduced is that of orthogonality.

Definition 5.6 Two function f and g in $L^2[a, b]$ are said to be **orthogonal** if $\langle f, g\rangle = 0$. A collection of functions $\{\phi_n\} \subset L^2[a, b]$ are said to form an **orthogonal set** if they are all nonzero functions and $\langle \phi_n, \phi_m\rangle = 0$ for all $m \neq n$. An orthogonal set $\{\phi_n\}$ is said to be **orthonormal** if $\|\phi_n\|_2 = 1$ for all n.

Example 5.4 The set of functions $\{\sin nx\}_{n=1}^{\infty}$ is an orthogonal set on the interval $[0, \pi]$, where the inner product is defined by $\langle f, g\rangle = \int_0^{\pi} f(x)g(x)dx$. Since $\int_0^{\pi} \sin^2 nx\, dx = \frac{\pi}{2}$ for all n, the functions $\{\sqrt{\frac{2}{\pi}} \sin nx\}$ form an orthonormal set. These facts follow from the computations given in Example 5.2. By similar computations, the functions $\{\cos nx\}_{n=0}^{\infty}$ form an orthogonal set in $L^2[0, \pi]$ with respect to the same inner product. Based on these examples, the collection of functions $\{\frac{1}{2}, \cos nx, \sin nx\}_{n=1}^{\infty}$ forms an orthogonal set in $L^2[-\pi, \pi]$ with respect to the inner product $\langle f, g\rangle = \int_{-\pi}^{\pi} f(x)g(x)dx$. (We have used the constant function $\frac{1}{2}$ instead of allowing $n = 0$ in the cosine term so that the norm of all the functions in this last collection is the same, namely π.)

1. In the literature, the notation $L^2[a, b]$ designates the class of square integrable functions in the sense of Lebesgué. In his dissertation, Henri Lebesgué introduced a generalization of the Riemann integral that overcame the weaknesses of the latter in regard to the problem of interchanging sequential limits of functions and integrals (the Riemann integral essentially requires uniform convergence to justify this interchange). The Lebesgué integral has had far-reaching consequences in the theory of Fourier expansions and partial differential equations and is a fundamental tool in more advanced courses. While we use the Lebesgué notation here, the theory of the Lebesgué integral is not needed in this text.

Example 5.5 The collection of complex-valued functions $\{e^{inx}\}_{n=-\infty}^{\infty}$ is an orthogonal set in $L^2[-\pi, \pi]$ with respect to the inner product $\langle f, g\rangle = \int_{-\pi}^{\pi} f(x)\overline{g(x)}dx$. The norm of each function in this collection is 2π.

Next we give a generalization of the classic theorem about the lengths of the sides of a right triangle.

Proposition 5.7 (Pythagorean Theorem) Let f and g be in $L^2[a, b]$ with $\langle f, g\rangle = 0$. Then $\|f + g\|_2^2 = \|f\|_2^2 + \|g\|_2^2$.

Proof This is immediate from the computation

$$\|f + g\|_2^2 = \langle f + g, f + g\rangle = \langle f, f\rangle + \langle f, g\rangle + \langle g, f\rangle + \langle g, g\rangle.$$

As the two middle terms are zero and the first and last are square norms, the result follows. □

We can now present a result that makes the formulas heuristically derived in the examples of the preceding section completely natural. Indeed, this result is a guiding principle of what to look for when applying the method of separation of variables.

Theorem 5.8 Let $\{\phi_n\}_{n=1}^{N}$ be an orthonormal set in $L^2[a, b]$. Then

$$\|f - \sum_{n=1}^{N} \langle f, \phi_n\rangle\, \phi_n\|_2 \leq \|f - \sum_{n=1}^{n} c_n\phi_n\|_2$$

for all choices of constants $\{c_n\}$.

Proof By adding and subtracting, we have

$$f - \sum_{n=1}^{n} c_n\phi_n = \left(f - \sum_{n=1}^{N} \langle f, \phi_n\rangle\, \phi_n\right) + \sum_{n=1}^{N} (\langle f, \phi_n\rangle - c_n)\phi_n.$$

Notice that $\left(f - \sum_{n=1}^{N} \langle f, \phi_n\rangle\, \phi_n, \phi_m\right) = 0$ for all m. Hence, in the above identity, the first term on the right is orthogonal to the second term on the right. By the Pythagorean theorem,

$$\|f - \sum_{n=1}^{n} c_n\phi_n\|_2^2 = \|f - \sum_{n=1}^{N} \langle f, \phi_n\rangle\, \phi_n\|_2^2 + \sum_{n=1}^{N} |\, \langle f, \phi_n\rangle - c_n|^2.$$

The last sum on the right is nonnegative and equals zero precisely when $c_n = \langle f, \phi_n\rangle$ for all n. This establishes the result. □

This result can be paraphrased: The linear combination

$$\sum_n \langle f, \phi_n \rangle \phi_n$$

gives the best approximation of f among any linear combination of the functions $\phi_1, \phi_2, \ldots \phi_N$. The function

$$P[f](x) = \sum_{n=1}^{N} \langle f, \phi_n \rangle \, \phi_n(x)$$

is called the orthogonal projection of f onto the subspace generated by $\{\phi_n\}$. By way of analogy, given a vector $\mathbf{v} \in \mathbb{R}^3$, its projection onto the plane generated by the unit vectors $\mathbf{e}_1$ and $\mathbf{e}_2$ (i.e., the x_1x_2-plane) is given by the formula $(\mathbf{v} \cdot \mathbf{e}_1)\mathbf{e}_1 + (\mathbf{v} \cdot \mathbf{e}_2)\mathbf{e}_2$.

In the realm of boundary value problems, we rarely work with orthonormal sets. Their use in the previous result makes for simplified formulas and has a direct analogy with finite dimensional vector algebra. If $\{\psi_n\}$ is an orthogonal set in $L^2[a, b]$, then the projection of $f \in L^2[a, b]$ onto the subspace generated by $\{\psi_n\}_{n=1}^{N}$ is given by

$$P[f](x) = \sum_{n=1}^{N} \frac{\langle f, \psi_n \rangle}{\|\psi_n\|^2} \psi_n(x).$$

As in the orthonormal case, this projection gives the best approximation of f in norm among all linear combinations of the functions $\{\psi_n\}_{n=1}^{N}$.

Definition 5.9 Let $\{\phi_n\}_{n=1}^{\infty}$be an orthogonal set in $L^2[a, b]$ with respect to the inner product (5.2.1). The sequence of numbers defined by

$$\widehat{f}(n) = \frac{\langle f, \phi_n \rangle}{\|\phi_n\|_2^2}, \ n = 1, 2, \ldots$$

is called the sequence of Fourier coefficients of f with respect to $\{\phi_n\}$. The series

$$\sum_n \widehat{f}(n) \phi_n(x)$$

is called the Fourier series of f with respect to this orthogonal set.

Example 5.6 Consider the orthogonal set $\{\sin nx\}_{n=1}^{\infty}$ in $L^2[0,\pi]$, the inner product given by $\langle f,g\rangle = \int_0^\pi f(x)g(x)dx$. The numbers given by the formula

$$\widehat{f}_s(n) = \frac{2}{\pi}\int_0^\pi f(x)\sin nx\,dx$$

form the sequence of Fourier coefficients with respect to this orthogonal set. This is the same formula derived heuristically in Example 5.2. Usually, these are called the Fourier sine coefficients (on the interval $[0,\pi]$). Likewise, the series

$$f \sim \sum_{n=1}^{\infty} \widehat{f}_s(n)\sin nx$$

is called the Fourier sine series of f. The projection of f onto the subspace generated by $\{\sin nx\}_{n=1}^{N}$ has the form

$$P_N[f](x) = \sum_{n=1}^{N} \widehat{f}_s(n)\sin nx.$$

Notice that $P_N[f](x)$ is just the Nth partial sum of the Fourier sine series of f. The latter are usually denoted $S_N f(x)$.

We finish this section with two results. The first provides an estimate of the Fourier coefficients in terms of the norm of the function. The second concerns convergence of Fourier series in norm and is a direct consequence of the calculation in the proof of the first.

Proposition 5.10 (Bessel's Inequality) Let $\{\phi_n\}$ be an infinite orthogonal set in $L^2[a,b]$ and let $f \in L^2[a,,b]$. Then

$$\sum_{n=1}^{\infty} |\widehat{f}(n)|^2 \|\phi_n\|_2^2 \leq \|f\|^2.$$

Proof This is based on the following computation using the properties of inner products and orthogonality.

$$
\begin{aligned}
0 \le \left\| f - \sum_{n=1}^{N} \widehat{f}(n)\phi_n \right\|_2^2 &= \left\langle f - \sum_{n=1}^{N} \widehat{f}(n)\phi_n, f - \sum_{n=1}^{N} \widehat{f}(n)\phi_n \right\rangle \\
&= \|f\|_2^2 - 2\,\mathrm{Re}\left\langle f, \sum_{n=1}^{N} \widehat{f}(n)\phi_n \right\rangle + \left\| \sum_{n=1}^{N} \widehat{f}(n)\phi_n \right\|_2^2 \\
&= \|f\|_2^2 - 2\sum_{n=1}^{N} |\widehat{f}(n)|^2 \|\phi_n\|_2^2 + \sum_{n=1}^{N} |\widehat{f}(n)|^2 \|\phi_n\|_2^2 \\
&= \|f\|_2^2 - \sum_{n=1}^{N} |\widehat{f}(n)|^2 \|\phi_n\|_2^2
\end{aligned}
$$

□

Corollary 5.11 Given the hypothesis of the preceding proposition, then the following two statements are equivalent:

1. For every $f \in L^2[a,b]$,

$$\lim_{N\to\infty} \|f - \sum_{n=1}^{N} \widehat{f}(n)\phi_n\|_2 = 0 \quad \text{(convergence in norm).} \tag{5.2.2}$$

2. For every $f \in L^2[a,b]$,

$$\|f\|_2^2 = \sum_{n=1}^{\infty} |\widehat{f}(n)|^2 \|\phi_n\|_2^2 \quad \text{(Parseval's equality).} \tag{5.2.3}$$

A set of orthogonal functions for which either of the above equivalent statements hold is said to be a **complete orthogonal set**, a property that has a nice geometric characterization as follows. Indeed, if $\phi(x)$ is an L^2 function such that $\langle \phi, \phi_n \rangle = 0$ for all n, then by statement 1 in Corollary 5.11, $\phi(x) = 0$. In other words, for norm convergence to hold for all functions f, then the only function orthogonal to every element in the orthogonal set $\{\phi_n\}$ is the zero function. In fact, the converse is true.

Proposition 5.12 Let $\{\phi_n\}$ be an orthogonal set in $L^2[a,b]$. Then statement 1 of Corollary 5.11 holds if and only if the only function in $L^2[a,b]$ orthogonal to every ϕ_n is the zero function.

Remark 5.13 An essential idea is that the Fourier coefficients relative to an orthogonal system provide a decomposition of a function f in terms of a sequence $\{\widehat{f}(n)\}$. Theorem 5.8 indicates that among all linear combinations of members of the orthogonal set, the combination of the Fourier components $\widehat{f}(n)\phi_n$ provide a best approximation. The question is: In what sense does the Fourier series represent f, i.e., how can we synthesize f from its Fourier components? Corollary 5.11 gives one meaning to the "$\sim$" that has been used in introducing the Fourier series of a function f:

$$f \sim \sum_n \widehat{f}(n)\phi_n.$$

Namely, it gives a necessary and sufficient condition for the partial sums of the Fourier series to converge in norm to the function. The condition, namely Parseval's equality, also enables us to compute the norm of the function in terms of the norm of the Fourier components $\widehat{f}(n)\phi_n$, a fact that is useful in energy calculations.

Remark 5.14 To reiterate a comment made at the beginning of this section, all of the results of this section hold in more general settings than $L^2[a,b]$. In particular, we will apply our results to $L^2(D)$ where D is a domain in $\mathbb{R}^d$ (Section 6.4) where the inner product has the form

$$\langle f, g\rangle_{L^2(D)} = \int_D f(x)\overline{g(x)}dV,$$

and to $L^2(S^2)$ where S^2 is the unit sphere in three dimensions (Section 8.1). The inner product in the latter case has the form

$$\langle f, g\rangle_{L^2(S^2)} = \int_{S^2} f(\omega)\overline{g(\omega)}d\omega,$$

where $d\omega$ is the surface element on the unit sphere.

Exercises

5.10 Verify that $\langle f, g\rangle = \int_a^b f(x)\overline{g(x)}\rho(x)dx$ satisfies the three defining properties of an inner product as listed in the beginning of this section. Assume $\rho \in C[a,b]$ and is strictly positive on $[a,b]$.

5.11 Verify that $\|\cdot\|_\infty$ defined in Remark 5.5 satisfies the properties of norm.

5.12 For $f, g \in L^2[a,b]$, show that $\|f+g\|_2^2 + \|f-g\|_2^2 = 2(\|f\|_2^2 + \|g\|_2^2)$. This is called the parallelogram law. Explain why.

5.13 Find two functions $f, g \in C[a,b]$ such that $\|\cdot\|_\infty$ does not satisfy the parallelogram law. This shows that there does not exist an inner product on $C[a,b]$ such that $\|f\|_\infty = \langle f, f\rangle^{1/2}$.

5.14 Demonstrate the triangle inequality by expressing $\|f+g\|_2^2$ as an inner product. Use the properties of the inner product and the Cauchy–Schartz inequality.

5.15 Let $\{\phi_n\}$ be an orthonormal set in $L^2[0,l]$ and set ϕ_n^e and ϕ_n^o be the even and odd extensions of ϕ_n, respectively. Show that $\{2^{-1/2}\phi_n^e\} \cup \{2^{-1/2}\phi_n^o\}$ forms an orthonormal set in $L^2[-l,l]$.

5.3 Classical Fourier Series

Definition and Examples

Given a function $f \in L^2[0,\pi]$, the Fourier sine and Fourier cosine series of f are, respectively,

$$\begin{array}{ll} f \sim \sum_{n=1}^{\infty} \widehat{f}_s(n)\sin nx, & \widehat{f}_s(n) = \frac{2}{\pi}\int_0^{\pi} f(x)\sin nx\,dx \\ f \sim \frac{\widehat{f}_c(0)}{2} + \sum_{n=1}^{\infty} \widehat{f}_c(n)\cos nx, & \widehat{f}_c(n) = \frac{2}{\pi}\int_0^{\pi} f(x)\cos nx\,dx. \end{array} \tag{5.3.1}$$

The right-hand side of both lines give the Fourier sine and Fourier cosine coefficients of f. The basis for these definitions is that the sets $\{\sin nx\}$ and $\{1/2, \cos nx\}$ are orthogonal sets of functions on the interval $[0,\pi]$ relative to the inner product $\langle f, g\rangle = \int_0^{\pi} f(x)g(x)dx$, i.e.,

$$\int_0^{\pi} \sin nx \sin mx\,dx = \int_0^{\pi} \cos nx \cos mx\,dx = 0, \quad \text{for } n \neq m.$$

These same sets of functions are orthogonal on the interval $[-\pi,\pi]$ relative to the inner product $\langle f, g\rangle = \int_{-\pi}^{\pi} f(x)g(x)dx$ and, furthermore, they are mutually orthogonal, i.e.,

$$\int_{-\pi}^{\pi} \cos nx \sin mx\,dx = 0, \quad \text{for all } n \text{ and } m.$$

Based on this, given a function f on $[-\pi,\pi]$, the Fourier series and Fourier coefficients of f are defined by

$$\begin{array}{l} f \sim \frac{\widehat{f}_c(0)}{2} + \sum_{n=1}^{\infty}(\widehat{f}_c(n)\cos nx + \widehat{f}_s(n)\sin nx) \\ \widehat{f}_c(n) = \frac{1}{\pi}\int_{-\pi}^{\pi} f(x)\cos nx\,dx, \quad n = 0, 1, \ldots \\ \widehat{f}_s(n) = \frac{1}{\pi}\int_{-\pi}^{\pi} f(x)\sin nx\,dx, \quad n = 1, 2, \ldots. \end{array} \tag{5.3.2}$$

The connection between (5.3.1) and (5.3.2) should be clear. A odd function f on $[-\pi, \pi]$ has $\widehat{f}_c(n) = 0$ and the Fourier series of f and the Fourier sine series of f restricted to $[0, \pi]$ are the same infinite series. Conversely, given a function f on $[0, \pi]$, we can consider its odd extension, f^o, to $[-\pi, \pi]$ defined by

$$f^o(x) = \begin{cases} f(x), & x \in [0, \pi] \\ -f(-x), & x \in [-\pi, 0). \end{cases}$$

Then the Fourier series of f^o and the Fourier sine series of f are the same infinite series. The same comments apply to even functions on $[-\pi, \pi]$ and functions f on $[0, \pi]$ with even extensions

$$f^e(x) = \begin{cases} f(x), & x \in [0, \pi] \\ f(-x), & x \in [-\pi, 0). \end{cases} \tag{5.3.3}$$

Example 5.7 Let $f(x) = 1$ on $[0, \pi]$. Then $f^e(x) = 1$ on $[-\pi, \pi]$ and odd extension is given by

$$f^o(x) = \begin{cases} 1, & x \in [0, \pi] \\ -1, & x \in [-\pi, 0). \end{cases}$$

Then the Fourier series of the even extension is simply $f^e \sim 1$, whereas the Fourier expansion of the odd extension is a true infinite series,

$$f^o \sim \sum_{n=1}^{\infty} \frac{2}{\pi n} [1 - (-1)^n] \sin nx.$$

The sine coefficients were computed in Example 5.2.

From a computational perspective, these definitions can be treated in a more concise form if we make use of complex-valued functions. In this case the inner product takes the form

$$\langle f, g \rangle = \int_{-\pi}^{\pi} f(x)\overline{g(x)}dx.$$

The orthogonal system of functions is then $\{e^{inx}\}_{n=-\infty}^{+\infty}$ and the orthogonality relations become

$$\int_{-\pi}^{\pi} e^{inx} e^{-imx} dx = \begin{cases} 0, & n \neq m \\ 2\pi, & \text{otherwise.} \end{cases}$$

The **complex form of the Fourier series** of a function f (possibly complex valued) is

$$\begin{aligned} f &\sim \textstyle\sum_{n=-\infty}^{+\infty} \hat{f}(n)e^{inx} \\ \hat{f}(n) &= \textstyle\frac{1}{2\pi}\int_{-\pi}^{\pi} f(x)e^{-inx}\,dx. \end{aligned} \tag{5.3.4}$$

In the case where f is real valued, we have the symmetry relation for the Fourier coefficients

$$\hat{f}(-n) = \overline{\hat{f}(n)}. \tag{5.3.5}$$

Further, from the Euler identity $e^{i\theta} = \cos\theta + i\sin\theta$ we have

$$\hat{f}(n) = \frac{1}{2\pi}\int_{-\pi}^{\pi} f(x)(\cos nx - i\sin nx)dx = \frac{1}{2}\Big(\widehat{f_c}(n) - i\widehat{f_s}(n)\Big). \tag{5.3.6}$$

This last formula gives a direct relation between the Fourier coefficients in real form (5.3.2) and those in complex form (5.3.4). Combining the symmetry relation and formula (5.3.6) we have for real-valued functions f

$$\widehat{f_c}(n) = \hat{f}(n) + \hat{f}(-n), \quad \widehat{f_s}(n) = i\left(\hat{f}(n) - \hat{f}(-n)\right). \tag{5.3.7}$$

The point of emphasis is: *properties of Fourier coefficients in the real and complex forms should be the same for real-valued functions; often it is easier to work with the complex form from a computational standpoint.* The last point is brought to bear in the following examples.

Example 5.8 On the interval $[-\pi, \pi]$, we can find the (real) Fourier series of the function

$$f(x) = \begin{cases} x, & x \in [0, \pi] \\ 0, & x \in [-\pi, 0) \end{cases}$$

by computing the complex Fourier coefficients of f using integration by parts:

$$\begin{aligned} \hat{f}(n) = \frac{1}{2\pi}\int_0^{\pi} xe^{-inx}\,dx &= \frac{1}{2\pi}\left[\left(-x\frac{e^{-inx}}{in}\right|_0^{\pi} + \frac{1}{in}\int_0^{\pi} e^{-inx}dx\right] \\ &= \frac{1}{2\pi}\left[-\pi\frac{e^{-in\pi}}{in} + \frac{1}{n^2}\left(e^{-in\pi} - 1\right)\right] \\ &= \frac{1}{2}\left[\frac{1}{\pi n^2}\left((-1)^n - 1\right) - i\frac{(-1)^{n+1}}{n}\right], \end{aligned}$$

and $\hat{f}(0) = \pi/4$. From (5.3.7), we have

$$\widehat{f_c}(0) = \frac{\pi}{2}, \quad \widehat{f_c}(n) = \frac{1}{\pi n^2}\left((-1)^n - 1\right), \quad \widehat{f_s}(n) = \frac{(-1)^{n+1}}{n}.$$

Example 5.9 The Fourier sine and cosine series of the function

$$f(x) = \begin{cases} 0, & x \in [0, \pi/2) \\ 1, & x \in [\pi/2, \pi] \end{cases}$$

can be found simultaneously by computing the integral

$$\begin{aligned} \int_0^{\pi} f(x)\, e^{inx}\, dx &= \int_{\pi/2}^{\pi} e^{inx} dx = \frac{1}{in} \left(e^{inx} \right|_{\pi/2}^{\pi} \\ &= -\frac{\sin \frac{n\pi}{2}}{n} - i \frac{\cos n\pi - \cos \frac{n\pi}{2}}{n} \\ &= -\frac{\sin \frac{n\pi}{2}}{n} - i \frac{(-1)^n - \cos \frac{n\pi}{2}}{n}. \end{aligned}$$

It follows that $\widehat{f_c}(n) = -\frac{2 \sin \frac{n\pi}{2}}{\pi n}$ and $\hat{f}_s(n) = -\frac{2}{\pi n} \left((-1)^n - \cos \frac{n\pi}{2} \right)$.

Fourier Series on Other Intervals In the preceding discussion, there is nothing special about working on the interval $[-\pi, \pi]$, save simplification of formula. Furthermore, a linear change of variable allows us to translate all formulas to intervals of the form $[-L, L]$. Indeed, if $f(x)$ is a function defined on $[-L, L]$, then $g(y) = f(\frac{L}{\pi} y)$ is defined on $[-\pi, \pi]$. Writing the Fourier series and coefficients of the function g and applying this change of variable leads to the definition of the complex Fourier series of the original function f:

$$\begin{aligned} & f \sim \textstyle\sum_n \hat{f}(n) e^{i \frac{n\pi}{L} x} \\ & \hat{f}(n) = \textstyle\frac{1}{2L} \int_{-L}^{L} f(x) e^{-i \frac{n\pi}{L} x}\, dx. \end{aligned} \tag{5.3.8}$$

Similarly, we can write the formulas for the real form of the Fourier series on $[L, L]$:

$$\begin{aligned} & f \sim \tfrac{\widehat{f_c}(0)}{2} + \textstyle\sum_{n=1}^{\infty} \Big(\widehat{f_c}(n) \cos \frac{n\pi}{L} x + \widehat{f_s}(n) \sin \frac{n\pi}{L} x \Big) \\ & \widehat{f_c}(n) = \textstyle\frac{1}{L} \int_{-L}^{L} f(x) \cos \frac{n\pi}{L} x\, dx \\ & \widehat{f_s}(n) = \textstyle\frac{1}{L} \int_{-L}^{L} f(x) \sin \frac{n\pi}{L} x\, dx. \end{aligned} \tag{5.3.9}$$

It is left as an exercise to write the formulas for the Fourier cosine and sine series on the interval $[0, L]$. The important fact to remember is that all convergence and differentiation results in the next subsection immediately carry over to the case of Fourier series on other intervals.

Convergence of Fourier Series

Let f be defined on $[-\pi, \pi]$ and consider its Fourier representation,

$$f \sim \sum_n \widehat{f}(n) e^{inx}.$$

This subsection endeavors to explore the recovery of the function from the partial sums of its Fourier series. In other words, under what conditions and in what sense does it hold that

$$\lim_{N=\infty} S_N(f, x) = f(x)?$$

In the study of partial differential equations, three modes[2] of convergence are natural and useful: (1) convergence at individual points, also called pointwise convergence; (2) uniform convergence on $[-\pi, \pi]$; and (3) convergence in L^2-norm. A closely related fundamental question is: How are the properties of f reflected in the properties of the Fourier coefficients $\{\widehat{f}(n)\}$? One example in this vein has already been given. If $f \in L^2[-\pi, \pi]$, then

$$2\pi \sum_n |\widehat{f}(n)|^2 \leq \|f\|_2^2.$$

In particular, the series on the left-hand side converges. In preparation for discussing convergence results this related question is elaborated further.

Properties of Fourier Coefficients The following classical result is fundamental to the theory and is an immediate consequence of Bessel's inequality.

Proposition 5.15 (Riemann–Lebesgué Lemma) Let $f \in L^2[-\pi, \pi]$. Then

$$\hat{f}(n) \to 0 \quad \text{as} \quad |n| \to \infty.$$

Remark 5.16 In particular, this holds if $f \in \mathrm{PC}[-\pi, \pi]$. Looking at the defining formula for Fourier coefficients, we see that Fourier coefficients can be defined for functions integrable on the interval $[-\pi, \pi]$. Please note that square integrability implies integrability, but not the other way around. Riemann proved the above result for Riemann integrable functions. Later on, Lebesgué proved this result for functions integrable in a more general sense. An alternative argument that works in these cases essentially goes as follows. By a change of variable,

$$\widehat{f}(n) = \frac{1}{2\pi} \int_{-\pi}^{\pi} \tilde{f}(x) e^{-inx}\, dx = \frac{1}{2\pi} \int_{-\pi-\pi/n}^{\pi-\pi/n} \tilde{f}(x + \frac{\pi}{n}) e^{-in(x+\pi/n)}\, dx,$$

2. Fourier series of distributions and convergence in distribution is considered separately and is useful in discussion of solutions of the wave equation and Green's functions.

where $\tilde{f}$ is the 2π-periodization of f. Since $e^{-i\pi} = -1$ and the integrand is 2π-periodic, this can be rewritten as

$$\widehat{f}(n) = -\frac{1}{2\pi}\int_{-\pi}^{\pi} \tilde{f}(x + \frac{\pi}{n})e^{-inx}\, dx.$$

Adding the two formulas for the Fourier coefficient and taking the modulus we obtain

$$|\widehat{f}(n)| \leq \frac{1}{2\pi}\int_{-\pi}^{\pi} |\tilde{f}(x + \frac{\pi}{n}) - \tilde{f}(x)|dx.$$

This is a fundamental inequality for the Fourier coefficients of the function f. Obtaining Proposition 5.15 for continuous functions is relatively easy if f is a continuous function. More general functions can be treated using suitable approximation arguments.[3]

Proposition 5.15 has implications regarding differentiation of Fourier series term by term. Indeed, the function $f(x) = 1$ on $[0, \pi]$ has Fourier sine series

$$\frac{4}{\pi}\sum_{n=0}^{\infty} \frac{1}{2n+1}\sin(2n+1)x. \tag{5.3.10}$$

Formally differentiating the series term by term, one might think that f' has Fourier cosine series

$$\frac{4}{\pi}\sum_{n=0}^{\infty}\cos(2n+1)x.$$

However, the coefficients of this series do not tend to zero and hence this cannot be a Fourier series of an integrable function. Cast differently and working with the complex form, the question becomes: Given a Fourier series

$$f \sim \sum_{n} \hat{f}(n)e^{inx},$$

what conditions on f ensure that $\widehat{f'}(n) = (in)\hat{f}(n)$?

Lemma 5.17 Let $f \in C[-\pi, \pi]$ with $f' \in \mathrm{PC}[-\pi, \pi]$. Then

$$\widehat{f'}(n) = \frac{(-1)^n}{2\pi}\left[f(\pi) - f(-\pi)\right] + (in)\hat{f}(n). \tag{5.3.11}$$

3. A general integrable function can be incredibly pathological in its local behavior. The recent book [6] provides a wealth of examples in a historical context. Lebesgué's theory of integration supplies the machinery in order to manage this behavior. A nice introduction to Lebesgué's theory in the context of Fourier series can be found in [27].

Proof This is an exercise in integration by parts.

$$\begin{aligned}\widehat{f'}(n) &= \frac{1}{2\pi}\int_{-\pi}^{\pi} f'(x)\,e^{-inx}\,dx \\ &= \frac{1}{2\pi}\left(f(x)e^{-inx}\right|_{\pi}^{\pi} + \frac{in}{2\pi}\int_{-\pi}^{\pi} f(x)\,e^{-inx}\,dx \\ &= \frac{(-1)^n}{2\pi}[f(\pi) - f(-\pi)] + (in)\widehat{f}(n).\end{aligned}$$

□

Complete understanding of formula (5.3.11) comes from a geometric interpretation of functions on the interval $[-\pi, \pi]$. The map $x \to (\cos x, \sin x)$ takes the interval $[-\pi, \pi]$ onto the unit circle $\mathbf{T}$ in the plane. As a consequence, a function F on the circle can be defined via the formula $F(\cos x, \sin x) = f(x)$. Continuity and/or smoothness properties of F then involves the respective property of f along with a condition at the point $(-1, 0)$ on the circle. For example, if $f(-\pi) \neq f(\pi)$, then the function F is discontinuous at $(-1, 0)$. These considerations in mind, we define

$$C(\mathbf{T}) = \{f \in C[-\pi, \pi] : \, f(\pi) = f(-\pi)\}.$$

The condition that $f \in C(\mathbf{T})$ forces the boundary terms in (5.3.11) to vanish and provides the result regarding differentiating Fourier series term by term. A second understanding of this formula comes from associating Fourier series with 2π-periodic functions, i.e., functions defined on $\mathbb{R}$ such that

$$f(x + 2\pi) = f(x), \quad \text{for all } x \in \mathbb{R}.$$

Given a function defined on $[-\pi, \pi)$, its 2π-periodic extension, or 2π-periodization, is defined by the formula

$$\tilde{f}(x + 2n\pi) = f(x), \quad n \in \mathbb{Z}, \quad x \in [-\pi, \pi).$$

It is then clear that Fourier series, as we have defined them, can be thought of as series representations of the 2π-periodic functions. We now see that $f \in C(\mathbf{T})$ if and only if the 2π-periodic extension of f is continuous on the real line.

Theorem 5.18 Let $f \in C(\mathbf{T})$ with $f' \in \mathrm{PC}[-\pi, \pi]$. Then $\widehat{f'}(n) = (in)\hat{f}(n)$. Moreover, there is a constant C such that $\left|\hat{f}(n)\right| \leq C/|n|$.

Proof We need only demonstrate the inequality. From the first part, $\hat{f}(n) = \widehat{f'}(n)/(in)$. By the Riemann–Lebesgué lemma, $\widehat{f'}(n) \to 0$ as $|n| \to \infty$. In particular, this sequence is bounded. □

This result may be inductively extended: If $f^{(k)} \in C(\mathbf{T})$ and $f^{(k+1)}$ is piecewise continuous, then

$$\widehat{f^{(k+1)}}(n) = (in)^{k+1}\hat{f}(n).$$

In the case of Fourier sine and cosine series, we can apply the previous results by using the odd and even extensions of f, respectively. The following example illustrates this idea.

Example 5.10 Consider $f(x) = \pi^2 x - x^3$ on $[0, \pi]$. Then $f^o(x) = \pi^2 x - x^3$ and $f^{o\prime} \in C(\mathbf{T})$ with $f''(x) = -6x$. Hence the Fourier sine series of f can be differentiated twice term by term. This seems reasonable since the Fourier sine series of f is

$$\sum_{n=1}^{\infty} \frac{12(-1)^{n+1}}{n^3} \sin nx.$$

Differentiating once would leave n^2 in the denominator and a uniformly convergent series; the second differentiation leaves n in the denominator. On the other hand, the even extension of f is given by $f^e(x) = \pi^2|x| - |x|^3$ and this is not differentiable at $x = 0$. Nonetheless, $f^e \in \mathrm{PC}[-\pi, \pi]$, hence we may differentiate the Fourier cosine series of f term by term once. Again this seems reasonable since the cosine series of f is

$$\frac{\pi^3}{4} + \sum_{n=1}^{\infty} \left[\frac{2\pi}{n^2} \left(2(-1)^{n+1} - 1\right) + \frac{12}{\pi n^4} \left((-1)^n - 1\right) \right] \cos nx.$$

The Convergence Theorems With the above preparations, we are now in a position to discuss convergence properties of Fourier series. Here is the pointwise convergence theorem.

Theorem 5.19 Let f and f' be piecewise continuous on $[-\pi, \pi]$. Then

$$\lim_{N\to\infty} S_N f(x) = \lim_{N\to\infty} \sum_{n=-N}^{N} \hat{f}(n) e^{inx} = \frac{\tilde{f}(x+0) + \tilde{f}(x-0)}{2},$$

for all $x \in [-\pi, \pi]$.

The essence of the argument demonstrating this is based on an analytic expression for the partial sums in terms of $\tilde{f}$. Indeed,

$$\begin{aligned} S_N f(x) &= \sum_{n=-N}^{N} \widehat{f}(n)\, e^{inx} = \frac{1}{2\pi}\int_{-\pi}^{\pi} f(s) \sum_{n=-N}^{N} e^{in(x-s)}\, ds \\ &= \frac{1}{2\pi}\int_{-\pi}^{\pi} \tilde{f}(s) \sum_{n=-N}^{N} e^{in(x-s)}\, ds \\ &= \frac{1}{2\pi}\int_{-\pi}^{\pi} \tilde{f}(x-s) \sum_{n=-N}^{N} e^{ins}\, ds = \frac{1}{2\pi}\int_{-\pi}^{\pi} \tilde{f}(x-s)\, D_N(s)\, ds, \end{aligned}$$

where $D_N(x)$ is the Dirichlet kernel given by

$$D_N(s) = \sum_{|n|\le N} e^{ins}.$$

Noting that the sum is geometric, apply the geometric sum formula along with algebraic and trigonometric manipulation to obtain a closed form for the kernel:

$$D_N(s) = \frac{\sin(N+\frac{1}{2})s}{\sin\frac{s}{2}}. \qquad (5.3.12)$$

The Dirichlet kernel is clearly an even function and

$$\frac{1}{2\pi}\int_{-\pi}^{\pi} D_N(x)dx = 1.$$

Figure 5.5 gives the plot of the Dirichlet kernel for three values of N; note that as N increases the kernel becomes more localized around the origin. Consequently, the integral formula for the partial sums is interpreted as a **localized average** of f around the point x. This intuitive view of the Dirichlet kernel is comparable to the view given for the Gaussian kernel related to solution of the heat equation (see Chapter 1). A key difference is the oscillations present in the Dirichlet kernel, which are controlled by the smoothness hypothesis $f' \in \mathrm{PC}[-\pi,\pi]$.

Proof (of Theorem 5.19) By a change of variable, the partial sums can be written

$$\begin{aligned} S_N f(x) &= \frac{1}{2\pi}\int_{-\pi}^{\pi} \tilde{f}(x+y)D_N(y)dy \\ &= \frac{1}{2\pi}\int_{0}^{\pi} \tilde{f}(x+y)D_N(y)dy + \frac{1}{2\pi}\int_{0}^{\pi} \tilde{f}(x-y)D_N(y)dy, \end{aligned}$$

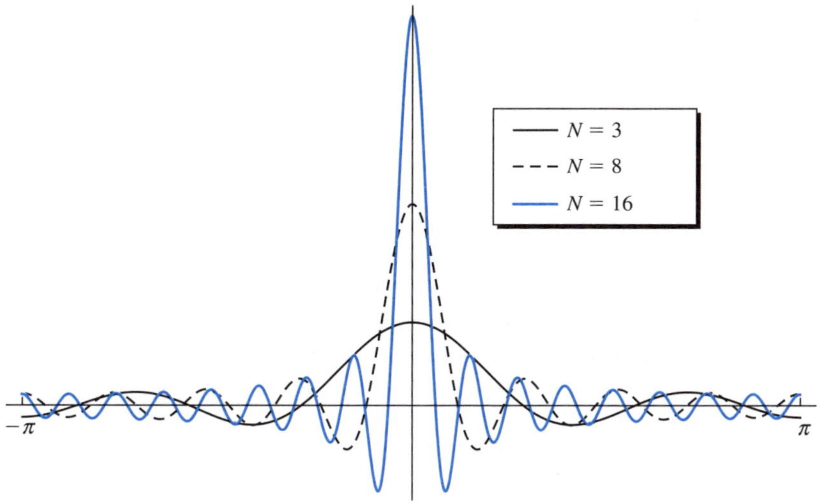

Figure 5.5 The Dirichlet Kernel, $D_N(x)$

where the original integral was split at $y = 0$. The proof will be concluded by showing

$$\lim_{N\to\infty} \frac{1}{2\pi} \int_0^\pi \tilde{f}(x \pm y) D_N(y) dy = \frac{1}{2} \tilde{f}(x \pm 0).$$

Both limits are demonstrated the same way; the details are given for the + case. Since $\int_0^\pi D_N(y)dy = \pi/2$, we need to show that

$$\lim_{N\to\infty} \frac{1}{2\pi} \int_0^\pi \left[\tilde{f}(x+y) - \tilde{f}(x+0)\right] D_N(y) dy = 0.$$

We write

$$\frac{1}{2\pi} \int_0^\pi \left[\tilde{f}(x+y) - \tilde{f}(x+0)\right] D_N(y) dy = \frac{1}{2\pi} \int_0^\pi g(y)\, Y(y) \sin(N+1/2)y\, dy, \tag{5.3.13}$$

where $g(y) = \left[\tilde{f}(x+y) - \tilde{f}(x+0)\right]/y$ and $Y(y) = y/\sin(y/2)$. It is easy to see that Y is continuous on $[0, \pi]$. Furthermore, applying l'Hôpital's rule gives

$$\lim_{y\to 0^+} g(y) = \tilde{f}'(x+0).$$

The latter limit implies that g is a piecewise continuous function on $[0, \pi]$. Hence, the product $g(y)Y(y)$ is piecewise continuous on $[0, \pi]$. We can interpret the right-hand side of (5.3.13) as the computation (up to a multiplicative constant) of the Fourier coefficients of the integrable function $g(y)Y(y)$ with respect to the orthogonal set $\{\sin(N+1/2)y\}$. The Riemann–Lebesgué lemma for this orthogonal system then gives that the right-hand side tends to zero as N tends to infinity, completing the proof. □

Historical Comment As mentioned before, Fourier series arose in the work of Joseph Fourier in his study of heat propagation in solids in the early 1800s. In the course of his investigations he made the brazen conjecture that such series can be defined for any function whatsoever and, furthermore, the resulting series converges to the function. This conjecture caused considerable controversy in the mathematical and scientific communities and the analysis of it did not occur until Riemann conceived the Riemann integral in the 1850s. The pointwise convergence theorem as stated above is essentially due to Dirichlet.

Here are two examples illustrating this theorem.

Example 5.11 Recall Example 5.7, where we found the Fourier sine series of the function $f(x) = 1$ on $[0, \pi]$. This series is the Fourier series of the odd extension of f. In Figure 5.3, representative partial sums are plotted along with the function f on the interval $[-\pi, \pi]$, illustrating the conclusion of the pointwise convergence theorem in a naive way as all of the functions $\sin nx$ vanish at $x = 0$. In a different light, consider the cosine series of $f(x) = x$ on $[0, \pi]$. This is the Fourier series of its even extension to $[-\pi, \pi]$. The plots of representative partial sums are in Figure 5.6, again illustrating convergence at every point due to the fact that f^e is continuous.

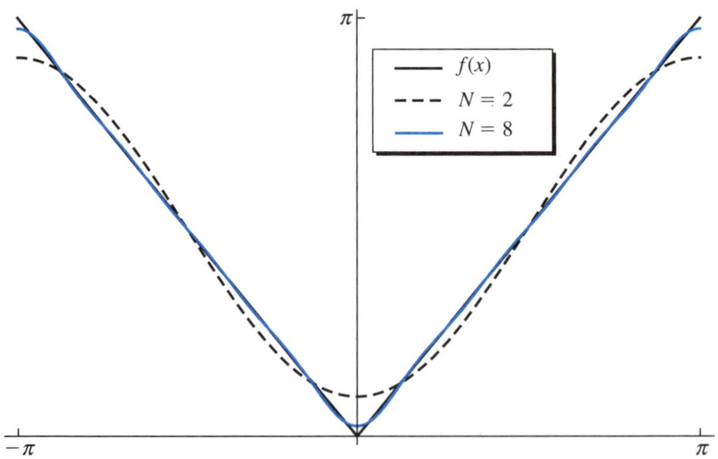

Figure 5.6 Partial Sums for $f(x) = |x|$

Remark 5.20 Note the behavior of the partial sums near the discontinuities of f^o in the first part of Example 5.11. The graphs indicate an overshoot on either side of a discontinuity; this overshoot phenomena is known as **Gibb's phenomena**, and is typical behavior for partial sums of Fourier series of piecewise continuous functions. The magnitude of the overshoot is approximately 9% of the absolute value of the jump. The following example further illustrates Gibb's phenomena.

Example 5.12 Consider the odd extension of the function in Example 5.9. By the result of that example its Fourier series would then be

$$f^o \sim \sum_{n=1}^{\infty} \frac{-2\left((-1)^n - \cos\frac{n\pi}{2}\right)}{\pi n} \sin nx.$$

The function is plotted in Figure 5.7 along with representative partial sums. According to the theorem, the series converges to $f(x)$ at every point in $[0,\pi]$ except $x = \frac{\pi}{2}$ and π where it converges to the respective values $\frac{1}{2}$ and 0. The same applies for the points $x = -\frac{\pi}{2}$ and $-\pi$ in the interval $[-\pi, 0]$.

Combining the differentiation formula and the pointwise convergence theorem gives a result on uniform convergence of Fourier series useful in some applications.

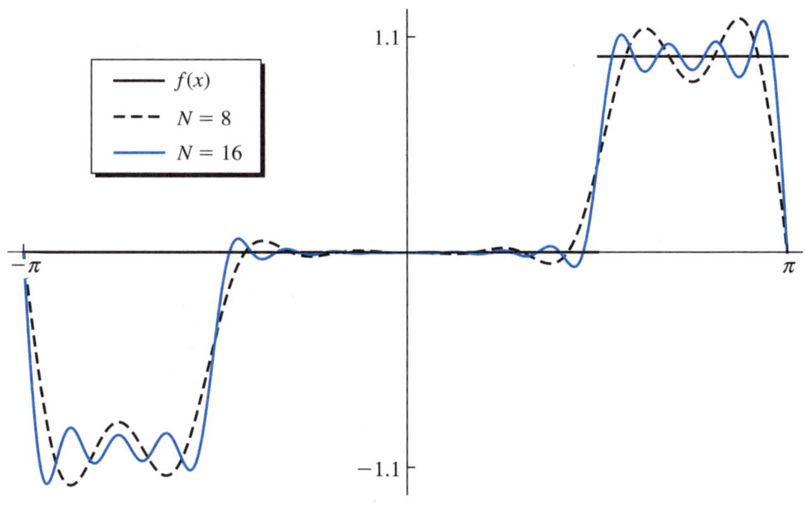

Figure 5.7 Plot for Example 5.12

Theorem 5.21 (Uniform Convergence) Let $f \in C(\mathbf{T})$ with f' piecewise continuous. Then the Fourier series of f converges uniformly to f on $[-\pi, \pi]$. In norm notation, $\|S_N f - f\|_\infty \to 0$ as $N \to \infty$.

Proof The hypothesis implies that $\hat{f}(n) = \widehat{f'}(n)/(in)$. Therefore, applying the Cauchy–Schwarz inequality, we have

$$\sum_{0<|n|\leq N} \left|\hat{f}(n)\right| \leq \left(\sum_{0<|n|\leq N} \frac{1}{n^2}\right)^{1/2} \left(\sum_{0<|n|\leq N} |\widehat{f'}(n)|^2\right)^{1/2}.$$

The first sum on the right-hand side converges as N tends to infinity, as it is a p series with $p > 1$; the second sum converges by Bessel's inequality. Hence, $\sum_n \left|\hat{f}(n)\right|$ converges and by the Weierstrass M-test, the Fourier series converges uniformly. □

Remark 5.22 The Weierstrass M-test immediately gives that $\sum_n |\widehat{f}(n)| < \infty$ implies that the Fourier series converges uniformly on $[-\pi, \pi]$. The beauty of Theorem 5.2.1 is that it gives a simple criterion in terms of the function f itself.

We conclude this section with the L^2-convergence result; this is a basic tool for energy calculations.

Theorem 5.23 (L^2 Convergence) Let $f \in L^2[-\pi, \pi]$. Then

$$\lim_{N\to\infty} \|S_N f - f\|_2 = 0$$

and Parseval's equality holds:

$$2\pi \sum_n |\widehat{f}(n)|^2 = \|f\|_2^2.$$

Proof We supply the proof in the case where $f \in \mathrm{PC}[-\pi, \pi]$ with piecewise continuous derivative; the general case follows from approximation arguments like those indicated in the proof of the Riemann–Lebesgué lemma. First, if $g \in C(T)$ with $g' \in \mathrm{PC}[-\pi, \pi]$, then we know from the previous result that $S_N g \to g$ uniformly. Now consider the estimate

$$\|S_N g - g\|_2 = \left[\int_{-\pi}^{\pi} |S_N g(x) - g(x)|^2\right]^{1/2} \leq (2\pi)^{1/2} \|S_N g - g\|_\infty.$$

The right-hand side converges to zero by uniform convergence. If f is piecewise continuous with piecewise continuous derivative, we approximate f with a function $g \in C(T)$ that has piecewise continuous derivative. Specifically, given $\varepsilon > 0$, there is a g with these properties such that

$$\|f - g\|_2 < \frac{\varepsilon}{3}.$$

To construct such a function g, we linearize f in a small interval around each of its jump discontinuities and at the endpoints (the latter to force $g(-\pi) = g(\pi)$). We can now write, using the triangle inequality,

$$\|S_N f - f\|_2 \le \|S_N f - S_N g\|_2 + \|S_N g - g\|_2 + \|g - f\|_2. \tag{5.3.14}$$

For the first term on the right, Bessel's inequality may be applied:

$$\|S_N f - S_N g\|_2^2 = \|S_N(f-g)\|_2^2 = 2\pi \sum_{|n|\le N} |\widehat{f-g}(n)|^2 \le \|f-g\|_2^2 < \left(\frac{\varepsilon}{3}\right)^2.$$

Hence, returning to (5.3.14), we have

$$\|S_N f - f\|_2 < \|S_N g - g\|_2 + 2\frac{\varepsilon}{3}.$$

Now we choose N_0 such that if $N \ge N_0$, $\|S_N g - g\|_2 < \varepsilon/3$. Then for such N, $\|S_N f - f\|_2 < \varepsilon$; this establishes the norm convergence. Parseval's equality now follows from Theorem 5.11. □

Fourier Series of the Dirac Delta

Given a distribution on $[-\pi, \pi]$, it is natural to question how to define its Fourier series. This is particularly of interest in using Fourier series to construct fundamental solutions of initial boundary value problems. On first thought, given $u \in \mathcal{D}'([-\pi, \pi])$, the candidate for its Fourier coefficients would be $\frac{1}{2\pi}(u, e^{-inx})$. However, the difficulty lies in the fact that the complex exponential is not a test function on the interval and u potentially can have singular behavior at the endpoints. However, in the case of the δ functional that acts by evaluation, this vein of thought presents no problem. Indeed, if we set

$$\widehat{\delta}(n) = \frac{1}{2\pi}(\delta, e^{-inx}) = \frac{1}{2\pi}$$

and form the series

$$\delta \sim \frac{1}{2\pi}\sum_n e^{inx} = \frac{1}{2\pi} + \frac{1}{\pi}\sum_{n=1}^{\infty} \cos nx, \tag{5.3.15}$$

it is easily seen that the series converges to δ in distribution. Indeed, if we take a test function ϕ and let $S_N\delta$ be the partial sums of the series (5.3.15), then

$$(S_N\delta, \phi) = \sum_{n=-N}^{N} \widehat{\phi}(n) \to \phi(0).$$

One way to make this rigorous is to generalize the notion of a distribution by considering other classes of test functions. In this case, consider the class $\mathcal{E} = C^\infty(\mathbf{T})$. A distribution would then be a continuous linear functional on $\mathcal{E}$. In other words, if $u \in \mathcal{E}'$ (the latter being the class of distributions in our new sense), and if $\{\phi_n\} \subset \mathcal{E}$ such that ϕ_n and all of its derivatives converge uniformly on $[-\pi, \pi]$, then $(u, \phi_n) \to (u, \phi)$ where $\phi = \lim_n \phi_n$. The Dirac delta clearly satisfies this continuity condition and since the complex exponentials e^{-inx} are test functions in the new sense, the above definition of Fourier coefficients is well defined. Notice that $\mathcal{D} \subset \mathcal{E}$ and, as a consequence, $\mathcal{E}' \subset \mathcal{D}'$. Furthermore, this inclusion is strict.[4] Intuitively, we have removed the aforementioned possible singular behavior.

For functions f that have pointwise convergent Fourier series, due to periodicity of the complex exponentials, the series converges on the real line to the 2π-periodization of the function f. The Dirac delta also has a 2π-periodization given by

$$\delta_{\text{per}}(x) = \sum_{n=-\infty}^{\infty} \delta(x + 2n\pi).$$

Furthermore, on any interval of length 2π, the series (5.3.15) converges in distribution to δ_{per}.

Example 5.13 (Fourier Sine Series) Consider the distribution on $[-\pi, \pi]$ given by $u(x) = \delta(x - \xi) - \delta(x + \xi)$, where $\xi \in (-\pi, \pi)$ is fixed. Its action on test functions ϕ is given by $(u, \phi) = \phi(\xi) - \phi(-\xi)$. Following the preceding discussion, the Fourier coefficients of u are given by

$$\begin{aligned} \widehat{u}(n) &= \frac{1}{2\pi}(\delta(x - \xi) - \delta(x + \xi), e^{-inx}) \\ &= \frac{1}{2\pi}(e^{-in\xi} - e^{in\xi}) \\ &= \frac{-i}{2\pi} \sin n\xi. \end{aligned}$$

Thus the distribution u has Fourier series

$$u \sim \frac{-i}{2\pi} \sum_n (\sin n\xi) e^{inx} = \frac{2}{\pi} \sum_{n=1}^{\infty} \sin n\xi \sin nx.$$

4. The function $u(x) = \frac{1}{x-\pi}$ defines an element of $\mathcal{D}'$, however u is not in $\mathcal{E}'$.

As a distribution, $(u, \phi) = 0$ for any even test function ϕ. Such distributions are said to be odd in analogy with the definition for functions. This example illustrates that odd distributions have Fourier series that can be written as sine series. The periodization of u is easily written as

$$u_{\text{per}}(x) = \delta_{\text{per}}(x - \xi) - \delta_{\text{per}}(x + \xi).$$

Remark 5.24 Assume that $\xi > 0$ in the example. Then the distribution u restricted to the interval $[0, \pi]$ (i.e., applied to test functions defined on this interval) simply becomes $u(x) = \delta(x - \xi)$. We can define its Fourier sine coefficients by $\widehat{u}_s(n) = \frac{2}{\pi}(u(x), \sin nx) = \frac{2}{\pi} \sin n\xi$ and the resulting Fourier sine series agrees with the result of the example. This is consistent with the definition of Fourier sine series for functions defined on $[0, \pi]$. Likewise, we can define the Fourier cosine series of u (Exercise 5.29).

Example 5.14 (Differentiation of Fourier Series) Consider the distribution δ' on $[-\pi, \pi]$. The Fourier coefficients and Fourier series are given by

$$\widehat{\delta'}(n) = \frac{1}{2\pi}(\delta'(x), e^{-inx}) = -\frac{1}{2\pi}(\delta(x), -ine^{-inx}) = \frac{in}{2\pi}$$

$$\delta' \sim \frac{1}{2\pi} \sum_n ine^{inx}.$$

If ϕ is a real-valued test function, then

$$(S_N\delta', \phi) = \sum_{|n| \leq N} in\widehat{\phi}(-n) = -\sum_{|n| \leq N} in\widehat{\phi}(n) \to -\phi'(0) = (\delta', \phi).$$

In this computation, we substitute $-n$ for n to obtain the second equality. The limit conclusion follows by the pointwise convergence theorem for Fourier series. This example illustrates the fact that Fourier series of distributions can be differentiated term by term.

Remark 5.25 The preceding discussion was meant to be a very brief introduction to Fourier series of distributions. In the following material, our primary application of the ideas will be in finding formulas for the fundamental solutions of initial boundary value problems. Thus we will really only require Fourier expansions for the Delta functional. For a more detailed look at Fourier series of distributions, the texts [10, 36] are recommended reading.

Exercises

5.16 Compute the real and complex form of the Fourier series of each of the following functions on the interval $[-\pi, \pi]$.

a. $f(x) = x$

b. $f(x) = |x|$

c. $f(x) = x^2$

d. $f(x) = \pi x - x^2$

Hint: use (a) and (c)!

5.17 Compute the real and complex form of the Fourier series of each of the following functions on the interval $[-\pi, \pi]$.

a. $f(x) = e^{ax}$

b. $f(x) = \sinh ax$, $a \neq 0$

c. $f(x) = \cos ax$, $a \neq 0, \pm 1, \pm 2, \ldots$

5.18 Compute the Fourier sine and cosine series of each of the following functions on the interval $[0, \pi]$.

a. $f(x) = x$

b. $f(x) = x^2$

c. $f(x) = 1$, if $0 < x < \pi/2$ and $= 0$ if $\pi/2 < x < \pi$

5.19 Let f be a real-valued function on $[-\pi, \pi]$.

a. If f is an odd function show that $\hat{f}(-n) = -\hat{f}(n)$.

b. If f is an even function, show that $\hat{f}(-n) = \hat{f}(n)$.

5.20 Work out the formulas for complex Fourier coefficients on $[-l, l]$, Fourier sine and cosine coefficients on $[0, l]$, and the relations between them [analogs of (5.3.6) and (5.3.7)]. You will first need to determine the orthogonality conditions for $\{1/2, \cos(n\pi x/l), \sin(n\pi x/l)\}$. This can be done by substituting $\pi x/l$ for x in the orthogonality conditions for $\{1/2, \cos nx, \sin nx\}$.

5.21 Derive (5.3.12).

5.22 State the analog of Theorem 5.19 for Fourier sine and cosine series. In each case determine $\lim_{N\to\infty} S_N f(0)$.

5.23 State the analog of Theorem 5.19 for functions defined on the interval $(-l, l)$.

5.24 Discuss the pointwise convergence properties of the Fourier series obtained in Exercises 5.16 and 5.17.

5.25 Show the following. HINT: Consider the Fourier series you computed in previous exercises.

a. $\sum_{n=1}^{\infty} \frac{1}{(2n-1)^2} = \frac{\pi^2}{8}$

b. $\sum_{n=1}^{\infty} \frac{(-1)^{n+1}}{n^2} = \frac{\pi^2}{12}$

c. $\sum_{n=1}^{\infty} \frac{1}{n^2} = \frac{\pi^2}{6}$

d. $\frac{a\pi}{\sin a\pi} = 1 + 2a^2 \sum_{n=1}^{\infty} \frac{(-1)^{n+1}}{n^2 - a^2}$, $a \neq 0, \pm 1, \pm 2, \ldots$

5.26 Find the Fourier cosine series of $f(x) = 0$, $0 < x < \pi/2$, $= x$, $\pi/2 < x < \pi$ and use this to find the Fourier sine series of f'.

5.27 Verify the formula for u_{per} given in Example 5.13.

5.28 An even distribution u on $[-\pi, \pi]$ has the defining property that $(u, \phi) = 0$ for every odd test function ϕ. Devise the definition of the Fourier cosine series of u and determine this explicitly for $\delta(x - \xi)$, where $0 < \xi < \pi$.

5.29 After making suitable definitions of Fourier sine and cosine series of distributions on the interval $[0, L]$, find these series for the distribution $\delta(x - \xi)$.

5.4 Eigenfunction Expansions

Sturm–Liouville Problems

The intent of this subsection is to complete the picture of some of the patterns observed and used in the examples of the first section such as the fact that the orthogonality conditions (called calculus formulas in Section 5.1) and the fact that the separation constants are real. For this we consider an ordinary differential equation of the form

$$(kX')' + qX + \mu\rho X = 0, \ a < x < b, \tag{5.4.1}$$

where the coefficients satisfy

A1 $k = k(x) \in C^1[a,b]$, $q \in C[a,b]$, and $\rho = \rho(x) \in C[a,b]$ with $k(x)$ and $\rho(x) > 0$ for all $x \in [a,b]$.

This equation occurs when applying, for example, separation of variables to the heat equation where the physical parameters are functions of x (see Exercise 5.30). Coupled with this equation are boundary conditions of one of two types. The first type are periodic boundary conditions of form

$$\begin{cases} X(a) = X(b) \\ X'(a) = X'(b). \end{cases} \tag{5.4.2}$$

This situation arose in Example 5.1 when applying separation of variables to Laplace's equation in the disk. The second type of boundary conditions take the form

$$\begin{cases} a_{11}X'(a) + a_{12}X(a) = 0 \\ a_{21}X'(b) + a_{22}X(b) = 0, \end{cases} \tag{5.4.3}$$

where we assume the coefficients a_{ij} satisfy the following conditions.

A2 The boundary conditions are nondegenerate, i.e., $a_{k1}^2 + a_{k2}^2 \neq 0$ for $k = 1, 2$.

A3 $a_{11}a_{12} \leq 0$ and $a_{21}a_{22} \geq 0$

Assumption A2 simply guarantees that both conditions are nontrivial. Assumption A3 is made on physical grounds, as was determined in Chapter 1.

Equation (5.4.1), which satisfies A1 coupled with either type of boundary condition, is called a **regular Sturm–Liouville problem**; in the case of periodic boundary conditions we further assume $k(a) = k(b)$. Singular Sturm–Liouville problems occur when k is allowed to be zero at one or both endpoints or when the function q is undefined at one or both of the endpoints. Examples occur when applying separation of variables to Laplace's equation in higher dimensions.

Notice that $X(x) \equiv 0$ is a solution for any value of μ. Hence, the fundamental question is: *do there exist values of μ for which the BVP has nontrivial solutions?* We have seen several examples where the answer to this question is affirmative, making the following definition appropriate.

Definition 5.26 Values of μ for which a Sturm–Liouville problem has nontrivial $C^1[a,b]$ solution are called **eigenvalues**. The corresponding solutions are **eigenfunctions**.

It should be noted that if X is an eigenfunction, then so is any multiple of X. Further, linear combinations of eigenfunctions with the same eigenvalue are again eigenfunctions. In other words, the set of eigenfunctions corresponding to a given eigenvalue forms a vector space, called the **eigenspace**.

Proposition 5.27 Let μ_1 and μ_2 be eigenvalues corresponding to eigenfunctions X_1 and X_2 of a regular Sturm–Liouville problem with boundary conditions as above. If $\mu_1 \neq \mu_2$, then

$$\int_a^b X_1(x)X_2(x)\rho(x)dx = 0. \tag{5.4.4}$$

In other words, the real-valued eigenfunctions are orthogonal with respect to the inner product $\langle f, g\rangle = \int_a^b f(x)g(x)\rho(x)dx$.

Proof The proof of this result is based on the following identity, known as Lagrange's identity, obtained via integration by parts:

$$(\mu_2 - \mu_1)\int_a^b X_1(x)X_2(x)\rho(x)dx = k(a)W[X_1, X_2](a) - k(b)W[X_1, X_2](b), \tag{5.4.5}$$

where $W[X_1, X_2](x) = X_1(x)X_2'(x) - X_1'(x)X_2(x)$, is the Wronskian of the two functions. Assume the boundary conditions have form (5.4.3). Then the Wronskian is zero at both $x = a$ and $x = b$. To see this, consider the boundary conditions at $x = a$,

$$a_{11}X_1'(a) + a_{12}X_1(a) = 0$$
$$a_{11}X_2'(a) + a_{12}X_2(a) = 0,$$

as a system of equations in a_{11} and a_{12}, at least one of which is not zero. Then this system has nondegenerate solution, which implies its coefficient matrix, the Wronskian at $x = a$, is zero. Similarly for the other endpoint $x = b$. In the periodic case, it is easy to see that $W[X_1, X_2](a) = W[X_1, X_2](b)$ from which the result follows given the extra condition on k. □

Remark 5.28 Notice that this proof did not use assumption A3. In fact the only place we will use this assumption is in showing that all the eigenvalues are nonnegative. For this reason, some authors leave out the assumption A3 in the definition of a regular Sturm–Liouville problem. In most applications of PDE, assumption A3 is valid, which is the reason we chose to include it in the definition.

An alternative formulation of the notions of eigenvalue and eigenfunction, and more in line with the terminology in linear algebra, can be given by considering the differential operator

$$L = -\frac{1}{\rho}\left[\frac{d}{dx}\left(k\frac{d}{dx}\right) + q\right].$$

Let $L^2([a,b], \rho(x)dx)$ be the space of all square integrable functions with respect to the weight function ρ and let $V_{bc} \subset L^2([a,b], \rho(x)dx)$ be the vector subspace of all C^2 functions that satisfy either periodic boundary conditions or those of form (5.4.3). Then restricting to this subspace, the Lagrange identity can be written in the form

$$\langle LX_1, X_2\rangle_\rho = \langle X_1, LX_2\rangle_\rho .$$

In other words, the operator L is **symmetric**[5] with respect to the underlying inner product on the subspace V_{bc}. Then an eigenvalue of the Sturm–Liouville problem is precisely a value of μ such that

$$LX = \mu X$$

has a nontrivial solution. Notice that the above equation has precisely the same form used in defining eigenvalues for linear transformations on a finite dimensional vector space. An interesting and useful observation to make is that orthogonality of eigenfunctions is a direct equivalent of the fact that L is a symmetric operator. In the rest of this text, eigenvalues of a BVP and eigenvalues of the associated differential operator will be terminology used interchangeably.

Example 5.15 Consider the equation $X'' + \mu X = 0$ on the interval $[0, \pi]$. Table 5.1 provides the eigenvalues and eigenfunctions for various boundary conditions met in the examples and exercises of the first section. In the last row, the numbers λ_n are solutions to the equation $\tan \pi\lambda = -\lambda$ (see Example 5.3). In all cases listed, the eigenspaces are one dimensional. The same equation on the interval $[-\pi, \pi]$ with periodic boundary conditions has eigenvalues n^2, $n = 0, 1, \ldots$ with corresponding eigenfunctions $\cos nx$ and $\sin nx$. Notice that except in the case of eigenvalue zero, the eigenspaces are two dimensional.

The other key properties of the eigenvalues are given in the following result.

Proposition 5.29 Consider a regular Sturm–Liouville problem. Then the following properties hold.

1. All eigenvalues are real.

5. Other terminology commonly used is to say the operator is *self-adjoint*.

Boundary Conditions	Eigenvalues	Eigenfunctions
$X(0) = X(\pi) = 0$	$n^2,\ n = 1, 2, \ldots$	$\sin nx$
$X'(0) = X'(\pi) = 0$	$n^2,\ n = 0, 1, 2, \ldots$	$\frac{1}{2}$, $\cos nx$
$X(0) = X'(\pi) = 0$	$(n + \frac{1}{2})^2,\ n = 0, 1, \ldots$	$\sin(n + \frac{1}{2})x$
$X'(0) = X(\pi) = 0$	$(n + \frac{1}{2})^2,\ n = 0, 1, \ldots$	$\cos(n + \frac{1}{2})x$
$X(0) = 0,\ X'(\pi) + X(\pi) = 0$	λ_n^2	$\sin \lambda_n x$

Table 5.1 Eigenvalues and Eigenfunctions of $\frac{d^2}{dx^2}$

2. All eigenvalues are nonnegative if $q(x) \leq 0$.

3. An eigenspace is at most two dimensional. In the case of boundary conditions (5.4.3), it is one dimensional.

Proof For property 1, if μ is an eigenvalue with $\mu \neq \bar{\mu}$ and corresponding eigenfunction X, then $\bar{\mu}$ is also an eigenvalue with corresponding eigenfunction $\bar{X}$. By the previous result $\int_a^b |X(x)|^2 \rho(x) dx = 0$. This implies that $X = 0$, a contradiction. To demonstrate property 2, we let μ be an eigenvalue with corresponding real-valued eigenfunction X. Then we derive the following identity by multiplying the ODE by X, integrating over the interval, and using integration by parts. This is known as the Rayleigh quotient.

$$\mu = \frac{-\left(k(x)X(x)X'(x)\right|_a^b + \int_a^b [X'(x)]^2 k(x) dx - \int_a^b q(x)[X(x]^2 dx}{\int_a^b X^2(x)\rho(x)dx} \tag{5.4.6}$$

In the case of periodic boundary conditions the result is immediate because the first term in the numerator is zero and $q \leq 0$. In the case of boundary conditions (5.4.3), the conclusion is obtained from the observation that $X(a)X'(a) \geq 0$ and $X(b)X'(b) \leq 0$. Indeed, multiply the first boundary condition by $X'(a)$ to get $a_{11}(X'(a))^2 + a_{12}X(a)X'(a) = 0$. Since a_{11} and a_{12} have opposite sign, it follows that $X(a)X'(a) \geq 0$. Similarly for the endpoint $x = b$.

We now turn to property 3. The set of solutions of any homogeneous linear second-order ODE is generated by a pair of linearly independent solutions, say X_1 and X_2, that can be taken to satisfy $X_1(a) = 1$, $X_1'(a) = 0$, $X_2(a) = 0$, and $X_2'(a) = 1$ (so their Wronskian is nonzero). This shows that any eigenspace has dimension at most two. If $c_1 X_1(x) + c_2 X_2(x)$ is the eigenfunction satisfying the boundary condition $a_{11}X'(a) + a_{12}X(a) = 0$, then $c_2 a_{11} + c_1 a_{12} = 0$. This linear relationship between c_1 and c_2 implies the eigenspace is one dimensional.
□

In the previous result, the conditions given in A2 on the coefficients in the boundary conditions was only used in proof that all the eigenvalues are nonnegative. It should be stressed that these conditions are sufficient in the case $q \leq 0$ for the eigenvalues to be nonnegative and are not necessary. The following example illuminates this point.

Example 5.16 Consider the equation $X'' + \mu X = 0$ on the interval $(0, L)$. Following Pinsky [23], write the boundary conditions in the form

$$(L \sin \alpha) X'(0) - (\cos \alpha) X(0) = 0$$
$$(L \sin \beta) X'(L) + (\cos \beta) X(L) = 0.$$

We may assume that $0 \leq \alpha, \beta < \pi$. The conditions A2 translate to imposing $\alpha, \beta \leq \pi/2$. Our focus here is to show that there is a larger region dependent on α and β where all eigenvalues of the problem are nonnegative. To determine this, consider the case $\mu = -\lambda^2 < 0$. Then any solution of the differential equation has form

$$X(x) = c_2 \cosh \lambda x + c_2 \sinh \lambda x.$$

Applying the boundary conditions yields the following equations for c_1 and c_2:

$$(-\cos \alpha) c_1 + (\lambda L \sin \alpha) c_2 = 0$$
$$(\lambda L \sin \beta \sinh \lambda L + \cos \beta \cosh \lambda L) c_1 + (\lambda L \sin \beta \cosh \lambda L + \cos \beta \sinh \lambda L) c_2 = 0.$$

For nontrivial solutions to exist, the determinant of the coefficient matrix must be zero. Computing this determinant, some algebra and a trig identity yield the following equation for this determinant to be zero:

$$\frac{\tanh \lambda L}{\lambda L} = -\frac{\sin(\alpha + \beta)}{\cos \alpha \cos \beta + (\lambda L)^2 \sin \alpha \sin \beta}. \tag{5.4.7}$$

For there to exist a negative eigenvalue, there must be positive solutions of the above equation in λ. There are four interesting cases to be considered: (i) $0 < \alpha, \beta < \pi/2$; (ii) $0 < \alpha < \pi/2 < \beta < \pi$; (iii) $0 < \beta < \pi/2 < \alpha < \pi$; and (iv) $\pi/2 < \alpha, \beta < \pi$. Case (i) is precisely the hypothesis A2 and we already know that all the eigenvalues are nonnegative. In case (ii), the left-hand side of the equation is less than or equal to one and the right-hand side has a vertical asymptote when $\lambda L = \sqrt{|\cot \alpha \cot \beta|}$. However, if α and β satisfy

$$\sin(\alpha + \beta) + \cos \alpha \cos \beta > 0, \tag{5.4.8}$$

then the right-hand side of (5.4.7) is greater than one at $\lambda = 0$ and is an increasing function of λ for $\lambda > 0$. Hence there are no solutions in λ of this equation and all eigenvalues of the BVP are nonnegative. By symmetry in α and β, the same is true in case (iii). Furthermore, in cases (ii) and (iii), if $\sin(\alpha+\beta) + \cos\alpha\cos\beta < 0$, then there is precisely one solution of equation (5.4.7), and hence one negative eigenvalue for the BVP. In the final case (iv), it turns out that there are two negative eigenvalues if (5.4.8) holds and only one if the opposite inequality holds. (The further analysis of the cases is left as an exercise.) Figure 5.8 provides plots of the curves $\sin(\alpha+\beta) + \cos\alpha\cos\beta = 0$ dividing the square $0 < \alpha, \beta < \pi$ into three regions. Regions I and III are where (5.4.8) holds; in region I all eigenvalues of the BVP are nonnegative and in region III there are precisely two negative eigenvalues. Region II is where there is precisely one negative eigenvalue.

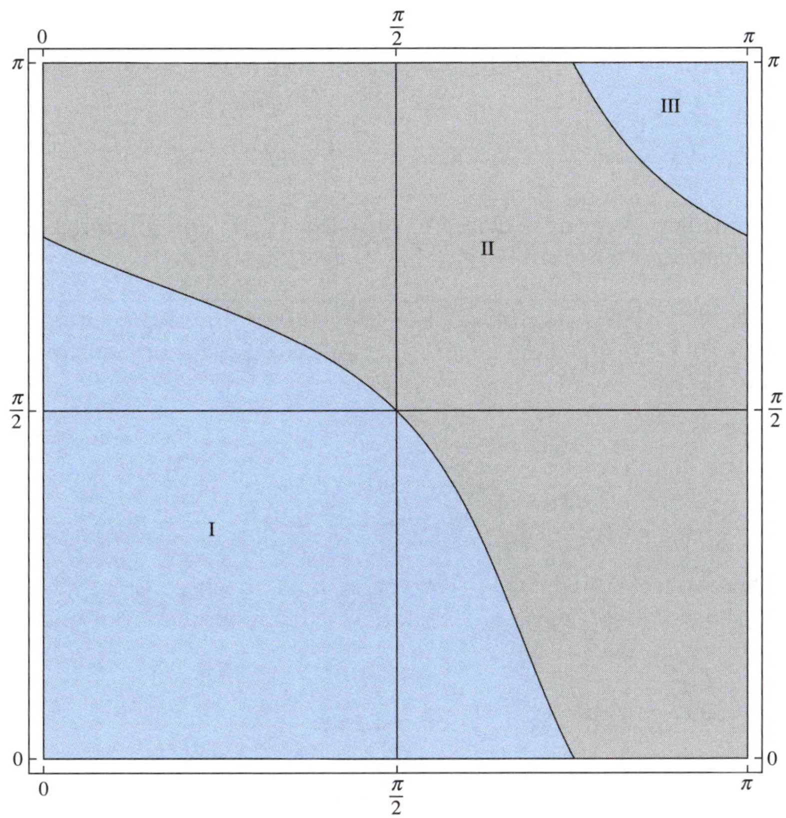

Figure 5.8 Regions for Example 5.16

We now turn to the question of existence of eigenvalues for Sturm–Liouville problems with the following classical result.

Theorem 5.30 Consider a regular Sturm–Liouville BVP with boundary conditions of the form (5.4.3). Then there exists a increasing sequence $\{\mu_n\}_{n=1}^{\infty}$ of eigenvalues with $\mu_n \to \infty$ such that the corresponding eigenfunctions satisfy Parseval's equality, i.e.,

$$\int_a^b |f(x)|^2 \rho(x) dx = \sum_{n=1}^{\infty} |\hat{f}_n|^2 \|X_n\|_2^2,$$

for every $f \in L^2([a, b], \rho(x)dx)$. Here $\{\widehat{f_n}\}$ is the sequence of Fourier coefficients with respect to the orthogonal set $\{X_n\}$. Moreover, writing $\mu_n = \lambda_n^2$, we have the asymptotic $\lambda_n = \left(\frac{n\pi}{b-a}\right) + O(\frac{1}{n})$. (The capital O means an expression bounded by a constant time $\frac{1}{n}$ as $n \to \infty$.)

This result will not be proved here; see [5, 10]. The Fourier expansion

$$f \sim \sum_n \widehat{f_n} X_n(x), \; \widehat{f_n} = \frac{1}{\|X_n\|} \int_a^b f(x) X_n(x) \rho(x) dx,$$

is often called the **eigenfunction expansion** of f. The following examples serve to illustrate the result.

Example 5.17 This example focuses on the asymptotic relation for the eigenvalue given Robin boundary conditions. Consider the Sturm–Liouville problem

$$\text{ODE}: X'' + \mu X = 0, \quad 0 < x < \pi$$
$$\text{BC}: \begin{cases} X(0) = 0 \\ X'(\pi) + X(\pi) = 0. \end{cases}$$

Then the eigenvalues are $\mu_n = -\lambda_n^2$, $n = 1, 2, \ldots$ where the λ_n's are solutions to the equation $\tan \pi\lambda = -\lambda$ and the corresponding eigenfunctions are $X_n(x) = \sin \lambda_n x$. Let

$$\Lambda_n = \|X_n\|_2^2 = \int_0^\pi \sin^2 \lambda_n x \, dx = \frac{\pi}{2} - \frac{\sin 2\lambda_n \pi}{4\lambda_n},$$

then Parseval's equality in this case becomes

$$\int_0^\pi |f(x)|^2 \, dx = \sum_{n=1}^{\infty} |\hat{f}_n|^2 \Lambda_n, \quad \text{where } \hat{f}_n = \frac{1}{\Lambda_n} \int_0^\pi f(x) \sin \lambda_n x dx.$$

The asymptotic behavior of the eigenvalues is illustrated as follows. Rewrite the λ equation in the form $\cot \pi\lambda = -1/\lambda$. The function $\cot \pi\lambda$ has zeros at the half-integers $n - 1/2, n = 1, 2, \ldots$ Apply Taylor's theorem about each of these zeros to obtain the estimate

$$\cot \pi\lambda = -\pi(\lambda - (n - 1/2)) + O((\lambda - (n - 1/2))^2).$$

Write $\varepsilon_n = \lambda_n - (n - \frac{1}{2})$, then $\varepsilon_n \to 0$ and in asymptotic form, the equation $\cot \pi\lambda = -1/\lambda$ becomes

$$-\pi\varepsilon_n + O(\varepsilon_n^2) = -\frac{1}{\varepsilon_n + n - 1/2} = O\left(\frac{1}{n}\right).$$

It follows that $\varepsilon_n = O(1/n)$ and, rewriting the above, we have

$$\lambda_n = n - \frac{1}{2} + O(1/n).$$

This corresponds with the asymptotic form given in Theorem 5.30.

Example 5.18 In this example, we focus on the Parseval equality and make use of the method of reflection. The Sturm–Liouville problem

$$X'' + \mu X = 0, \quad 0 < x < \pi$$
$$X(0) = X'(\pi) = 0$$

has eigenvalues $(n + \frac{1}{2})^2$ and corresponding eigenfunctions $\sin(n + \frac{1}{2})x$. Given $f \in L^2[0, \pi]$, the Fourier expansion has the form

$$f \sim \sum_{n=0}^{\infty} \widehat{f}(n)\sin(n + \frac{1}{2})x$$
$$\widehat{f}(n) = \frac{2}{\pi} \int_0^{\pi} f(x)\sin(n + \frac{1}{2})x\, dx. \tag{5.4.9}$$

We want to show that $\int_0^{\pi} f^2(x)dx = \frac{\pi}{2}\sum_n [\widehat{f}(n)]^2$. This will be accomplished using a technique known as the method of reflection; a more direct analytical approach is given in Exercise 5.33. Let $\mathcal{O}_1$ be the collection of functions $\{\sin(n + \frac{1}{2})x\}_{n=0}^{\infty}$. On the interval $[0, 2\pi]$, the set of functions $\mathcal{O}_2 = \{\sin \frac{m}{2}x\}_{m=1}^{\infty}$ form an orthogonal set, indeed these are the building blocks of Fourier sine series on this interval. Given $g \in L^2[0, 2\pi]$, its Fourier sine coefficients have the form

$$\widehat{g}_s(m) = \frac{1}{\pi} \int_0^{2\pi} g(x) \sin \frac{m}{2} x \, dx$$

and the Parseval equality holds (by results of Section 5.3):

$$\int_0^{2\pi} g^2(x) dx = \pi \sum_m [\widehat{g}_s(m)]^2.$$

The set $\mathcal{O}_1$ is a subset of $\mathcal{O}_2$, it contains precisely the functions in $\mathcal{O}_2$ for which m is an odd integer, i.e., $m = 2n + 1$, $n = 0, 1, \ldots$. Furthermore, the functions in $\mathcal{O}_1$ as functions on $[0, 2\pi]$ are even about the line $x = \pi$. The functions in $\mathcal{O}_2$ that are not in $\mathcal{O}_1$ are odd about $x = \pi$. For the function f, consider its even extension about $x = \pi$ defined by

$$f^e(x) = \begin{cases} f(x), & 0 < x < \pi \\ f(2\pi - x), & \pi < x < 2\pi. \end{cases}$$

It is easy to see that $f^e \in L^2[0, 2\pi]$. Also, since f^e is even about $x = \pi$ and $\sin \frac{m}{2} x$ is odd about $x = \pi$ when m is even, $\widehat{f^e}_s(2n) = 0$, $n = 1, 2, \ldots$. We now compute the other sine coefficients:

$$\begin{aligned} \widehat{f^e}_s(2n+1) &= \frac{1}{\pi} \left\{ \int_0^{\pi} f(x) \sin(n + \frac{1}{2}) x dx + \int_{\pi}^{2\pi} f(2\pi - x) \sin(n + \frac{1}{2}) x \, dx \right\} \\ &= \frac{2}{\pi} \int_0^{\pi} f(x) \sin(n + \frac{1}{2}) x \, dx = \widehat{f}(n), \end{aligned}$$

where we have replaced x by $2\pi - x$ in the integral over the interval $[\pi, 2\pi]$. We now have

$$\sum_{n=0}^{\infty} [\widehat{f}(n)]^2 = \sum_{m=1}^{\infty} [\widehat{f^e_s}(m)]^2 = \frac{1}{\pi} \int_0^{2\pi} [f^e(x)]^2 \, dx = \frac{2}{\pi} \int_0^{\pi} f^2(x) dx,$$

the last step using the same substitution as before. This completes the demonstration that Parseval's equality holds for this Sturm–Liouville problem.

Pointwise convergence of eigenfunction expansions is much more subtle than the case of classical Fourier series. Here is a classical result.

Theorem 5.31 Let f and $f' \in \mathrm{PC}[a,b]$. Let $\{\mu_n, X_n\}$ be a complete orthogonal set of eigenfunctions of a regular Sturm–Liouville problem with boundary conditions of the form (5.4.3). Let

$$f \sim \sum_n \hat{f}_n X_n(x)$$

be the Fourier expansion of f. Then for $x \in (a,b)$,

$$\lim_{N\to\infty} S_N f(x) = \frac{1}{2}[f(x+) + f(x-)],$$

where $S_N f(x)$ is the N partial sum of the Fourier expansion. Moreover, if $f \in C^2[a,b]$ and satisfies the boundary conditions, then the series converges uniformly on $[a,b]$.

Example 5.19 Consider the regular Sturm–Liouville problem of Example 5.17. As noted previously, the eigenvalues have form $\mu = \lambda^2$ where λ is a solution to the equation $\tan \pi\lambda = -\lambda$. These can be computed numerically, e.g., the first four decimal places are

$$\lambda_1 = 0.7876, \quad \lambda_2 = 1.6716, \quad \lambda_3 = 2.6162, \quad \lambda_4 = 3.5865.$$

The corresponding eigenfunctions are $X_n(x) = \sin \lambda_n x, n = 1, 2, \ldots$. A function $f \in \mathrm{PC}[0,\pi]$ has Fourier expansion of the form

$$f \sim \sum_{n=1}^{\infty} \hat{f}_n \sin \lambda_n x,$$

where

$$\hat{f}_n = \frac{1}{\Lambda_n}\int_0^{\pi} f(x) \sin \lambda_n x \, dx, \quad \Lambda_n = \int_0^{\pi} \sin^2 \lambda_n x \, dx = \frac{\pi}{2} - \frac{\sin 2\pi\lambda_n}{4\lambda_n}.$$

The results of these computations for two examples are given in the following table.

Function	Coefficients
$f(x) = \begin{cases} 1, & 0 < x, \pi/2 \\ -1, & \pi/2 < x < \pi \end{cases}$	$\frac{4(1-2\cos\lambda_n \frac{\pi}{2} + \cos\lambda_n\pi)}{2\pi\lambda_n - \sin 2\lambda_n\pi}$
$g(x) = -\frac{\pi}{1+\pi}x + \pi$	$\frac{4\pi}{2\pi\lambda_n - \sin 2\lambda_n\pi}$

Typical partial sums for each of these are given in Figure 5.9, effectively illustrating Theorem 5.31.

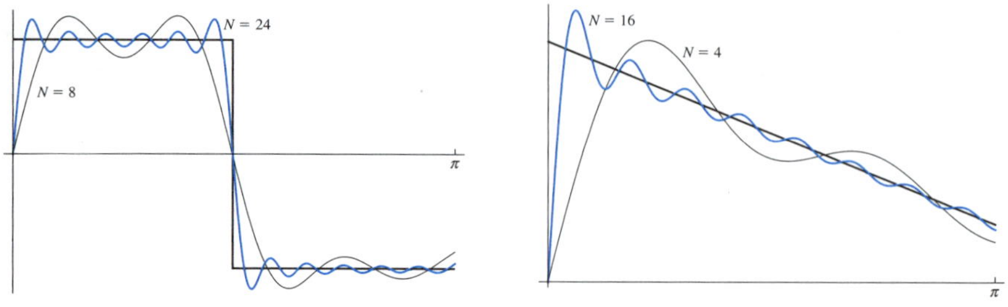

(a) Partial Sums for $f(x)$ (b) Partial Sums for $g(x)$

Figure 5.9 Plots for Example 5.19

Application to Green's Functions for ODEs

The Fredholm Alternative In Section 4.2 we gave a method for constructing Green's functions for a class of boundary value problems associated with a second-order ordinary differential equation. The method applies to equations of the form

$$(kX')' + qX = f, \quad a < x < b \tag{5.4.10}$$

with boundary conditions (5.4.3). A limitation of the method was that no solution of the corresponding homogeneous equation could satisfy both boundary conditions. In other words, the method applies provided zero is not an eigenvalue of the associated Sturm–Liouville problem

$$(kX')' + qX + \mu X = 0$$

coupled with the same boundary conditions. In this latter case, a solution will exist if the function f is orthogonal to the corresponding eigenfunction; however, it is not unique. The following example illustrates this idea.

Example 5.20 Consider the equation $X'' + X = f(x)$ on the interval $[0, \pi]$ with boundary conditions $X(0) = X(\pi) = 0$. We assume $f \in C[0, \pi]$. Since $\mu = 1$ is an eigenvalue of the BVP

$$\begin{aligned} X'' + \mu X &= 0, \quad 0 < x < \pi \\ X(0) = X(\pi) &= 0, \end{aligned}$$

we assume that $\langle f, \sin x\rangle = \int_0^\pi f(x) \sin x \, dx = 0$. The fundamental set for the equation $X'' + X = 0$ is $\{\sin x, \cos x\}$. Since $\sin x$ satisfies both boundary conditions, we seek a solution of the nonhomogeneous equation of the form $X(x) = v(x)\sin x$. Computing the derivatives and substituting into the equation gives

$$v'' \sin x + 2v' \cos x = f(x),$$

which can be rewritten as $(v' \sin^2 x)' = f(x)\sin x$. Integrating and rearranging leads to the formula

$$v' = \csc^2 x \int_0^x f(\xi)\sin \xi \, d\xi + C \csc^2 x,$$

where C is the integration constant. Integrating again we have

$$v(x) = \int_0^x \csc^2 \xi \int_0^\xi f(y)\sin y \, dy \, d\xi - C \cot x + D.$$

There is no problem with the existence of the integral on the right-hand side as the inner integral is of order ξ^2 for small ξ, which cancels out the singular behavior of $\csc^2 \xi$. We can now write

$$X(x) = \sin x \int_0^x \csc^2 \xi \int_0^\xi f(y)\sin y \, dy \, d\xi - C \cos x + D \sin x.$$

The first boundary condition implies $C = 0$. Integrating by parts and rearranging yields the formula

$$X(x) = -\cos x \int_0^x f(\xi)\sin \xi \, d\xi + \sin x \int_0^x f(\xi)\cos \xi \, d\xi + D \sin x.$$

Because of the orthogonality condition, $X(\pi) = 0$ for any choice of D. (This constant can be determined if a suitable third condition is imposed; the usual condition is that $X(x)$ is orthogonal to the eigenfunction $\sin x$.)

The preceding discussion and example leads to a dichotomy regarding the nonhomogeneous ODE (5.4.10) with boundary conditions (5.4.3) known as the **Fredholm alternative**:

- If zero is not an eigenvalue of the associated Sturm–Liouville problem, then there exists a unique solution to the nonhomogeneous equation satisfying the boundary conditions. In fact, a Green's function exists for the problem.
- If zero is an eigenvalue of the associated Sturm–Liouville problem with corresponding eigenfunction X_0, then the nonhomogeneous equation has a unique solution satisfying the boundary conditions if and only if the function f satisfies the orthogonality condition $\int_a^b f(x)X_0(x)dx = 0$. The solution is determined up to a constant.

Eigenfunction Expansions of Distributions and Green's Functions Recall that a distribution on a finite interval (a, b) is a continuous linear functional on the space of test functions. In this case, the test functions are smooth functions that vanish near the endpoints of the interval. Given a complete orthogonal set of real valued eigenfunctions $\{\phi_n\}$ associated with a Sturm–Liouville problem, it is generally the case that $\{\phi_n\} \subset C^1[a, b]$; however, they are not test functions. Nonetheless, for some distributions a Fourier expansion with respect to $\{\phi_n\}$ can be defined and the series will converge in distribution to u. The important case here is the distribution $\delta_\xi(x)$ where ξ is a fixed point in (a, b). As the delta functional acts by evaluation, the Fourier series takes the form

$$\delta(x - \xi) \sim \sum_{n=1}^{\infty} \frac{\phi_n(\xi)}{\|\phi_n\|_2^2} \phi_n(x).$$

To show this converges in distribution, let S_N denote the Nth partial sum, then for any test function ψ,

$$(S_N, \psi) = \sum_{n=1}^{N} \frac{\phi_n(\xi)}{\|\phi_n\|_2^2} (\phi_n, \psi) = \sum_{n=1}^{N} \phi_n(\xi)\widehat{\psi}(n) \to \psi(\xi) = (\delta(x - \xi), \psi(x)),$$

where we have applied Theorem 5.31.

Recall that the Green's function, $G(x, \xi)$, was the weak solution of the ordinary differential equation

$$(kX')' + q(x)X = \delta(x - \xi)$$

that satisfied the boundary conditions (5.4.3). We are assuming that $\mu = 0$ is not an eigenvalue of the associated Sturm–Liouville problem. Let $\{\phi_n(x)\}_{n=1}^{\infty}$ be a complete orthogonal set of eigenfunctions of this problem with corresponding eigenvalues $\{\mu_n\}$. Then the Green's function will have expansion given by

$$G(x, \xi) \sim \sum_{n=1}^{\infty} \frac{\phi_n(\xi)}{\mu_n \|\phi_n\|_2^2} \phi_n(x),$$

the series convergent in the sense of distribution. The following example illustrates this idea and other features of the expansion.

Example 5.21 Consider the problem $u'' = f$ on $[0, \pi]$ with boundary conditions $u(0) = u(\pi) = 0$. A complete set of orthogonal eigenfunctions for this problem is $\{\sin nx\}_{n=1}^{\infty}$ with eigenvalues $\{n^2\}$. Hence the Green's function has Fourier expansion

$$G(x, \xi) \sim \frac{2}{\pi} \sum_{n=1}^{\infty} \frac{\sin n\xi \sin nx}{n^2}.$$

Notice that the series is actually uniformly convergent by the Weierstrass M-test. The series also directly reflects the fact that the Green's function is symmetric in its arguments, i.e., $G(x, \xi) = G(\xi, x)$.

Remark 5.32 The uniform convergence of the eigenfunction expansion of the Green's function is typical in the case of ordinary differential equations of the type we have considered. This is in sharp contrast to what we will find in the case of partial differential equations in subsequent chapters.

Exercises

5.30 Consider the heat equation in the form $\rho u_t = (ku_x)_x$, where $\rho = \rho(x)$ and $k = k(x)$ are both positive functions and satisfy the hypothesis assumed in this section. By seeking separated solutions of form $u(x, t) = e^{-\mu t} X(x)$, show that X must satisfy the ODE (5.4.1).

5.31 Reconsider Exercise 5.30 for the wave equation $\rho u_{tt} = (ku_x)_x$ and the separated solutions of the form $u(x, t) = e^{i\mu t} X(x)$. Give a physical argument for seeking solutions using these complex exponentials.

5.32 Demonstrate Lagrange's identity (5.4.5).

5.33 Verify the conclusion of Example 5.18 by expanding $\sin(n + \frac{1}{2})x$ using the addition theorem for the sine function and applying Parseval's equality for cosine and sine series on $[0, \pi]$.

5.34 Work out the remaining details of Example 5.16.

5.35 Show that Parseval's equality holds for the BVP $X'' + \mu X = 0,\ 0 < x < \pi$, $X'(0) = X(\pi) = 0$.

5.36 Let $f(x) = -\frac{1+\pi}{2+\pi} x^2 + \pi x$. Note that $f(0) = 0$ and $f'(\pi) + f(\pi) = 0$. Compute the Fourier expansion of f relative to the eigenfunctions of Example 5.19. Plot the second and fourth partial sums along with the graph of f.

5.37 Let $f \in C^1[0,\pi]$ with $f'' \in \mathrm{PC}[0,\pi]$ and suppose f satisfies the boundary conditions of Example 5.19. Expanding f in terms of the eigenfunctions of this example, show $\sum |\hat{f}_n| < \infty$. Further, the Fourier expansion of f converges uniformly on $[0,\pi]$. Relate this to Exercise 5.36. HINT: Calculate $\widehat{f''}_n$ using integration by parts.

5.38 Find the Fourier eigenfunction expansion of the Green's function for the equation $X'' = f$ on the interval $(0,\pi)$ with boundary conditions $X(0) = X'(\pi) = 0$. Show that it converges in distribution to the Green's function.

5.39 Reconsider Exercise 5.38 with boundary conditions $X(0) = 0$ and $X'(\pi) + X(\pi) = 0$.

Chapter 6

Diffusion and Waves With Boundaries

Throughout this chapter we will apply the eigenfunction expansions of the previous chapter—along with other suitable techniques—to initial boundary value problems associated with the heat and wave equations where the spatial domain is rectangular. After we provide an overview of some useful techniques, we will focus the one-dimensional heat and wave equations, respectively in Sections 6.2 and 6.3. In both sections, solution formulas take the form of infinite series, and discussion of convergence is always at the forefront. We will apply the resulting series to a variety of qualitative problems for the equations, e.g., asymptotics and energy considerations. In the final section, we introduce multiple Fourier series and their application to heat flow and waves in higher dimensions.

6.1 Problem Types and Useful Techniques

Let $\mathcal{L}$ denote either the wave operator $\frac{\partial^2}{\partial t^2} - c^2\frac{\partial^2}{\partial x^2}$ or the heat operator $\frac{\partial}{\partial t} - \alpha^2\frac{\partial^2}{\partial x^2}$. One focus in this chapter is to develop techniques for solving initial boundary value problems of the form

$$\begin{aligned}
&\text{PDE: } \mathcal{L}(u) = Q(x,t), \quad 0 < x < L, \quad t > 0\\
&\text{BC: } \begin{cases} B_o[u] \stackrel{def}{=} a_{11}u_x(0,t) + a_{12}u(0,t) = g_0(t) \\ B_L[u] \stackrel{def}{=} a_{21}u_x(L,t) + a_{22}u(L,t) = g_L(t) \end{cases}\\
&\text{IC: } \begin{cases} u(x,0) = f(x), & \text{for heat} \\ \begin{cases} u(x,0) = f(x) \\ u_t(x,0) = g(x) \end{cases}, & \text{for wave.} \end{cases}
\end{aligned}$$

Our approach to this general problem is to use the linearity of the equation and break the problem into three essential subproblems of the following types.

- **Type I** Homogeneous equation and boundary conditions ($Q = 0$ and $g_0 = g_L = 0$). Separation of variables and eigenfunction expansions is the technique of choice here. The separated solutions that satisfy the boundary conditions have the form $u_n(x,t) = T_n(t)X_n(x)$, where the X_n's are orthogonal eigenfunctions of $\frac{d^2}{dx^2}$ on the interval $[0, L]$. The solution to the problem is constructed using superposition and involves coefficients determined via a Fourier-type formula. The resulting series solution is called the **Fourier representation of the solution.** (Recall the heat flow examples in the beginning of Chapter 5.)

- **Type II** Homogeneous boundary conditions and zero for initial conditions ($f = g = 0$). Here, there are two subcases to consider.

  - The function Q is time independent, i.e., $Q(x,t) = Q(x)$. There is a steady-state solution $U(x)$ of the partial differential equation that satisfies the boundary conditions. Writing $u(x,t) = U(x) + v(x,t)$, the function $v(x,t)$ then satisfies an initial boundary value problem of type I.

  - The function Q is time dependent. Here we employ a technique akin to the method of variation of parameters. Specifically, if $\{X_n\}$ are the eigenfunctions mentioned in type I, we expand Q as

$$Q(x,t) \sim \sum_n q_n(t)X_n(x)$$

    and seek a solution of the nonhomogeneous partial differential equation of the form

$$u(x,t) = \sum_n a_n(t)X_n(x).$$

    Upon substitution into the equation this leads to an ordinary differential equation for the coefficients $a_n(t)$ that can be solved using a variety of known techniques.

- **Type III** Homogeneous equation and zero for the initial state. Again, there are two subcases.

  - Time-independent boundary conditions, i.e., the functions g_0 and g_L are constants. This can be handled just like the first subcase of type II.

– Time-dependent boundary conditions. One way to treat this case is to find a simple function $w(x,t)$ that satisfies the boundary conditions and write $v(x,t) = u(x,t) - w(x,t)$. Simple computation then shows that $v(x,t)$ satisfies an initial boundary value problem that can be broken into two separate problems of types I and II.

Undoubtedly, this breakdown into types is an oversimplification. Nonetheless, it gives a bag of tricks useful as a starting point for solving general problems. Further, although the discussion was focused on the wave and heat equations in one dimension, the basic ideas can be modified and applied to equations in higher dimensions and to time-independent equations such as Laplace's equation where the role of the time variable is played by one of the spatial variables (see Chapter 7).

6.2 Heat Flow Problems in One Dimension

Homogeneous Problems

The general initial boundary value problem considered here takes the form

$$\begin{aligned} &\text{PDE: } u_t = \alpha^2 u_{xx}, \quad 0 < x < L, \quad t > 0 \\ &\text{BC: } \begin{cases} a_{11}u_x(0,t) + a_{12}u(0,t) = 0 \\ a_{21}u_x(L,t) + a_{22}u(L,t) = 0 \end{cases} \\ &\text{IC: } u(x,0) = f(x), \end{aligned} \tag{6.2.1}$$

where the coefficients a_{ij} are constants and the initial data f is assumed to be piecewise continuous with piecewise continuous derivative. Assuming separated solutions of the form $u(x,t) = T(t)X(x)$ leads to the BVP for $X(x)$,

$$\begin{aligned} &\text{ODE: } X'' + \mu X = 0, \quad 0 < x < L \\ &\text{BC: } \begin{cases} a_{11}X'(0) + a_{12}X(0) = 0 \\ a_{21}X'(L) + a_{22}X(L) = 0. \end{cases} \end{aligned} \tag{6.2.2}$$

We make the following assumption: *all eigenvalues of this boundary value problem are nonnegative.* Writing $\mu = \lambda^2$, it follows that the separated solutions have the form

$$u_n(x,t) = e^{-(\lambda_n \alpha)^2 t} X_n(x),$$

where the function $X_n(x)$ is the eigenfunction corresponding to eigenvalue λ_n. The collection $\{X_n(x)\}$ forms an orthogonal set on the interval $[0, L]$; hence, pursuing the ideas of Example 5.2, the conjecture for the solution to this problem has the form

$$u(x,t) = \sum_n \widehat{f}(n) e^{-(\lambda_n \alpha)^2 t} X_n(x), \tag{6.2.3}$$

where the coefficients $\widehat{f}(n)$ are the Fourier coefficients of the initial data f with respect to the orthogonal set $\{X_n(x)\}$. Our purpose here is to present the justification of this solution formula, describe its asymptotics as $t \to \infty$, and obtain and estimate on the energy gauge defined by

$$E(t) = \int_0^L u^2(x,t)dx.$$

The solution formula will also be rewritten in heat kernel form as

$$u(x,t) = \int_0^L f(\xi)\mathcal{S}_{\text{bc}}(x,\xi,t)dx,$$

where

$$\mathcal{S}_{\text{bc}}(x,\xi,t) = \sum_n e^{-(\lambda_n\alpha)^2 t}\frac{X_n(\xi)X_n(x)}{\|X_n\|^2}. \tag{6.2.4}$$

Formally, this is derived from (6.2.3) by substituting the formula for the Fourier coefficients and interchanging the sum and integral. The function $\mathcal{S}_{\text{bc}}$ is the heat kernel for the problem; the subscript bc denotes the boundary conditions, e.g., *dd* will mean homogeneous Dirichlet conditions at each end. We begin by exploring two examples.

Example 6.1 (Homogeneous Dirichlet Boundary Conditions) Consider the IBVP (6.2.1) with boundary conditions

$$u(0,t) = u(L,t) = 0.$$

Physically, this corresponds to heat flow in a uniform thin rod or slab with both boundaries held at temperature zero. In Example 5.2 this problem was considered in the case $L = \pi$ as motivation for the method of separation of variables. In this slightly more general problem, the separated solutions take the form

$$u_n(x,t) = e^{-(\lambda_n\alpha)^2 t}\sin\lambda_n x,$$

where $\lambda_n = \frac{n\pi}{L}$, $n = 1, 2, \ldots$. Following that example, the conjectured solution has form

$$u(x,t) = \sum_{n=1}^{\infty} \widehat{f}_s(n)e^{-(\lambda_n\alpha)^2 t}\sin\lambda_n x, \tag{6.2.5}$$

where $\widehat{f}_s(n)$ is the nth-Fourier sine coefficient of f on the interval $[0, L]$. Toward justification of the solution, we make the following claims: (a) the series is uniformly convergent on $[0, L]$ and $t \geq t_0 > 0$; (b) the series can be differentiated term by term in either variable and to any order; (c) u is a solution to the heat equation; (d) $u(x, 0) = f(x)$ at points of continuity; i.e., the initial condition is satisfied. Proof of these claims is as follows.

Claim (a). There is a constant M such that $|\widehat{f}_s(n)| \leq M$ for all n by the Riemann–Lebesgué Lemma. Hence, for $t \geq t_0$ and $0 \leq x \leq L$, we have $|\widehat{f}_s(n)e^{-(\lambda_n \alpha)^2 t} \sin nx| \leq Me^{-(\lambda_n \alpha)^2 t_0} = M_n$. Since $\sum_n M_n$ converges, the Weierstrass M-test (Theorem A.3) applies, proving the claim.

Claim (b). The series of term-by-term derivatives in x is

$$\sum_n \lambda_n \widehat{f}_s(n) e^{-(\lambda_n \alpha)^2 t} \cos \lambda_n x.$$

Similar to claim (a), we have that

$$|\lambda_n \widehat{f}_s(n) e^{-(\lambda_n \alpha)^2 t} \cos \lambda_n x| \leq M \lambda_n e^{-(\lambda_n \alpha)^2 t_0} = M_n.$$

Since $\sum_n M_n$ converges, the Weierstrass M-test implies that the series of term-by-term derivatives is uniformly convergent. The result on differentiating series, Theorem A.2, then implies that $u_x(x,t)$ exists and

$$u_x(x,t) = \sum_n n \widehat{f}_s(n)\, e^{-(\alpha n)^2 t} \cos nx.$$

A similar argument shows that the series for $u(x,t)$ can be differentiated term by term with respect to t. Furthermore, this process can be iterated to any order since the series $\sum_n \lambda_n^m e^{-(\lambda_n \alpha)^2 t_0}$ is convergent for any $m > 0$.

Claim (c). This claim follows by claim (b) because the terms in the series for $u(x,t)$ individually are solutions of the heat equation.

Claim (d). The initial condition is satisfied by the pointwise convergence theorem for Fourier series, Theorem 5.19.

We can easily determine the asymptotics of the solution because of the rapid decay of the exponential function $e^{-(\lambda_n \alpha)^2 t}$. To obtain this, we break off the first term of the series

$$u(x,t) = \widehat{f}_s(1) e^{-(\lambda_1 \alpha)^2 t} \sin \lambda_1 x + \sum_{n=2}^{\infty} \widehat{f}_s(n) e^{-(\lambda_n \alpha)^2 t} \sin \lambda_n x.$$

Using the previous notation, we can estimate the error term by:

$$\begin{aligned}\left| \sum_{n=2}^{\infty} \widehat{f}_s(n) e^{-(\lambda_n \alpha)^2 t} \sin \lambda_n x \right| &\leq M \sum_{n=2}^{\infty} e^{-n^2 (\alpha\pi/L)^2 t} \\ &\leq M \sum_{n=2}^{\infty} \left(e^{-(\alpha\pi/L)^2 t} \right)^n = \frac{M e^{-2(\alpha\pi/L)^2 t}}{1 - e^{-(\alpha\pi/L)^2 t}},\end{aligned}$$

where the geometric sum formula ($\sum_{n=m}^{\infty} r^n = \frac{r^m}{1-r}$, provided $r < 1$) has been applied in the last step. Let $t_0 > 0$ and assume $t \geq t_0$, then $1 - e^{-(\alpha\pi/L)^2 t} \geq 1 - e^{-(\alpha\pi/L)^2 t_0}$. It follows that our error term is $O(e^{-2(\alpha\pi/L)^2 t})$ as $t \to \infty$, where O means "bounded by a constant times the exponential function in parenthesis." Notice that this error estimate is uniform in the spatial variable x. Summarizing, we have the asymptotic estimate

$$u(x,t) = \widehat{f}_s(1) e^{-(\lambda_1 \alpha)^2 t} \sin \lambda_1 x + O(e^{-2(\lambda_1 \alpha)^2 t}).$$

It is important to note that the first term is dominant, i.e., tends to zero slower than the second. We can obtain an estimate for the energy gauge by simply squaring both sides and integrating:

$$\begin{aligned} E(t) &= \frac{1}{2}\left[\widehat{f}_s(1)\right]^2 (1) e^{-2(\lambda_1\alpha)^2 t} \int_0^l \sin^2 \lambda_1 x \, dx + O(e^{-3(\lambda_1\alpha)^2 t}) \\ &= \frac{L}{4}\left[\widehat{f}_s(1)\right]^2 e^{-2(\lambda_1\alpha)^2 t} + O(e^{-3(\lambda_1\alpha)^2 t}). \end{aligned}$$

A physical interpretation of these asymptotics is that the first eigenvalue λ_1 governs the decay of the solution and the energy. In the case where $\widehat{f}_s(1) = 0$, larger eigenvalues must considered to obtain the asymptotics.

Finally, to obtain the heat kernel for the problem, we substitute the formula for the Fourier sine coefficients into (6.2.5) and interchange the sum and integral to obtain

$$u(x,t) = \int_0^L f(\xi) \mathcal{S}_{\mathrm{dd}}(x,\xi,t) d\xi,$$

where

$$\mathcal{S}_{\mathrm{dd}}(x,\xi,t) = \frac{2}{L} \sum_{n=1}^{\infty} e^{-(\lambda_n\alpha)^n t} \sin \frac{n\pi\xi}{L} \sin \frac{n\pi x}{L}.$$

This is justified as the preceding series is uniformly convergent for $\xi \in [0, L]$ and any $t > 0$. When $t = 0$ the series becomes

$$\frac{2}{L} \sum_{n=1}^{\infty} \sin \frac{n\pi\xi}{L} \sin \frac{n\pi x}{L},$$

which diverges pointwise but converges in distribution to $\delta(x - \xi)$ and is the Fourier sine series of the latter.

Example 6.2 (Homogeneous Mixed Boundary Conditions) Consider the initial boundary value problem (6.2.1), where the boundary conditions are

$$\text{BC:} \begin{cases} u(0,t) = 0 \\ u_x(L,t) + hu(L,t) = 0, \end{cases}$$

where $h > 0$. Physically the left end is held at constant temperature zero and the right end convects into a medium whose temperature is zero. The separated solutions in this case have the form

$$u_n(x,t) = e^{-(\lambda_n \alpha)^2 t} \sin \lambda_n x,$$

where λ_n^2 are the eigenvalues of the BVP

$$\text{ODE: } X'' + \mu X = 0, \quad 0 < x < L$$
$$\text{BC:} \begin{cases} X(0) = 0 \\ X'(0) + hX(0) = 0. \end{cases}$$

Explicitly, λ_n are the solutions of the transcendental equation $\tan \lambda L = -\frac{h}{\lambda}$. (See Example 5.17 for the case $L = \pi$ and $h = 1$.) The solution to this problem takes the form

$$u(x,t) = \sum_{n=1}^{\infty} \widehat{f}(n) e^{-(\lambda_n \alpha)^2 t} \sin \lambda_n x,$$

where the coefficients $\widehat{f}(n)$ are the Fourier coefficients of the initial data f with respect to the orthogonal set $\{\sin \lambda_n x\}$; i.e.,

$$\widehat{f}(n) = \frac{1}{\Lambda_n} \int_0^L f(x) \sin \lambda_n x \, dx$$
$$\Lambda_n = \int_0^L \sin^2 \lambda_n x \, dx = \frac{L}{2} - \frac{\sin 2\lambda_n L}{4\lambda_n}.$$

The justification of the solution proceeds just as in the previous case where the fact that $\lambda_n \approx \frac{\pi n}{L}$ is used (Theorem 5.30) in the application of the Weierstrass M-test and Theorem 5.31 is applied to deduce that the initial condition is satisfied. To derive the asymptotic for the solution we write

$$u(x,t) = \widehat{f}(1) e^{-(\lambda_1 \alpha)^2 t} \sin \lambda_1 x + R.$$

The estimate for the remainder is carried out as before, with the small change of using the inequality $\left(n-\frac{1}{2}\right)\frac{\pi}{L} < \lambda_n < \frac{n\pi}{L}$:

$$|R| \le M\sum_{n=2}^{\infty} e^{-(\lambda_n\alpha)^2 t} \le M\sum_{n=2}^{\infty} e^{-(n-1/2)^2(\alpha\pi/L)^2 t}$$
$$\le e^{\frac{1}{2}\left(\frac{\alpha\pi}{L}\right)^2 t}\sum_{n=2}^{\infty}\left(e^{-\left(\frac{\alpha\pi}{L}\right)^2 t}\right)^n = \frac{e^{-\frac{3}{2}\left(\frac{\alpha\pi}{L}\right)^2 t}}{1-e^{-\left(\frac{\alpha\pi}{L}\right)^2 t}}.$$

Since $\lambda_1 \le \frac{\pi}{L}$, we have $e^{-\frac{3}{2}\left(\frac{\alpha\pi}{L}\right)^2 t} \le e^{-\frac{3}{2}(\lambda_1\alpha)^2 t}$ and it follows that

$$u(x,t) = \widehat{f}(1)e^{-(\lambda_1\alpha)^2 t}\sin\lambda_1 x + O\left(e^{-\frac{3}{2}(\lambda_1\alpha)^2 t}\right).$$

Likewise, the energy can be estimated as before:

$$E(t) = \Lambda_1\left[\widehat{f}(1)\right]^2 e^{-2(\lambda_1\alpha)^2 t} + O\left(e^{-\frac{5}{2}(\lambda_1\alpha)^2 t}\right).$$

Again notice that the first eigenvalue governs the asymptotics of both the solution and energy. The heat kernel in this example takes the form

$$\mathcal{S}_{\mathrm{dr}}(x,\xi,t) = \sum_{n=1}^{\infty} e^{-(\lambda_n\alpha)^2 t}\frac{\sin\lambda_n\xi\sin\lambda_n x}{\Lambda_n}.$$

At $t=0$, the series is the Fourier series with respect to the orthogonal set $\{\sin\lambda_n x\}$ of the Dirac delta $\delta(x-\xi)$. A contour plot of the heat kernel is given in Figure 6.1 for small and large values of t, indicating the fact that the kernel approximates a delta functional for t small and diffuses for t large, respectively. (The darker color indicates values closer to zero.) Also note the contour behavior in the lower-left and upper-right corners, indicative of the boundary conditions.

The previous two examples illustrate the following theorem whose proof is a careful adaptation of the methods we employed.

Theorem 6.1 Under the preceding assumptions, (6.2.3) gives the solution to the IBVP (6.2.1). Furthermore, if $\widehat{f}(1) \neq 0$ and all eigenvalues are positive, there is a constant $k > 1$ such that as $t \to \infty$,

$$u(x,t) = \widehat{f}(1)e^{-(\lambda_1\alpha)^2 t}X_1(x) + O\left(e^{-k(\lambda_1\alpha)^2 t}\right)$$

and

$$E(t) = \Lambda_1\left[\widehat{f}(1)\right]^2 e^{-2(\lambda_1\alpha)^2 t} + O\left(e^{-(k+1)(\lambda_1\alpha)^2 t}\right).$$

Figure 6.1 Contour Plot of Heat Kernel: Example 6.2

In the case where $\lambda_0 = 0$ is an eigenvalue of (6.2.2) (Neumann boundary conditions, $\lambda_n = \frac{n\pi}{L}$, $n = 0, 1, \ldots$), then the solution has the form

$$u(x,t) = \frac{\widehat{f_c}(0)}{2} + \sum_{n=1}^{\infty} \widehat{f_c}(n) e^{-(\lambda_n \alpha)^2 t} \cos \lambda_n x$$

with asymptotics given by

$$u(x,t) = \frac{\widehat{f_c}(0)}{2} + O\left(e^{-(\lambda_1 \alpha)^2 t}\right)$$

and

$$E(t) = \int_0^L \left[u(x,t) - \widehat{f_c}(0)/2\right]^2 dx = O\left(e^{-2(\lambda_1 \alpha)^2 t}\right). \tag{6.2.6}$$

In all cases, the heat kernel is given by (6.2.4).

Remark 6.2

1. In the case of Neumann boundary conditions, the quantity $E(t)$ represents the square norm of the difference between the solution and the possibly nonzero steady state.

2. In the case of Dirichlet boundary conditions, Theorem 6.1 sharpens the energy estimates obtained in Chapter 3 using the Poincaré inequality; see Example 3.5.

The Nonhomogeneous Equation

In this subsection, techniques are presented solve IBVPs for the nonhomogeneous heat equation. The problem of interest takes the form

$$\begin{aligned} \text{PDE: } & u_t = \alpha^2 u_{xx} + Q(x,t), \quad 0 < x < L, \quad t > 0 \\ \text{BC: } & B_0[u] = B_L[u] = 0 \\ \text{IC: } & u(x,0) = 0. \end{aligned} \tag{6.2.7}$$

Here $Q(x,t)$ represents a heat source inside the rod due to a chemical reaction or some other mechanism. It has units of degrees per unit time so represents a rate of heat generation.

If the source term is independent of time, $Q(x,t) = Q(x)$, then the problem has a steady-state solution, denoted $U(x)$. The latter is a solution of the ordinary differential equation

$$\alpha^2 U'' + Q = 0,$$

which satisfies the same boundary conditions as u. Hence the steady state can be obtained by integration and algebraic manipulation. Substituting $u(x,t) = v(x,t) + Q(x)$ into (6.2.7), it follows that $v(x,t)$ satisfies the IBVP

$$v_t = \alpha^2 v_{xx}, \quad B_0[v] = B_L[v] = 0, \quad v(x,0) = -U(x).$$

The reduced problem for v was considered in the previous subsection. In fact we can employ the asymptotics for v to obtain such for u. The following example illustrates this case.

Example 6.3 Assume $Q(x,t) = Q_0$, a constant, and the boundary conditions are $u(0,t) = u(L,t) = 0$. Integrating twice, we obtain the steady-state solution of the form

$$U(x) = -\frac{1}{2\alpha^2} Q_0 x^2 + c_1 x + c_2.$$

We can determine the integration constants from the boundary conditions as $c_2 = 0$ and $c_1 = \frac{Q_0 L}{2\alpha^2}$. Thus the steady state can be written

$$U(x) = \frac{Q_0 x}{2\alpha^2}(L - x).$$

Now the Fourier sine coefficients of $-U(x)$ are given by $\frac{2Q_0L^2}{(n\pi)^3\alpha^2}(1-(-1)^n)$ and hence the transient solution $v(x,t)$ is given by

$$\begin{aligned} v(x,t) &= -\sum_{n=1}^{\infty} \frac{2Q_0L^2}{(n\pi)^3\alpha^2}(1-(-1)^n)e^{-(\lambda_n\alpha)^2t}\sin\lambda_n x \\ &= -\frac{4Q_0L^2}{\pi^3\alpha^2}\sum_{n=1,3,\ldots} \frac{1}{n^3}e^{-(\lambda_n\alpha)^2t}\sin\lambda_n x, \end{aligned}$$

where $\lambda_n = \frac{n\pi}{L}$. The solution to the nonhomogeneous problem is then $U(x)+v(x,t)$. From the previous subsection, we have the asymptotic

$$v(x,t) = -\frac{4Q_0L^2}{\pi^3\alpha^2}e^{-(\lambda_1\alpha)^2t}\sin\lambda_1 x + O\Big(e^{-2(\lambda_1\alpha)^2t}\Big).$$

This implies an asymptotic for $u(x,t)$ of the form

$$u(x,t) = U(x,t) + O\Big(e^{-(\lambda_1\alpha)^2t}\Big).$$

This solution is plotted in Figure 6.2 where $\alpha=1$, $Q_0=1$, and $L=\pi$.

Turning now to the general nonhomogeneous heat equation, we use an approach akin to the method of variation of parameters for ordinary differential equations. As above, we let $\{X_n(x)\}$ be the eigenfunctions of the BVP (6.2.2) with corresponding eigenvalues $\{\lambda_n\}$, and seek a solution of the equation of the

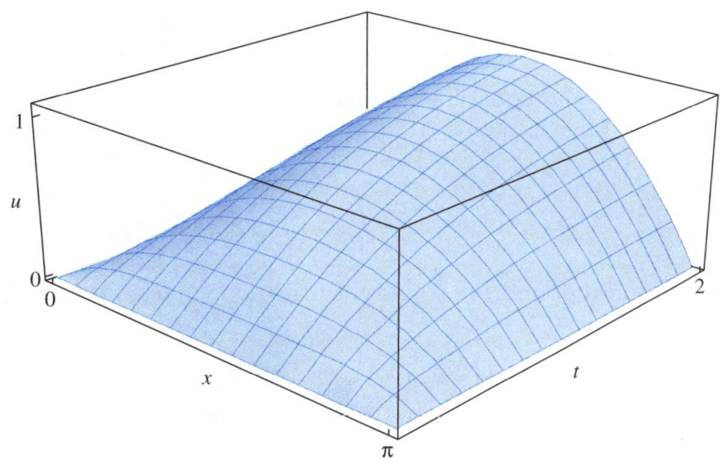

Figure 6.2 Heat Equation: Time-Independent Source Term

form

$$u(x,t) = \sum_n a_n(t)X_n(x). \tag{6.2.8}$$

Notice that the boundary conditions are automatically satisfied, so the coefficients $a_n(t)$ must be determined so as to satisfy the nonhomogeneous equation and the initial condition. The latter is easy: $a_n(0) = 0$ for all n. Proceeding formally, we expand the source term $Q(x,t)$ in terms of the eigenfunctions

$$Q(x,t) = \sum_n q_n(t)X_n(x),$$

where $q_n(t)$ is the nth Fourier coefficient of $Q(x,t)$ with respect to the eigenfunction X_n. Substituting this and the assumed form of the solution (6.2.8) into the equation, we arrive at

$$\sum_n a_n'(t)X_n(x) = -\alpha^2 \sum_n \lambda_n^2 a_n(t) + \sum_n q_n(t)X_n(x).$$

Collecting terms this can be written

$$\sum_n \left[a_n' + (\lambda_n\alpha)^2 a_n - q_n\right] X_n = 0.$$

Thus, for the idea to work, for each n the coefficient $a_n(t)$ must be chosen to satisfy the following initial value problem:

$$a_n' + (\lambda_n\alpha^2)a_n = q_n, \quad a_n(0) = 0. \tag{6.2.9}$$

This is a first-order linear ODE with integrating factor $e^{(\lambda_n\alpha)^2 t}$. When we multiply by this factor the equation becomes $\left[e^{(\lambda_n\alpha)^2 t}a_n\right]' = e^{(\lambda_n\alpha)^2 t}q_n$, which upon integration supplies the solution:

$$a_n(t) = e^{-(\lambda_n\alpha)^2 t}\int_0^t e^{(\lambda_n\alpha)^2\tau}q_n(\tau)d\tau. \tag{6.2.10}$$

Conceivably, we have found the solution to the nonhomogeneous IBVP (6.2.7) as

$$u(x,t) = \sum_n \left[\int_0^t e^{-(\lambda_n\alpha)^2(t-\tau)}q_n(\tau)d\tau\right]X_n(x). \tag{6.2.11}$$

Convergence questions and verification of the solution remains, issues that are most easily handled on an individual problem basis.

Example 6.4 In this example we take as boundary conditions $u(0,t) = u(L,t) = 0$ and a source term of the form

$$Q(x,t) = \begin{cases} Q_0 x(1-x), & 0 < t < t_0 \\ 0, & t \geq t_0, \end{cases}$$

where $t_0 > 0$ is fixed. From the previous example, $\lambda_n = \frac{n\pi}{L}$, $X_n(x) = \sin\frac{n\pi x}{L}$, and

$$Q(x,t) = \frac{4L^2 Q_0}{\pi^3 \alpha^2} \sum_{n=1,3,\dots} \frac{1}{n^3} \sin \lambda_n x, \quad \text{if } 0 < t < t_0.$$

The initial value problem for the coefficient $a_n(t)$ becomes

$$a_n' + (\lambda_n \alpha)^2 a_n = \begin{cases} \frac{4L^2 Q_0}{\pi^3 \alpha^2 n^3}, & 0 < t < t_0,\ n = 1, 3, \dots \\ 0, & t \geq t_0 \text{ or } n = 2, 4, \dots, \end{cases}$$

and $a_n(0) = 0$ for all n. It is immediate that $a_n(t) = 0$ for all t if n is even. Computing the integral in (6.2.10) yields

$$a_n(t) = \frac{4L^4 Q_0}{\pi^5 \alpha^4 n^5} \begin{cases} \left[1 - e^{-(\lambda_n \alpha)^2 t}\right], & 0 < t < t_0 \\ \left[e^{-(\lambda_n \alpha)^2 (t - t_0)} - e^{-(\lambda_n \alpha)^2 t}\right], & t > t_0. \end{cases}$$

Notice that for either realm $0 < t < t_0$ or $t > t_0$, the coefficients and their derivatives with respect to t have rapid decay. For $0 < t < t_0$, the coefficients decay like a power of n and for $t > t_0$, they have exponential decay. Hence, we can apply the Weierstrass M-test to conclude that the series for $u(x,t)$ can be differentiated in either variable term by term. Thus, in both realms the series for $u(x,t)$ provides a strong solution to the nonhomogeneous heat equation. However, $t = t_0$ is problematic. The function $a_n(t)$ satisfies $a_n(t_0-) = a_n(t_0+)$, so that $a_n(t)$ is continuous at $t = t_0$. However $a_n'(t)$ has a jump discontinuity at $t = t_0$. Nonetheless, viewing $u(x,t)$ as a distribution, the claim is that for any test function $\phi = \phi(x,t)$ on $[0,L] \times (0,\infty)$,

$$\langle u_t - \alpha^2 u_{xx}, \phi \rangle = \langle Q, \phi \rangle. \tag{6.2.12}$$

In other words, $u(x,t)$ as defined by the series provides a weak solution to the equation.

The essential step in justifying this stems from the formula

$$\begin{aligned}\langle u_t - \alpha^2 u_{xx}, \phi\rangle &= -\langle u, \phi_t - \alpha^2 \phi_{xx}\rangle \\ &= -\int u(x,t)\big[\phi_t(x,t) + \alpha^2 \phi_{xx}(x,t)\big]\,dx\,dt.\end{aligned}$$

The integral on the right is a double integral over the support of the test function ϕ. The heuristic reason for the equality in (6.2.12) is that this double integral essentially ignores the behavior near the line $t = t_0$ (a lower-dimensional set). The technical details are as follows; they involve the notion of a partition of unity.[1] It suffices to consider the case where the support of ϕ intersects the line $t = t_0$. Let R be a rectangle in $(0, L) \times (0, \infty)$ that contains the support of ϕ. Divide R into two subrectangles R_l and R_u, the parts of R below and above the line $t = t_0$, respectively. Let $\delta > 0$ be a small number, chosen below, and let R_δ be the subrectangle of R for which $t_0 - \delta < t < t_0 + \delta$ (see Figure 6.3). Let ψ_1, ψ_2, and ψ_3 be test functions that are positive inside R_l, R_δ, and R_u, respectively, and further vanish on the respective boundaries. Let $\phi_i = \psi_i / \sum_{k=1}^{3} \psi_k$. Then $\sum_i \phi_i = 1$; the collection ϕ_1, ϕ_2, and ϕ_3 form a partition of unity of the rectangle R. We can now write

$$(u_t - \alpha^2 u_{xx} - Q, \phi) = \sum_{i=1}^{3} (u_t - \alpha^2 u_{xx} - Q, \phi_i \phi). \tag{6.2.13}$$

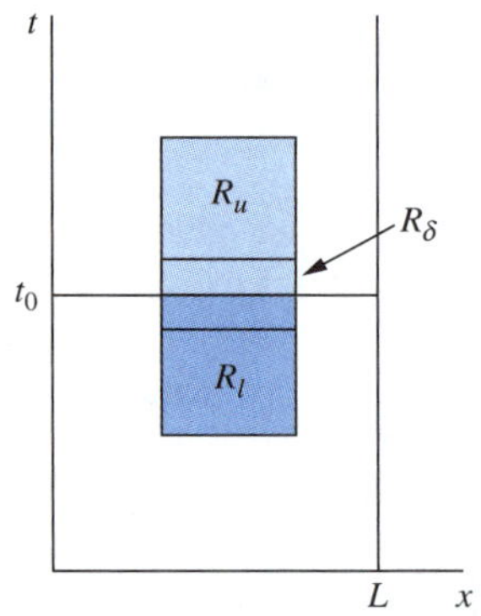

Figure 6.3 Partition of Unity Argument

1. The notion of a partition of unity is a fundamental tool in the theory of distributions. The interested reader could consult the book by Strichartz [30] for an overview of its role in distribution theory.

The first and third terms on the right-hand side are zero because u is a strong solution inside the rectangles R_l and R_u. To handle the second term we write

$$(u_t - \alpha^2 u_{xx}, \phi_2\phi) = -\int_{R_\delta} u(x,t)\left[(\phi_2(x,t)\phi(x,t))_t \right. \\ \left. +\alpha^2 \left(\phi_2(x,t)\phi(x,t)\right)_{xx}\right] dx\,dt,$$

where we have applied the definition of the derivative in distribution. Now the area of R_δ is less than $L\delta$ and the integrand above is bounded, hence $\left|(u_t - \alpha^2 u_{xx}, \phi_2\phi)\right| \le M\delta$ for some constant M. Likewise, $|(Q, \phi_2\phi)| \le M\delta$. From the decomposition (6.2.13) it follows that

$$\left|(u_t - \alpha^2 u_{xx} - Q, \phi)\right| \le M\delta,$$

for suitable constant M. The desired conclusion follows as δ can be made arbitrarily small.

For the choices $Q_0 = 1$, $\alpha = 1$, $t_0 = 1$, and $L = \pi$, a plot of the solution is given in Figure 6.4. From a physical perspective, this plot indicates the expected behavior. For $t < 1$, the solution is smooth and depicts the growth in temperature in the rod due to the source term. For $t > 1$ the solution is also smooth and the plot represents the decay of the solution (the equation is a homogeneous heat equation at this point). At $t = 1$, the solution depicts nonsmoothness; this is not surprising as the source term is switched off at this point.

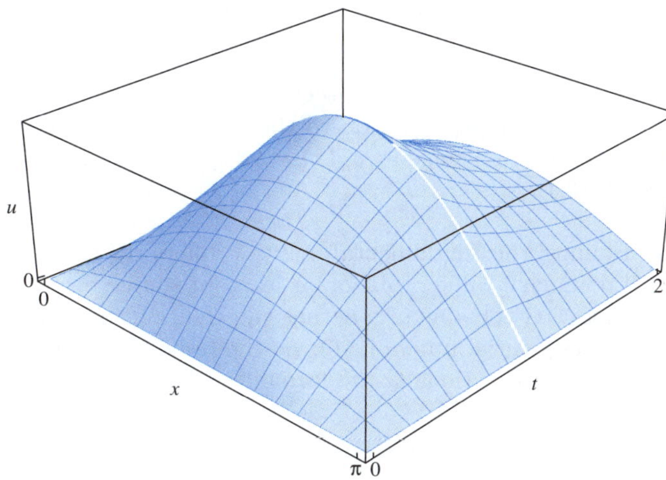

Figure 6.4 Solution of Nonhomogeneous Heat Equation: Example 6.4

The methodology we employed can be used to find the fundamental solution for the IBVP (6.2.7), i.e., the weak solution of the problem for the case where $Q(x,t) = \delta(t-\tau)\delta(x-\xi)$, $\tau > 0$ and $0 < \xi < L$. The initial value problem for $a_n(t)$ becomes

$$a_n' + (\lambda_n \alpha)^2 a_n = \delta(t-\tau)\frac{X_n(x)}{\|X_n\|^2}, \qquad a_n(0) = 0,$$

where we used the Fourier expansion $\delta(x-\xi) = \sum_n \frac{X_n(\xi)X_n(x)}{\|X_n\|^2}$. Using the method from Example 6.4, we can write

$$\left[e^{(\lambda_n\alpha)^2 t} a_n\right]' = e^{(\lambda_n\alpha)^2 t}\delta(t-\tau)\frac{X_n(\xi)}{\|X_n\|^2}.$$

As a distribution $e^{(\lambda_n\alpha)^2 t}\delta(t-\tau) = e^{(\lambda_n\alpha)^2 \tau}\delta(t-\tau)$ and since $[H(t-\tau)]' = \delta(t-\tau)$, the solution becomes

$$a_n(t) = H(t-\tau)e^{-(\lambda_n\alpha)^2(t-\tau)}\frac{X_n(\xi)}{\|X_n\|^2}.$$

Thus the fundamental solution in this case has the form

$$K_{\text{bc}}(x,\xi,t,\tau) = H(t-\tau)\sum_{n=1}^{\infty} e^{-(\lambda_n\alpha)^2(t-\tau)}\frac{X_n(\xi)X_n(x)}{\|X_n\|^2}.$$

The series converges in distribution and if $Q(x,t)$ is a test function, the solution to the IBVP (6.2.7) can be written

$$u(x,t) = \int_0^\infty \int_0^L K_{\text{bc}}(x,\xi,t,\tau)Q(\xi,\tau)d\xi\, d\tau. \tag{6.2.14}$$

Substituting the series for K_{bc} and integrating term by term leads back to (6.2.11) and provides a strong solution to the heat equation. (Because Q is a test function, the Fourier coefficients $q_n(\tau) = O(\lambda_n^{-2m})$ for any positive integer m; see Exercise 6.13.)

Remark 6.3 Finding the fundamental solution offers an alternative way to solve the initial boundary value problem, at least for source terms that are test functions. The question remains of whether the solution formula (6.2.14) can be extended to more general functions Q. As the previous example illustrates, for reasonably behaved functions Q the formula will provide a weak solution.

Exercises

6.1 Verify the conclusions of Theorem 6.1 for the case of homogeneous Neumann boundary conditions.

6.2 Consider the heat equation $u_t = \alpha^2 u_{xx}$ on the interval $0 < x < L$ with boundary conditions $u(0,t) = u_x(L,t) = 0$ and initial state $u(x,0) = f(x)$.

a. Determine the Fourier series representation of the solution, and verify the asymptotics given in Theorem 6.1.

b. Explicitly determine the solution in the case where $f(x) = L - x$.

c. In part (b) with $L = \pi$ and $\alpha = 1$, plot an approximation to the solution in space–time.

6.3 Assume $\alpha = 1$ and $L = \pi$. Consider a homogeneous heat flow problem with mixed boundary conditions $u_x(0,t) = u(\pi,t) = 0$. Determine the solution, plot approximations in space–time, and determine the asymptotics for large t for each of the following initial conditions.

a. $f(x) = 1$

b. $f(x) = x$

6.4 For the homogeneous heat equation with boundary conditions $u(0,t) = 0$ and $u(\pi,t) = 1$ with initial condition $u(x,0) = 1$, show that the solution satisfies $u(x,t) = \frac{x}{\pi} + \frac{4}{\pi}e^{-\alpha^2 t}\sin x + O(e^{-4\alpha^2 t})$ as $t \to \infty$. HINT: The steady-state solution is $U(x) = x/\pi$. Write $v(x,t) = u(x,t) - U(x)$ and determine the IBVP satisfied by v.

6.5 Find the Fourier representation of the solution to the IBVP: $u_t = \alpha^2 u_{xx}$, $0 < x < L$, $t > 0$, $u(0,t) = 0$, $u_x(L,t) = h[T - u(L,t)]$, and $u(x,0) = 0$. Here $h > 0$ and T are constants. Determine the asymptotics of the solution. HINT: Find the steady-state solution $U(x)$ and write $u(x,t) = U(x) + v(x,t)$.

6.6 Repeat Exercise 6.5 with the boundary conditions replaced by $u_x(0,t) = 0$ and $u_x(L,t) = h[T - u(L,t)]$.

6.7 Let $u(x,t)$ and $U(x)$ denote the solution and steady-state solution of a heat flow problem. (The latter is assumed to exist.) The relaxation time is defined as the quantity

$$\tau = -\lim_{t\to\infty} \frac{1}{t} \log |u(x,t) - U(x)|.$$

The relaxation time is a gauge in the time domain of how quickly a solution tends to steady state. Determine the relaxation time for the solutions found in Exercises 6.4, 6.5, and 6.6.

6.8 Consider the initial boundary value problem: $u_t = \alpha^2 u_{xx} - bu$, $0 < x < L$, $t > 0$; $u_x(0,t) = u_x(L,t) = 0$; $u(x,0) = T$.

a. Determine the steady-state solution.

b. Find the solution to the IBVP.

HINT: Set $w(x,t) = e^{bt}u(x,t)$, what problem does w solve?

6.9 Consider the initial boundary value problem $u_t = \alpha^2 u_{xx} - \beta u_x$, $0 < x < L$, $t > 0$; $u(0,t) = u(L,t) = 0$; $u(x,0) = f(x)$.

a. Write $u(x,t) = e^{ax+bt}v(x,t)$ and determine the constants a and b so that v satisfies the homogeneous heat equation $v_t = \alpha^2 v_{xx}$.

b. Use part (a) to find the solution of the original IBVP. (A particular case of this IBVP occurs in diffusion models for transistor design; see [35].)

6.10 In Example 6.4 replace the source function by

$$Q(x,t) = \begin{cases} Q_0 x(L-x)(t_0 - t), & 0 < t < t_0 \\ 0, & t \geq t_0. \end{cases}$$

Determine the solution of the IBVP in this case. Is the resulting series solution a strong solution in this case? Explain. As in the example, plot an approximation to the solution.

6.11 Find the solution to the nonhomogeneous problem $u_t = \alpha^2 u_{xx} + Q_0 e^{-at} \sin \frac{\pi x}{L}$, $0 < x < L$, $t > 0$; $u(0,t) = u(L,t) = 0$; $u(x,0) = 0$. Here Q_0 and a are constants.

6.12 Repeat Exercise 6.11 but replace the boundary conditions by $u(0,t) = u_x(0,t) = 0$.

6.13 Let $D = \frac{d^2}{dx^2}$, let $\{\lambda_n\}$ and $\{X_n\}$ be the eigenvalues and eigenfunctions of D on $[0,L]$ for a given set of boundary conditions, and let f be a test function on $[0,L]$. For any positive integer m, show that $\widehat{D^m f}(n) = \lambda_n^{2m} \widehat{f}(n)$ where $\widehat{f}(n)$ is the nth Fourier coefficient of f with respect to the orthogonal set $\{X_n\}$. Conclude that $\widehat{f}(n) = O(\lambda_n^{-2m})$ as $n \to \infty$. HINT: This is an exercise in integration by parts.

6.14 Show that $K_{\text{bc}}(x,\xi,t,\tau) = H(t-\tau)S_{\text{bc}}(x,\xi,t-\tau)$. (This is a formula relating the fundamental solution to the heat kernel.)

6.15 Write out explicitly the Fourier expansion of the fundamental solutions for the following boundary conditions.

a. $u(0,t) = u(L,t) = 0$

b. $u(0,t) = u_x(L,t) = 0$

c. $u(0,t) = 0$ and $u_x(L,t) + hu(L,t) = 0$, where $h > 0$ is a constant.

In each case, verify that the series converges in distribution.

6.3 The Wave Equation in One Dimension

The Vibrating String

The initial boundary value problem for a clamped string takes the following form.

$$\begin{aligned} \text{PDE} &: u_{tt} = c^2 u_{xx}, \quad 0 < x < L, \quad t > 0 \\ \text{BC} &: u(0,t) = u(L,t) = 0 \\ \text{IC} &: \begin{cases} u(x,0) = f(x) \\ u_t(x,0) = g(x) \end{cases} \end{aligned} \tag{6.3.1}$$

We have assumed that the density and tension are constant. The separated solutions of the wave equation that satisfy the boundary conditions have the form

$$\cos\frac{n\pi ct}{L}\sin\frac{n\pi x}{L}, \quad \sin\frac{n\pi ct}{L}\sin\frac{n\pi x}{L}, \quad n = 1, 2, \ldots,$$

or we may form the linear combination

$$u_n(x,t) = \left(a_n \cos\frac{n\pi ct}{L} + b_n \sin\frac{n\pi ct}{L}\right)\sin\frac{n\pi x}{L}, \; n = 1, 2, \ldots . \tag{6.3.2}$$

These are called the **fundamental modes of vibration**. In Figure 6.5, plots are given for $a_n = 1$, $b_n = 0$, at $t = 0$. Notice that for each fixed x, the amplitude is simple harmonic in time with frequency $\frac{nc}{L}$. These frequencies are called the natural frequencies of the string; the case $n = 1$ is the first harmonic. The fundamental modes are standing waves characterized by the presence of points on the string, called nodes, that do not move in time. Another useful expression for the fundamental modes is obtained by applying the addition formulas for sine and cosine as

$$\begin{aligned} u_n(x,t) = \frac{a_n}{2}&\left[\sin\frac{n\pi}{L}(x+ct) + \sin\frac{n\pi}{L}(x-ct)\right] \\ &+ \frac{b_n}{2}\left[\cos\frac{n\pi}{L}(x-ct) - \cos\frac{n\pi}{L}(x+ct)\right]. \end{aligned} \tag{6.3.3}$$

To determine a solution to the initial value problem we seek a solution by applying superposition of the form

$$u(x,t) = \sum_{n=1}^{\infty}\left(a_n \cos\frac{n\pi ct}{L} + b_n \sin\frac{n\pi ct}{L}\right)\sin\frac{n\pi x}{L}.$$

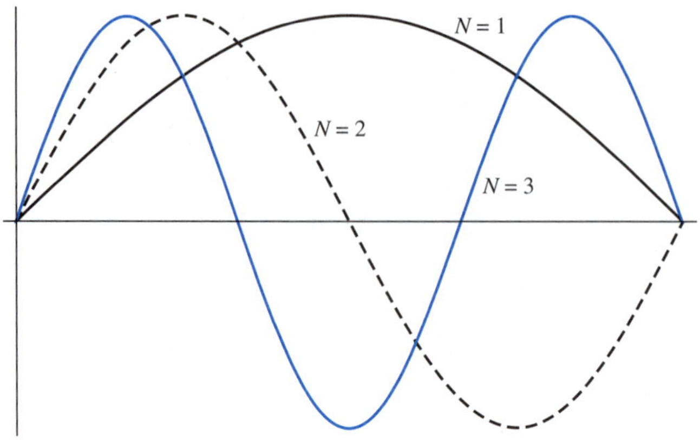

Figure 6.5 Fundamental Modes for a Vibrating String

At this point computations will be formal and break into two natural cases.

Case 1. **The Plucked String.** Here the string is given an initial position defined by $f(x)$ and released from rest, $g(x) = 0$. Substituting $t = 0$ into the formula for u gives

$$f(x) = \sum_{n=1}^{\infty} a_n \sin \frac{n\pi x}{L}$$

and, hence, $a_n = \widehat{f}_s(n)$, the nth Fourier sine coefficient of f on $[0, L]$. Differentiating the series for u with respect to time and substituting $t = 0$ yields

$$0 = \sum_{n=1}^{\infty} \frac{n\pi c}{L} b_n \sin \frac{n\pi x}{L},$$

and it follows that $b_n = 0$ for all n. Thus, the conjectured solution for the plucked sting is

$$u(x,t) = \sum_{n=1}^{\infty} \widehat{f}_s(n) \cos \frac{n\pi ct}{L} \sin \frac{n\pi x}{L}. \tag{6.3.4}$$

Case 2. **The Struck String.** The initial position in this case is $f(x) = 0$ and the string is given an initial velocity defined by the function $g(x)$. The computations are similar to case 1 and lead to a solution of the form

$$u(x,t) = \sum_{n=1}^{\infty} \frac{L}{n\pi c} \widehat{g}_s(n) \sin \frac{n\pi ct}{L} \sin \frac{n\pi x}{L}. \tag{6.3.5}$$

By the linearity of the wave equation, the conjectured solution for the string initial value problem is simply the sum of the results in the two cases, i.e.,

$$u(x,t) = \sum_{n=1}^{\infty} \left[\widehat{f}_s(n) \cos \frac{n\pi ct}{L} + \frac{L}{n\pi c} \widehat{g}_s(n) \sin \frac{n\pi ct}{L} \right] \sin \frac{n\pi x}{L}. \qquad (6.3.6)$$

This is commonly called the Fourier representation of the solution. In contrast to the Fourier representation of the solution of the homogeneous heat equation, essentially the only decay of the general term of the preceding series comes from the Fourier sine coefficients of the initial data. Hence for this series to provide a strong solution of the wave equation, suitable conditions will be needed on the initial data to justify differentiating the series term by term. In the case of the plucked string, the series (6.3.4) can be differentiated twice term by term in either variable provided the series $\sum |\widehat{f}_s(n)| n^2$ converges. Appealing to the theory of Fourier series Theorem 5.18, a condition on f that would ensure this would be that the odd extension $f^o \in C^2(\mathbf{T})$ with $f''' \in \mathrm{PC}[0,L]$. Similarly, a condition for (6.3.5) to define a strong solution is that $\sum |\widehat{g}_s(n)| n$ converges, which is implied by the conditions $g^o \in C^1(\mathbf{T})$ with $g'' \in \mathrm{PC}[0,L]$. Such smoothness conditions on f or g are rather artificial in that there are physically realizable initial data that do not satisfy them. The following example is illustrative.

Example 6.5 Consider a plucked string where

$$f(x) = \begin{cases} x, & 0 < x < L/2 \\ L - x, & L/2 < x < L. \end{cases}$$

Computation of the Fourier sine coefficients is straightforward with result $\widehat{f}_s(n) = \frac{4L}{(n\pi)^2} \sin \frac{n\pi}{2}$. Now $\widehat{f}_s(n) = 0$ in the case where n is even. This follows geometrically as the function $f(x)$ is even about the line $x = L/2$, whereas if n is even, $\sin \frac{n\pi x}{L}$ is odd about this line. Thus the solution can be written as

$$u(x,t) = \sum_{n=1,3,\ldots}^{\infty} \frac{4L(-1)^{n+1}}{(n\pi)^2} \cos \frac{n\pi ct}{L} \sin \frac{n\pi x}{L}.$$

Alternatively, we can rewrite the series solution using (6.3.3) as

$$u(x,t) = \sum_{n=1,3,\ldots} \frac{2L(-1)^{n+1}}{(n\pi)^2} \left[\sin \frac{n\pi}{L}(x+ct) + \sin \frac{n\pi}{L}(x-ct) \right],$$

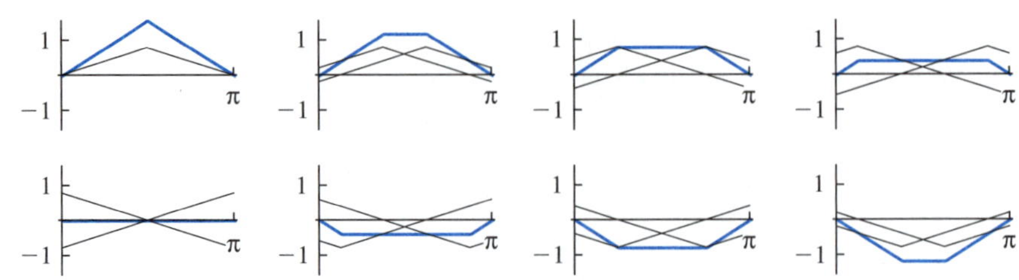

Figure 6.6 Time Snapshots of String Vibrations

which is a sum of a sum of a left and a right moving wave. In either form, it can be shown that the series can be differentiated once term by term in either variable. However, there is not sufficient decay in the coefficients to permit a second differentiation. We will see shortly that the series does define a weak solution of the wave equation. Truncating the series appropriately, we can plot snapshots of both forms of the weak solution. See Figure 6.6, where we have set $L = \pi$ and $c = 1$; t varies from 0 to $7\pi/8$ in multiples of $\pi/8$. The solution u is the blue curve, the black curves are the moving waves decomposing u.

To discuss the solution (6.3.6) rigorously, let's assume that $f \in C[0, L]$ with $f(0) = f(L) = 0$ and has a piecewise continuous derivative. It follows that the odd extension $f^o(x)$ is continuous on the circle $\mathbf{T}$ with piecewise continuous derivative. These conditions imply $\sum |\widehat{f}_s(n)| < \infty$ (Theorem 5.21), and it follows by the Weierstrass M-test that the series

$$\sum \widehat{f}_s(n) \cos \frac{n\pi ct}{L} \sin \frac{n\pi x}{L}$$

is uniformly convergent for $(x, t) \in [0, L] \times [0, \infty)$. (This handles the case of a plucked string.) For the initial velocity, assume $g \in \mathrm{PC}[0, \pi]$. Then $\sum |\widehat{g}_s(n)|^2$ converges and by the Cauchy–Schwarz inequality, $\sum \frac{|\widehat{g}_s(n)|}{n}$ converges. Again by the Weierstrass M-test,

$$\sum \frac{L}{n\pi c} \widehat{g}_s(n) \sin \frac{n\pi ct}{L} \sin \frac{n\pi x}{L}$$

converges uniformly. It follows that, under our assumptions on the initial data, the series defining $u(x, t)$ is uniformly convergent and u is a continuous function of x and t. As immediate consequence, $u(x, t)$ is a distribution on $[0, L] \times [0, \infty)$. Further, the series can be differentiated twice term by term in the distributional sense (Proposition 4.3) and hence $u(x, t)$ is a weak solution to the wave equation. These observations are worth summarizing in the following result.

Proposition 6.4 Let the initial data satisfy $f \in C[0, L]$ with $f(0) = f(L) = 0$ and $f' \in \mathrm{PC}[0, L]$, and $g \in \mathrm{PC}[0, L]$. Then the series (6.3.6) defines a weak solution of the initial value problem (6.3.1) for the clamped string.

As we did for Example 6.5, we can express the weak solution in d'Alembert form as follows. We write $u(x,t) = u_f(x,t) + u_g(x,t)$, breaking the sum into the pieces dependent on $\widehat{f}_s$ and $\widehat{g}_s$. Using (6.3.3), we have

$$u_f(x,t) = \frac{1}{2}\sum \widehat{f}_s(n)\left[\sin\frac{n\pi}{L}(x+ct) + \sin\frac{n\pi}{L}(x-ct)\right]$$
$$= \frac{1}{2}\left[\widetilde{f^o}(x+ct) + \widetilde{f^o}(x-ct)\right],$$

where in the second line the pointwise convergence theorem for Fourier series has been applied and $\widetilde{f^o}(x)$ is the $2L$–periodic extension of f^o. Likewise, for u_g,

$$\begin{aligned} u_g(x,t) &= \frac{1}{2}\sum \frac{L}{n\pi c}\widehat{g}_s(n)\left[\cos\frac{n\pi}{L}(x-ct) - \cos\frac{n\pi}{L}(x+ct)\right] \\ &= \frac{1}{2c}\sum \widehat{g}_s(n)\int_{x-ct}^{x+ct} \sin\frac{n\pi s}{L}\,ds \\ &= \frac{1}{2c}\int_{x-ct}^{x+ct} \widetilde{g^o}(s)ds. \end{aligned}$$

The last step is justified by the fact that we can integrate Fourier series term by term. Putting the two expressions together, we have derived d'Alembert's formula for the vibrating clamped string, namely:

$$u(x,t) = \frac{1}{2}\left[\widetilde{f^o}(x+ct) + \widetilde{f^o}(x-ct)\right] + \frac{1}{2c}\int_{x-ct}^{x+ct} \widetilde{g^o}(s)ds. \tag{6.3.7}$$

Energy Computations for the Vibrating String

Keeping the same assumptions on the initial data, d'Alembert's formula for the weak solution implies that the weak derivatives $u_x(x,t)$ and $u_t(x,t)$ are piecewise continuous functions of x and hence both are square integrable on $[0,L]$. Recall that the total kinetic and potential energy of the string are respectively given by

$$\mathrm{KE}(t) = \frac{1}{2}\int_0^L u_t^2(x,t)dx \quad \text{and} \quad \mathrm{PE}(t) = \frac{c^2}{2}\int_0^L u_x^2(x,t)dx.$$

Both of these can be calculated by an application of Parseval's identity as follows. Differentiating the series solution (6.3.6) for $u(x,t)$ term by term with respect to t and applying Parseval's equality (for sine series on $[0, L]$) and trigonometric identities gives

$$\begin{aligned}\mathrm{KE}(t) &= \frac{L}{4}\sum_{n=1}^{\infty}\Big(-\frac{n\pi c}{2}\hat{f}_s(n)\sin\frac{n\pi c}{L}t + \hat{g}_s(n)\cos\frac{n\pi c}{L}t\Big)^2 \\ &= \frac{\pi^2c^2}{4L}\sum_{n=1}^{\infty} n^2|\hat{f}_s(n)|^2\sin^2\frac{n\pi c}{L}t - \frac{\pi c}{4}\sum_{n=1}^{\infty} n\hat{f}_s(n)\hat{g}_s(n)\sin\frac{2n\pi c}{L}t \\ &\quad + \frac{L}{4}\sum_{n=1}^{\infty}|\hat{g}_s(n)|^2\cos^2\frac{n\pi c}{L}t.\end{aligned}$$

Similarly, for the potential energy,

$$\begin{aligned}\mathrm{PE}(t) = \frac{\pi^2c^2}{4L}\sum_{n=1}^{\infty} n^2|\hat{f}_s(n)|^2\cos^2\frac{n\pi c}{L}t + \frac{\pi c}{4}\sum_{n=1}^{\infty} n\hat{f}_s(n)\hat{g}_s(n)\sin\frac{2n\pi c}{L}t \\ + \frac{L}{4}\sum_{n=1}^{\infty}|\hat{g}_s(n)|^2\sin^2\frac{n\pi c}{L}t.\end{aligned}$$

Adding the two results together provides an alternative derivation of conservation of energy, i.e.,

$$E(t) = \mathrm{KE}(t) + \mathrm{PE}(t) = \frac{\pi^2c^2}{4L}\sum_{n=1}^{\infty} n^2|\hat{f}_s(n)|^2 + \frac{L}{4}\sum_{n=1}^{\infty}|\hat{g}_s(n)|^2 = E(0).$$

Henceforth we write $E(t) = E$.

As a mechanical process, the vibration of a string occurs because of an exchange between the kinetic and potential energies in time. It is natural to inquire about the balance of time between the two forms of energy. Observing that both the kinetic and potential energy are periodic functions, with period L/c, we define the **averaged kinetic and potential energies** as

$$\mathcal{A}_{\mathrm{KE}} = \frac{c}{L}\int_0^{L/c}\mathrm{KE}(t)\,dt \quad\text{and}\quad \mathcal{A}_{\mathrm{PE}} = \frac{c}{L}\int_0^{L/c}\mathrm{PE}(t)dt,$$

respectively. It is clear that $\mathcal{A}_{\mathrm{KE}} + \mathcal{A}_{\mathrm{PE}} = E$.

Taking the difference between kinetic and potential energies and using the identity $\cos 2u = \cos^2 u - \sin^2 u$ yields

$$\begin{aligned}\mathrm{KE}(t) - \mathrm{PE}(t) = -\frac{\pi^2c^2}{4L}\sum_{n=1}^{\infty} n^2|\hat{f}_s(n)|^2\cos\frac{2n\pi c}{L}t - \frac{\pi c}{2}\sum_{n=1}^{\infty} n\hat{f}_s(n)\hat{g}_s(n)\sin\frac{2n\pi c}{L}t \\ + \frac{L}{4}\sum_{n=1}^{\infty}|\hat{g}_s(n)|^2\cos\frac{2n\pi c}{L}t.\end{aligned}$$

All three series on the right-hand side are uniformly convergent in t and so can be integrated term by term. Averaging the formula over a time cycle of period L/c, we obtain $\mathcal{A}_{\text{KE}} - \mathcal{A}_{\text{PE}} = 0$. The two formulas for the averaged energies now imply that

$$\mathcal{A}_{\text{KE}} = \frac{E}{2} = \mathcal{A}_{\text{PE}}.$$

In physical terms this can be expressed as: *averaged on a time cycle, the total energy of the string is equally split between kinetic and potential energy.* In physics, this is an example of an **energy equipartition principle**.

Remark 6.5 The ideas in this subsection can also be applied to the wave equation with Neumann or mixed Dirichlet and Neumann boundary conditions. The application of Parseval's equality breaks down in the case of Robin boundary conditions. The conservation law in this situation also involves terms from an energy exchange at the boundary (see Exercise 1.16).

The Fundamental Solution

Employing the method used to find the fundamental solution for diffusion problems, here we find the weak solution of the problem:

$$\begin{aligned} &\text{PDE: } u_{tt} = c^2 u_{xx} + \delta(t-\tau)\delta(x-\xi), \quad 0 < x < L, \quad t > 0 \\ &\text{BC: } u(0,t) = u(L,t) = 0 \\ &\text{IC: } \begin{cases} u(x,0) = 0 \\ u_t(x,0) = 0. \end{cases} \end{aligned} \tag{6.3.8}$$

The idea is to seek a solution of the form

$$\sum_{n=1}^{\infty} a_n(t) \sin \lambda_n x,$$

where $\lambda_n = \frac{n\pi}{L}$. Of course, coefficients a_n also depend on the parameters $\tau < t$ and $\xi \in (0, L)$. The methodology can be employed for other types of boundary conditions (see the Exercises). Using the expansion $\delta(x-\xi) = \frac{2}{L}\sum_n \sin \lambda_n \xi \sin \lambda_n x$, differentiating the series for the solution term by term, and substituting into the wave equation leads to the following initial value problem for the coefficients:

$$a_n'' + (\lambda_n c)^2 a_n = \frac{2}{L}\delta(t-\tau) \sin \lambda_n \xi, \quad a_n(0) = a_n'(0) = 0.$$

This is a second-order nonhomogeneous ODE that can be solved using the method of variation of parameters. The homogeneous equation $a_n'' + (\lambda_n c)^2 a_n = 0$ has fundamental set $\{\cos \lambda_n ct, \sin \lambda_n ct\}$. We seek a solution of the nonhomogeneous equation of the form

$$a_n(t) = v(t)\cos \lambda_n ct + w(t)\sin \lambda_n ct.$$

Variation of parameters then gives that the coefficients v and w must satisfy the system of equations

$$\begin{gathered} v' \cos \lambda_n ct + w' \sin \lambda_n ct = 0 \\ -\lambda_n cv' \sin \lambda_n ct + \lambda_n cw' \cos \lambda_n t = \tfrac{2}{L}\delta(t-\tau)\sin \lambda_n \xi \end{gathered}$$

with initial conditions $v'(0) = w'(0) = 0$. Eliminating v' between the equations, we arrive at

$$\begin{aligned} w'(t) &= \frac{2}{L\lambda_n c} \cos \lambda_n ct\, \delta(t-\tau)\sin \lambda_n \xi \\ &= \frac{2}{L\lambda_n c} \cos \lambda_n c\tau\, \delta(t-\tau)\sin \lambda_n \xi, \end{aligned}$$

where we used the fact that $\cos \lambda_n ct\, \delta(t-\tau) = \cos \lambda_n c\tau\, \delta(t-\tau)$ as distributions. It follows that

$$w(t) = \frac{2}{L\lambda_n c} H(t-\tau)\cos \lambda_n c\tau \sin \lambda_n \xi,$$

and similarly

$$v(t) = -\frac{2}{L\lambda_n c} H(t-\tau)\sin \lambda_n c\tau \sin \lambda_n \xi.$$

Putting these into the formula for a_n and using the addition formula for the sine function yields

$$a_n(t) = \frac{2}{L\lambda_n c} H(t-\tau)\sin \lambda_n c(t-\tau)\sin \lambda_n \xi.$$

Thus, the fundamental solution is given by

$$K(x,\xi,t,\tau) = H(t-\tau)\sum_{n=1}^{\infty} \frac{2}{n\pi c} \sin \frac{n\pi c(t-\tau)}{L} \sin \frac{n\pi \xi}{L} \sin \frac{n\pi x}{L},$$

where the series converges in distribution. (The series is one of locally integrable functions on space–time; to show it converges in distribution, simply apply it to a test function and demonstrate that the resulting series converges.)

If $Q(x,t)$ is a test function on $[0,L]\times[0,\infty)$, then $u(x,t) = \langle K(x,\xi,t,\tau), Q(\xi,\tau)\rangle$ gives a weak solution of the equation $u_{tt} = c^2u_{xx} + Q(x,t)$ satisfying the boundary conditions and initial conditions of (6.3.8). In fact, this is a strong solution of the nonhomogeneous wave equation, which can be seen as follows. Let

$$q_n(\tau) = \frac{2}{L}\int_0^L Q(\xi,\tau)\sin\frac{n\pi\xi}{L}\,d\xi$$

be the Fourier sine coefficients of Q. We know that $q_n(\tau) = O(n^{-m})$ for every integer $m \geq 0$ and uniformly in τ (since $Q(\xi,\tau)$ is bounded in τ and the odd extension of Q in the ξ-variable is $C^\infty(\mathbf{T})$). Hence, we can write

$$\begin{aligned} u(x,t) &= \sum_{n=1}^{\infty}\frac{L}{n\pi c}\left\langle H(t-\tau)\sin\frac{n\pi c(t-\tau)}{L}, q_n(\tau)\right\rangle \sin\frac{n\pi x}{L} \\ &= \sum_{n=1}^{\infty}\frac{L}{n\pi c}\left[\int_0^t q_n(\tau)\sin\frac{n\pi c(t-\tau)}{L}d\tau\right]\sin\frac{n\pi x}{L}. \end{aligned} \tag{6.3.9}$$

This last series and the series of term-by-term derivatives in either variable and any order are uniformly convergent by the aforementioned growth property on $q_n(\tau)$.

Formula (6.3.9) can be applied to more general source terms $Q(x,t)$ to solve the nonhomogeneous wave equation. However, justification of the solution needs to be done on an individual problem basis.

Example 6.6 In this example we find the solution to $u_{tt} = c^2u_{xx} + u_0\cos\omega t$, $0 < x < L$, $t > 0$ satisfying homogeneous Dirichlet boundary conditions and with $u(x,0) = u_t(x,0) = 0$. It is assumed that $\omega \neq \frac{n\pi}{L}$ for all $n = 1, 2, \ldots$. We could apply (6.3.9) but the resulting trigonometric integrals are tedious to compute. Instead, we carry out a direct approach seeking a solution of form $u(x,t) = \sum a_n(t)\sin\frac{n\pi x}{L}$. Substituting into the equation and equating coefficients yields

$$a_n'' + \left(\frac{n\pi c}{L}\right)^2 a_n = \frac{2}{n\pi}[1-(-1)^n]u_0\cos\omega t \tag{6.3.10}$$

with initial conditions $a_n(0) = a_n'(0) = 0$. The equation has a particular solution of the form $a_n(t) = C_n\cos\omega t$. Substituting and solving for C_n we arrive at

$$C_n = \frac{2[1-(-1)^n]u_0}{\pi n\left[\left(\frac{n\pi c}{L}\right)^2 - \omega^2\right]}.$$

The solution of the differential equation satisfying the initial conditions then takes the form

$$a_n(t) = C_n \cos \omega t - C_n \cos \frac{n\pi ct}{L}.$$

(The general solution of (6.3.10) has the form of the particular solution plus a linear combination of $\cos \frac{n\pi ct}{L}$ and $\sin \frac{n\pi ct}{L}$; the coefficients of the latter are computed from the initial conditions.) Thus, the solution we seek takes the form

$$\begin{aligned} u(x,t) = \cos \omega t \sum_{n=1,3,\ldots}^{\infty} & \frac{4u_0}{\pi n \left[\left(\frac{n\pi c}{L}\right)^2 - \omega^2\right]} \sin \frac{n\pi x}{L} \\ & - \sum_{n=1,3,\ldots} \frac{4u_0}{\pi n \left[\left(\frac{n\pi c}{L}\right)^2 - \omega^2\right]} \cos \frac{n\pi ct}{L} \sin \frac{n\pi x}{L}. \end{aligned}$$

By an easy application of the Weierstrass M-test, this series as well as its series of term-by-term derivatives in either variable converge uniformly for all $t > 0$ and $0 \le x \le L$. With more effort, for any small number $\delta > 0$, the series of second derivatives converges uniformly for $\delta \le x \le L - \delta$. Hence, the series provides a strong solution to the nonhomogeneous wave equation.

Exercises

6.16 Under the conditions imposed on the functions f and g, show that the series for $u(x,t)$ given by (6.3.6) can be differentiated.

6.17 Let $0 < a < L$, $h > 0$, and set

$$f(x) = \begin{cases} hx/a, & 0 \le x \le a \\ h(L-x)/(L-a), & a \le x \le L. \end{cases}$$

a. Determine the Fourier representation of the solution to (6.3.1) where $g(x) = 0$.

b. Let $L = \pi$ and $h = 1$. Show that the terms in the series for $u(x,t)$ for larger n contribute more to $u(x,t)$ as a becomes closer to L.

HINT: Explore the Fourier coefficients of f for $a = L/2$, $a = 3L/4$, $a = 7L/8$, etc. (An interpretation of this is that plucking a string near its end gives a tone with higher harmonics.)

6.18 Replace the boundary conditions in (6.3.1) with $u_x(0,t) = u_x(L,t) = 0$.

a. Determine the Fourier representation of the solution and d'Alembert's formula.

b. What energy equipartition principle holds in this case?

6.19 Repeat Exercise 6.18 but with the boundary conditions $u(0,t) = u_x(L,t) = 0$.

6.20 Determine the Fourier representation of the wave kernel for the clamped string (replace the initial conditions in (6.3.1) by $u(x,0) = 0$ and $u_t(x,0) = \delta(x-\xi)$). Verify that it is a weak solution of the wave equation.

6.21 Determine the Fourier representation of the modified wave kernel for the clamped string and verify that it defines a weak solution of the wave equation.

6.22 Find the Fourier representation of the fundamental solution to the wave equation on $[0, L]$ where the boundary conditions are $u(0,t) = u_x(L,t) = 0$.

6.23 Find the Fourier representation of the wave kernel and modified wave kernel for the boundary conditions of Exercise 6.22.

6.24 Find the Fourier representation of the solution of the problem: $u_{tt} + 2au_t + a^2u = c^2u_{xx}$, $0 < x < L$, $t > 0$, $u(0,t) = u(L,t) = 0$, $u(x,0) = 0$ and $u_t(x,0) = g(x)$. Assume that $g \in \mathrm{PC}[0,L]$. HINT: Use the substitution $v(x,t) = e^{at}u(x,t)$ and determine an IBVP for v.

6.25 Find the Fourier solution of the problem: $u_{tt} = c^2u_{xx} - a^2u$, $0 < x < L$, $t > 0$, $u(0,t) = u(L,t) = 0$, $u(x,0) = 0$ and $u_t(x,0) = g(x)$. Assuming $g \in \mathrm{PC}[0,L]$, justify the solution. HINT: This must be approached using separation of variables.

6.26 Repeat Exercise 6.25 for the equation $u_{tt} + 2au_t = c^2u_{xx}$.

6.4 Multiple Fourier Series

In one dimension, the Laplace operator is just the second derivative operator and as we saw in Section 3.1, the method of separation of variables for the heat or wave equations leads to a boundary value problem for the equation

$$X'' + \mu X = 0$$

on a finite interval. Boundary conditions take one of three forms at each end of the interval, leading to nine possible eigenfunction expansions. For example, homogeneous Dirichlet conditions at each end leads to expansions in a series of sines.

In higher dimensions the natural analog of this framework is the equation

$$\nabla^2 X + \lambda X = 0 \tag{6.4.1}$$

in a domain $D \subset \mathbb{R}^d$. Called Helmholtz equation, it arises in a variety of PDE problems, for example, the vibrations of a rectangular membrane. Boundary conditions can take any of the three forms or a mixture of them. In the case of a Robin boundary condition, we assume it has the form

$$\frac{dX}{dn} + aX = 0, \ x \in \partial D,$$

where $a(x)$ is a function on the boundary assumed to be nonnegative.

Definition 6.6 Consider Helmholtz equation on a domain $D \subset \mathbb{R}^d$ with boundary conditions being homogeneous Dirichlet, Neumann, Robin, or a mixture of these. A value of μ for which the BVP has a nontrivial solution of class $C^1(\overline{D})$ is called an eigenvalue of $-\nabla^2$; the corresponding solution is called an eigenfunction.

Example 6.7 Consider Helmholtz equation on the rectangle $R = (0, L_1) \times (0, L_2)$ with a homogeneous Dirichlet condition on each segment of the boundary. The separated solutions of the equation satisfying the boundary conditions have the form

$$X_{n_1,n_2}(x) = \sin\left(\frac{n\pi x_1}{L_1}\right)\sin\left(\frac{m\pi x_2}{L_2}\right),$$

where $x = (x_1, x_2) \in D$ and $n, m = 1, 2, \ldots$. These are eigenfunctions of ∇^2 on R with eigenvalue $\lambda_{n,m} = \left(\frac{n\pi}{L_1}\right)^2 + \left(\frac{m\pi}{L_2}\right)^2$. This can be analytically derived by writing $X(x) = X_1(x_1)X_2(x_2)$, substituting into the Helmholtz equation and exploring the separated ordinary differential equations (Exercise 6.27). By direct computation, this double-indexed family of functions forms an orthogonal set on R:

$$\int_D X_{n,m}(x)X_{n',m'}(x)dA(x) = \begin{cases} 0, & n \neq n'' \text{ or } m \neq m' \\ \frac{L_1L_2}{4}, & \text{otherwise.} \end{cases}$$

Given a square integrable function $f(x)$ on R, we can expand it in a **double Fourier sine series**,

$$f \sim \sum_{n_1,n_2=1}^{\infty} \widehat{f}(n,m)X_{n,m}(x)$$

$$\widehat{f}(n,m) = \frac{4}{L_1L_2}\int f(x)X_{n,m}(x)dA(x).$$

Example 6.17 is typical of what one can expect for separated solutions of the Helmholtz equation on the rectangle with any of the combinations of boundary conditions on the sides. Indeed, letting $X(x) = X_1(x_1)X_2(x_2)$, substituting into Helmholtz equation, and rearranging yields

$$\frac{X_1''}{X_1} = -\frac{X_2''}{X_2} - \mu.$$

Introducing a separation constant $-\lambda$, we obtain a pair of ordinary differential equations:

$$X_1'' + \lambda X_1 = 0,\ 0 < x_1 < L_1, \text{ and } X_2'' + (\mu - \lambda)X_2 = 0,\ 0 < x_2 < L_2.$$

Each of these ODEs is associated with boundary conditions (coming from those on the rectangle) and hence the eigenvalues and eigenfunctions for the ODE problems can be determined. The eigenvalues and eigenfunctions of the Laplacian on D are then the sum and product of the eigenvalues and eigenfunctions, respectively, of the ODE boundary value problems. We will let D and N denote Dirichlet and Neumann, respectively, and B, T, L, and R stand for the bottom, top, left, and right sides of the rectangle D, respectively. Then Table 6.1 provides some possible examples of orthogonal solutions of the Helmholtz equation on the rectangle D using separation of variables as outlined.

Given any of the orthogonal sets of eigenfunctions on the Laplacian on rectangles generated in this fashion, a multiple Fourier expansion can be defined using the definition given in Section 3.2 and as was done in Example 6.17. A fundamental question concerns the convergence properties of multiple Fourier expansions built from these sets of orthogonal functions. Pointwise convergence of multiple Fourier expansions is a very involved topic for which many unsolved problems remain. One case where pointwise convergence is clear is for test functions $\phi(x)$, which automatically satisfy the same boundary conditions as the eigenfunctions $X_{n,m}(x)$. Using integration by parts we can show that $|\widehat{\phi}(n,m)| \leq C\,\lambda_{n,m}^{-N}$ for any integer N. This is more than sufficient to show that

B	T	L	R	Orthogonal Solutions	Eigenvalues
D	D	D	D	$\sin\frac{n\pi x_1}{L_1}\sin\frac{m\pi x_2}{L_2}$	$\left(\frac{n\pi}{L_1}\right)^2 + \left(\frac{m\pi}{L_2}\right)^2$
D	D	D	N	$\sin\frac{n\pi x_1}{L_1}\sin(m+\frac{1}{2})\frac{\pi x_2}{L_2}$	$\left(\frac{n\pi}{L_1}\right)^2 + \left((m+\frac{1}{2})\frac{\pi}{L_2}\right)^2$
D	D	N	D	$\sin\frac{n\pi x_1}{L_1}\cos(m+\frac{1}{2})\frac{\pi x_2}{L_2}$	$\left(\frac{n\pi}{L_1}\right)^2 + \left((m+\frac{1}{2})\frac{\pi}{L_2}\right)^2$
N	N	N	N	$\cos\frac{n\pi x_1 \Pi}{L_1}\cos\frac{m\pi x_2}{L_2}$	$\left(\frac{n\pi}{L_1}\right)^2 + \left(\frac{m\pi}{L_2}\right)^2$

Table 6.1 Separated Orthogonal Solutions of the Helmholtz Equation

the Fourier expansion of ϕ and of any term-by-term partial derivative converges pointwise, in fact uniformly.

The following result resolves the question relative to L^2–convergence.

Theorem 6.7 Let $\{X_{1n}\}_{n=1}^{\infty}$ and $\{X_{2m}\}_{m=1}^{\infty}$ be complete sets of eigenfunctions of the one-dimensional Laplace operator on the intervals $(0, L_1)$ and $(0, L_2)$, respectively, with homogeneous Dirichlet, Neumann, Robin, or mixed boundary conditions placed on the endpoints of the intervals (possibly different sets of boundary conditions on each interval). Let the corresponding eigenvalues be $\{\lambda_{1n}\}$ and $\{\lambda_{2m}\}$, respectively. Then the eigenvalues of $-\nabla^2$ on the rectangle $R = (0, L_1) \times (0, L_2)$ with boundary conditions induced from the those on the intervals, are precisely all sums of the form $\lambda_{nm} = \lambda_{1n} + \lambda_{2m}$ with corresponding eigenfunctions $X_{n,m}(x) = X_{1n}(x_1)X_{2m}(x_2)$. This set of eigenfunctions forms a complete set of orthogonal functions in $L^2(R)$ with respect to the inner product

$$\langle f, g\rangle = \int_R f(x)\overline{g(x)}dA(x), \ \ f, g \in L^2(R),$$

where dA is the area element.

Proof First of all,

$$\begin{aligned}\nabla^2 X_{n,m} &= (\partial_{11}X_{1n})X_{2n} + X_{1n}(\partial_{22}X_{2,n}) \\ &= -(\lambda_{1n} + \lambda_{2m})X_{1n}X_{2m} = -(\lambda_{1n} + \lambda_{2m})X_{n,m}.\end{aligned}$$

Furthermore, each function $X_{n,m}$ satisfies a set of boundary conditions on the boundary of the rectangle determined by the boundary conditions on the intervals. For example, if X_{1n} satisfies homogeneous Dirichlet conditions at $x = 0$, then $X_{n,m}(0, x_2) = 0$, which is a Dirichlet condition on the left side of the rectangle. Thus the function $X_{n,m}$ is an eigenfunction of ∇^2 on D with eigenvalue $\lambda_{1n} + \lambda_{2m}$. To verify orthogonality we have

$$\begin{aligned}\langle X_{n,m}, X_{n',m'}\rangle &= \int_0^{L_2}\int_0^{L_1} X_{1n}(x_1)X_{2m}(x_2)\overline{X_{1n'}(x_1)X_{2m'}(x_2)}dx_1dx_2 \\ &= \left(\int_0^{L_1} X_{1n}(x_1)\overline{X_{1n'}(x_1)}dx_1\right)\left(\int_0^{L_2} X_{2m}(x_2)\overline{X_{2m'}(x_2)}dx_2\right).\end{aligned}$$

Hence $\langle X_{n,m}, X_{n',m'}\rangle = 0$ if $n \neq n'$ or $m \neq m'$, which is the desired orthogonality. Completeness of the orthogonal set can be shown using Proposition 5.12 by showing that if $\phi \in L^2(R)$ is such that $\langle \phi, X_{n,m}\rangle = 0$ for all n and m, then $\phi = 0$. We give the argument in the case of $\phi \in C(R) \cap L^2(R)$, which carries the essential ideas of the general case. Define $g_m(x_1) = \int_0^{L_2} \phi(x_1, x_2)\overline{X_{2m}(x_2)}dx_2$.

Then $g_m \in L^2(0, L_1)$ since by the Cauchy–Schwarz inequality

$$\int_0^{L_1} |g_m(x_1)|^2 dx_1 \le \int_0^{L_1} \left(\int_0^{L_2} |\phi(x_1, x_2)|^2 dx_2 \right) \|X_{2m}\|_2^2 dx_1$$
$$= \|X_{2m}\|_2^2 \int_R |\phi(x)|^2 dA(x).$$

Now we simply compute $\int_0^{L_1} g_m(x_1)\overline{X_{1n}(x_1)}dx_1 = \langle \phi, X_{n,m} \rangle = 0$ for all n. Since $\{X_{1n}\}$ is a complete orthogonal set on $(0, L_1)$, it follows that $g_m(x_1) = 0$ for all m and all x_1. But then the completeness of the orthogonal set $\{X_{2m}\}$ on $(0, L_2)$ implies $\phi(x_1, x_2) = 0$ as desired. □

A nice corollary of this is that all eigenfunctions of ∇^2 on a rectangle are constructed using separation of variables. The proof is left to the reader.

Corollary 6.8 Let $\{X_{1n}\}, \{X_{2m}\}$, and $\{X_{n,m}\}$ be as in the previous result. If X is an eigenfunction of $-\nabla^2$ satisfying the same boundary conditions as the $X_{n,m}$ with λ its corresponding eigenvalue, then $\lambda = \lambda_n + \lambda_m$ for some n and m, and X is in the corresponding eigenspace.

Remark 6.9 Theorem 6.7 and its corollary can be extended to higher dimensions; for example, in three dimensions this would lead to triple Fourier expansions. Constructing orthogonal sets of eigenfunctions on rectangular regions in three dimensions uses the method of separation of variables; some examples are given in the exercises.

We conclude this section with a result concerning eigenfunctions of the Laplacian on general domains in $\mathbb{R}^d$; the multiple Fourier expansions previously considered are one set of examples. Finding explicit formulas for the eigenvalues and eigenfunctions on general domains is analytically impossible; even the problem of proving their existence is highly technical and requires methodologies beyond our current scope.

Theorem 6.10 Let $D \subset \mathbb{R}^3$ (the case $d = 2$ is similar) be a domain where we impose homogeneous Dirichlet, Neumann, Robin, or mixed conditions on the boundary for the Helmholtz equation. Then the following statements hold.

1. The operator $-\nabla^2$ restricted to the class of $C^2(D)$ functions that satisfy the boundary conditions is a symmetric operator with respect to the inner product

$$\langle u, v \rangle = \int_D u\bar{v}dV.$$

   Consequently, eigenfunctions corresponding to distinct eigenvalues are orthogonal with respect to this inner product.

2. All eigenvalues are real.

3. All eigenvalues are nonnegative.

Proof The proof of these statements is analogous to the corresponding statements for regular Sturm–Liouville problems. Here we give the proof for item 3; the others are left as exercises.

Let μ be an eigenvalue of the Laplacian with corresponding real eigenfunction X. Multiplying Helmholtz equation by X and integrating over the domain gives

$$\int_D X\nabla^2 X dV + \mu \int_D X^2 dV = 0.$$

Applying Green's first identity to the first integral on the left-hand side and rearranging terms then gives

$$\mu \int_D X^2 dV = -\int_{\partial D} X\boldsymbol{\nabla} X \cdot \mathbf{n}_o dS + \int_{\partial D} |\nabla X|^2 dS.$$

The result will follow if we can show that the first term on the right-hand side is nonnegative. This is immediate in the case of a Dirichlet or Neumann condition—the first term is zero. For a Robin condition, $\boldsymbol{\nabla} X \cdot \mathbf{n}_o = -aX$, where we have assumed $a > 0$. It then follows that the first term on the right-hand side is also nonnegative in this case. □

Application to Diffusion and Waves

Multiple Fourier expansions can be applied to boundary value problems in higher dimensions in much the same way as was they are in one-dimensional BVPs. Rather than work out general results, we provide two examples to illustrate the ideas.

Example 6.8 (Diffusion in Two Variables) In this example we consider the diffusion equation $u_t = \alpha^2\nabla^2 u$ in the rectangle $R = \{0 < x_1 < L_1,\ 0 < x_2 < L_2\}$ with homogeneous Dirichlet boundary conditions. We will determine the heat kernel and then solve a particular IBVP. We apply separation of variables by seeking solutions of the diffusion equation of the form $u(x,t) = T(t)X(x)$. Substituting and introducing a separation constant $-\lambda$ generates the ODEs

$$T' + \lambda\alpha^2 T = 0, \quad \nabla^2 X + \lambda X = 0.$$

Thus, we need the eigenfunctions of $-\nabla^2$ that satisfy the Dirichlet boundary conditions. From Example 6.7, the eigenvalues and eigenfunctions are given by

$$\lambda_{n,m} = \left(\frac{n\pi}{L_1}\right)^2 + \left(\frac{m\pi}{L_2}\right)^2, \; X_{n,m}(x) = \sin\frac{n\pi x_1}{L_1}\sin\frac{m\pi x_2}{L_2}, \; n, m = 1, 2, \ldots.$$

Solving the equation for T, the separated solutions satisfying the boundary conditions have the form

$$u_{n,m}(x) = e^{-\lambda_{n,m}\alpha^2 t}X_{n,m}(x).$$

To determine the heat kernel, we consider an initial condition of the form $u(x,0) = \delta(x-\xi)$ where $\xi \in R$. Proceeding as in the one-dimensional case, we first determine the expansion of the delta functional in terms of the eigenfunctions. The Fourier coefficients are easily computed as

$$\frac{4}{L_1L_2}\left(\delta(x-\xi), X_{n,m}(x)\right) = \frac{4}{L_1L_2}X_{n,m}(\xi)$$

and thus

$$\delta(x-\xi) \sim \frac{4}{L_1L_2}\sum_{n,m=1}^{\infty} X_{n,m}(\xi)X_{n,m}(x).$$

We can easily check that the series converges in distribution to $\delta(x-\xi)$. So to find the heat kernel, we seek a solution of the form

$$\sum_{n,m=1}^{\infty} a_{n,m}e^{-\lambda_{n,m}\alpha^2 t}X_n(x),$$

which satisfies the initial condition. Comparing the preceding series with that for $\delta(x-\xi)$ provides the formula for the heat kernel as

$$\mathcal{S}(x,\xi,t) = \frac{4}{L_1L_2}\sum_{n,m=1}^{\infty} e^{-\lambda_{n,m}\alpha^2 t}X_{n,m}(\xi)X_{n,m}(x).$$

As long as $t > 0$, this series is uniformly convergent for $x \in R$ and $\xi \in R$ due to the rapid decay of the exponential function.

Now let's consider an initial condition of the form $u(x,0) = f(x)$ and suppose for the moment that f is a test function on R. Then the solution of the corresponding IBVP is given by

$$\begin{aligned} u(x,t) &= \int_R f(\xi)\mathcal{S}(x,\xi,t)dA(\xi) \\ &= \sum_{n,m=1}^{\infty} \widehat{f}(n,m)e^{-\lambda_{n,m}\alpha^2 t}X_{n,m}(x), \end{aligned} \tag{6.4.2}$$

where $\widehat{f}(n,m)$ are the Fourier coefficients of f with respect to the eigenfunctions $X_{n,m}$. For more general functions f, say continuous on R, this formula provides a strong solution of the heat equation due to the rapid decay of the exponentials. The initial condition may fail in the pointwise sense but can be interpreted as being valid in the sense of L^2−convergence by Theorem 6.7. An explicit computational example is left to Exercise 6.32.

Remark 6.11 Two remarks are pertinent regarding Example 6.8.

1. Let $\mathcal{S}_{\mathrm{dd},L}$ denote the one-dimensional heat kernel on the interval $[0,L]$ with homogeneous Dirichlet conditions at the endpoints (this was denoted $\mathcal{S}_{\mathrm{dd}}$ in Section 6.2). Then by multiplying the series together we can show that the heat kernel determined in Example 6.8 can be expressed $\mathcal{S}(x,\xi,t) = \mathcal{S}_{\mathrm{dd},L_1}(x_1,\xi_1,t)\mathcal{S}_{\mathrm{dd},L_2}(x_2,\xi_2,t)$, a product of one-dimensional heat kernels. This is a general phenomena for heat kernels in higher dimension regardless of the particular homogeneous boundary conditions.

2. The methodology employed in the example will work just as easily in three dimensions.

Example 6.9 (The Vibrating Rectangular Membrane) Let R be the rectangle of Example 6.8 and consider the wave equation $u_{tt} = c^2\nabla^2 u$ in R with homogeneous Dirichlet boundary conditions and initial conditions $u(x,0) = f(x)$ and $u_t(x,0) = g(x)$, where the requirements on f and g will be discussed shortly. Here we are interested in constructing a solution to the problem and discussing the fundamental modes of vibration. Proceeding as in the previous example, the separated solutions have the form

$$u_{n,m}(x,t) = (a_{n,m}\cos\sqrt{\lambda_{n,m}}ct + b_{n,m}\sin\sqrt{\lambda_{n,m}}ct)X_{n,m}(x).$$

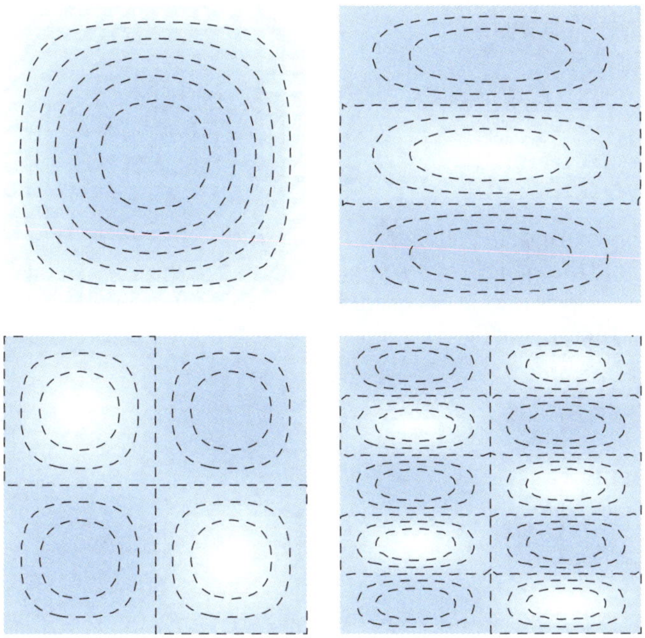

Figure 6.7 Membrane Nodal Contours

The fundamental modes of vibration have graphical profile given by the eigenfunctions $X_{n,m}(x)$. Contour plots of a few of these are given in Figure 6.7. The plots indicate the presence of nodal lines for the fundamental modes, i.e., lines in the rectangle R that do not move under vibration. The fundamental modes vibrate with frequency $\sqrt{\lambda_{n,m}}c$. Relative to the vibrating string, a contrasting feature of the vibrating rectangular membrane is that the set of allowable frequencies $\{\sqrt{\lambda_{n,m}}c\}$ are not integer multiples of a fundamental frequency. For this reason, the rectangular membrane does not produce what we would associate with musical sound.

Turning now to the initial boundary value problem, a solution will take the form

$$u(x,t) = \sum_{n,m=1}^{\infty} (a_{n,m} \cos \sqrt{\lambda_{n,m}}ct + b_{n,m} \sin \sqrt{\lambda_{n,m}}ct) X_{n,m}(x),$$

where the coefficients are determined from the initial data in complete analogy with the vibrating string. The resulting computation provides the conjecture for the solution formula

$$u(x,t) = \sum_{n,m=1}^{\infty} \left(\widehat{f}(n,m) \cos \sqrt{\lambda_{n,m}} ct + \frac{\widehat{g}(n,m)}{\lambda_{n,m} c} \sin \sqrt{\lambda_{n,m}} ct \right) X_{n,m}(x). \tag{6.4.3}$$

It can be shown that if f is continuous on the closure of the rectangle and satisfies the natural condition $f(x) = 0$ for $x \in \partial R$ and g is continuous on R, formula (6.4.3) gives a weak solution to the wave equation. As a computational example, we take $L_1 = L_2 = \pi$, $c = 1$, $g = 0$, and

$$f(x) = \begin{cases} x_1 x_2, & 0 \le x_1 \le \pi/2,\ 0 \le x_2 \le \pi/2 \\ x_1(\pi - x_2), & 0 \le x_1 \le \pi/2,\ \pi/2 \le x_2 \le \pi \\ (\pi - x_1)x_2, & \pi/2 \le x_1 \le \pi,\ 0 \le x_2 \le \pi/2 \\ (\pi - x_1)(\pi - x_2), & \pi/2 \le x_1 \le \pi,\ \pi/2 \le x_2 \le \pi. \end{cases}$$

The function f is continuous on R and is zero on the boundary of R. Computing the Fourier coefficients of f with respect to the eigenfunctions we have

$$\widehat{f}(n,m) = \frac{4}{\pi^2} \int_0^\pi \int_0^\pi f(x_1,x_2) X_{n,m}(x_1,x_2) dx_2 dx_1 = \frac{16}{\pi^2 n^2 m^2} \sin \frac{n\pi}{2} \sin \frac{m\pi}{2}.$$

Thus, the solution reads (note that $\widehat{g}(n,m) = 0$ for all n and m)

$$u(x,t) = \sum_{n,m=1}^{\infty} \widehat{f}(n,m) \cos \sqrt{n^2 + m^2} t \sin n x_1 \sin m x_2,$$

where we have substituted $\lambda_{n,m} = n^2 + m^2$. Truncating the series for the range $1 \le n, m \le 25$ snapshots of the approximate solution for various values of t are given in Figure 6.8. Note the rectangular ridge lines in the plots, indicating the propagation of the singularities in the initial data.

Remark 6.12 We can use Parseval's identity for the multiple sine expansion in Example 6.9 and deduce conservation of energy by expressing the total energy in terms of the energy of the fundamental modes. However, the the lack of harmonics precludes deducing an equipartition principle, at least in the form found for the vibrating string.

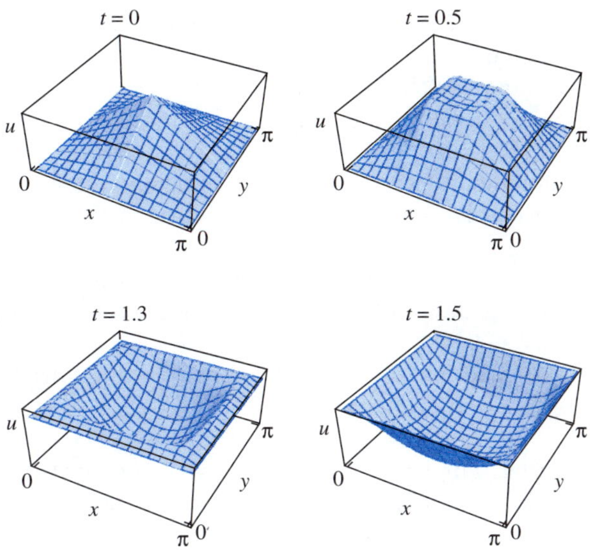

Figure 6.8 Vibrating Membrane Snapshots.

Exercises

6.27 Using separation of variables, demonstrate all the claims made in Table 6.1. Determine the range on n and m.

6.28 Find the eigenvalues and eigenfunctions of ∇^2 on the cube $(0, \pi) \times (0, \pi) \times (0, \pi)$ in $\mathbb{R}^3$ given the following homogeneous boundary conditions.

a. Dirichlet conditions on all sides

b. Neumann conditions on the vertical sides and Dirichlet conditions on the top and bottom

6.29 Let $\lambda_{n,m}$ and $X_{n,m}(x)$ be as in Theorem 6.7.

a. Write down Parseval's equality.

b. For each entry in Table 6.1, write out Parseval's equality explicitly.

6.30 Supply the proof of Corollary 6.8.

6.31 Supply the proof of Theorem 6.10 part (1). HINT: Write down Green's second identity and show that under the boundary conditions, the surface integrals are zero.

6.32 In Example 6.8, let $L_1 = L_2 = \pi$, $\alpha = 1$, and let $f(x) = 1$ on R. Compute the double Fourier sine coefficients of f and form the series (6.4.2). Truncating the sum for u by summing the terms for $n, m = 1, 2, \ldots 15$, plot the resulting approximation for $t = 0$, 0.02, 0.05, 0.1.

6.33 Find the fundamental solution of the diffusion equation in a rectangle, assuming homogeneous Dirichlet boundary conditions.

6.34 Using Parseval's equality, demonstrate conservation of energy for the solution (6.4.3) under the assumptions for f and g made in Example 6.9.

6.35 Compute the solution (6.4.3) explicitly for the initial data $f(x) = x_1(L_1 - x_1)x_2(L_2 - x_2)$ and $g(x) = 0$.

6.36 Find the Fourier representation of the wave kernel for the vibrating membrane on a rectangle with homogeneous Dirichlet boundary conditions.

Chapter 7

Laplace's Equation

Harmonic functions are solutions of Laplace's equation. Our focus in this chapter is the study of qualitative properties of harmonic functions and the construction of solution formulas for boundary value problems. The first section begins with a list of examples of harmonic functions (many that we derived in previous chapters), and we explore transformations that preserve the property of being harmonic. Particular attention is given to the transformations called inversion in a circle and sphere, which are useful in electrostatic problems and the study of Laplace's equation in exterior domains. The second and third sections are devoted to qualitative properties of harmonic functions and the uniqueness problem for boundary value problems. In the fourth section, we focus on Green's functions for the Dirichlet problem for Laplace's equation, including their construction where the domain is a ball or a half-space. A fundamental representation formula (Theorem 7.4) is the basis for these three sections. In the final section, we apply Fourier methods to a variety of boundary value problems and Green's function calculation.

7.1 Examples of Harmonic Functions

A strong solution to Laplace's equation is called a **harmonic function**. In one dimension, harmonic functions are solutions to the equation $u'' = 0$ and are simply linear functions. In higher dimensions, linear functions are easily checked to be harmonic; however, a plethora of other examples can be given. For example, in three dimensions the functions

$$x_1x_2, \quad x_1x_3, \quad x_2x_3, \quad x_1^2 - x_2^2, \quad x_1^2 - x_3^2, \quad x_2^2 - x_3^2 \tag{7.1.1}$$

are all harmonic. These are all homogeneous polynomials of degree 2 (a polynomial $p(x)$ is said to be homogeneous of degree n if $p(\lambda x) = \lambda^n p(x)$). We leave it to the reader to determine other examples of homogeneous harmonic polynomials.

In spherical coordinates, the radial part of the Laplace operator is the differential operator

$$\frac{d^2}{d\rho^2}+\frac{2}{\rho}\frac{d}{d\rho}=\rho^{-2}\frac{d}{d\rho}\left(\rho^2\frac{d}{d\rho}\right).$$

Hence all radial harmonic functions in $\mathbb{R}^3$, $u=u(\rho)$, satisfy the differential equation $(\rho^2 u')'=0$. Two linearly independent solutions are

$$u(\rho)=1 \quad \text{and} \quad u(\rho)=\frac{1}{\rho}.$$

The connection of the latter with physics is clear: A unit charge at the origin produces a potential at a point $x\neq 0$ proportional to $1/\rho$, where $\rho=|x|$. Similarly, radial harmonic functions, $u=u(r)$, in two dimensions are solutions of the equation $(r\,u')'=0$. Here the linearly independent solutions are

$$u(r)=1 \quad \text{and} \quad u(r)=\log r.$$

In Chapter 3, Example 3.4, and in the first section of Chapter 5 we learned that the functions $r^{|n|}e^{in\theta}$ and $r^{-|n|}e^{in\theta}$, where $n\in\mathbb{Z}$, are all solutions to Laplace's equation in $\mathbb{R}^2$ (for the second function, clearly we must avoid the origin). The complete collection of solutions arrived at by considering the Laplace operator in polar form can be written in unified form:

$$u_n(r,\theta)=\begin{cases}1,\ r^n\cos n\theta,\ r^n\sin n\theta, & n=1,2,\ldots\\ \log r,\ r^{-n}\cos n\theta,\ r^{-n}\sin n\theta, & n=1,2,\ldots.\end{cases} \tag{7.1.2}$$

Here we have written the functions as real valued; the first line is harmonic in the entire plane, the second line harmonic in the plane except for the origin. By the superposition principle, linear combinations of harmonic functions are also harmonic, e.g.,

$$2+r^2\cos 2\theta-r^5\sin 5\theta$$

is a harmonic function in the plane. Three-dimensional analogs of these harmonic functions in the plane will be determined in the next chapter.

New harmonic functions can be generated from known ones by applying certain transformations in the independent variables. Let D and D' be two domains in $\mathbb{R}^d$, and suppose there is an invertible map given by $x'=x'(x)$ of D onto D'. In three dimensions, this map would have component form:

$$x_1'=x_1'(x_1,x_2,x_3),\quad x_2'=x_2'(x_1,x_2,x_3),\quad x_3'=x_3'(x_1,x_2,x_3).$$

The inverse of this transformation of coordinates would be written $x = x(x')$, a map from D' onto D. Let $u = u(x)$ be a function defined on D and $u(x')$ be defined on D' by the composition formula $u(x') = u(x(x'))$. In Chapter 6 we used such transformations to covert a second-order PDE in two variables to its canonical form. The question to be addressed here is: If $u(x)$ is harmonic in D, for what transformations does $u(x')$ satisfy Laplace's equation in D' with respect to the variables x'? In other words, if $\partial_{11}u + \partial_{22}u + \partial_{33}u = 0$ in D, for what transformations does it follow that

$$\partial'_{11}u + \partial'_{22}u + \partial'_{33}u = 0$$

in D'? (Recall that the prime means differentiation in the primed variables.) A transformation for which this is true is said to preserve harmonicity.

We saw in Chapter 6 that the Laplacian is invariant under translations ($x' = x + h$) and rotations ($x' = Ax$). Another way to state this invariance is that translations and rotations preserve harmonicity. Two other elementary transformations that preserve harmonicity are the following.

1. *Reflections.* In $\mathbb{R}^2$, this means reflection about any fixed line in the plane. For example,

   $$x'_1 = x_1, \quad x'_2 = -x_2$$

   represents reflection in the x_1-axis,

   $$x'_1 = -x_1, \quad x'_2 = x_2$$

   represents reflection in the x_2-axis, and

   $$x'_1 = x_2, \quad x'_2 = x_1$$

   represents reflection in the line $x_2 = x_1$. In $\mathbb{R}^2$, a reflection about any given line can be obtained as a combination of one of these with translation and a rotation. In $\mathbb{R}^3$, reflections would be relative to a fixed plane. That reflections preserve harmonicity follows by the chain rule.

2. *Scaling.*

   $$x' = \lambda x,$$

   where λ is a nonzero constant.

The procedure for obtaining new harmonic functions from known ones under a transformation is as follows. Let $u(x)$ be harmonic in D. Find the set $D' = \{x' \in \mathbb{R}^d : x(x') \in D\}$ and then write the function $u(x') = u(x(x'))$. As a last step, drop the prime notation.

Example 7.1 The function $\log(x_1^2 + x_2^2)^{1/2}$ is harmonic in the plane except at the origin. Therefore, the function

$$\log((x_1 - h_1)^2 + (x_2 - h_2)^2)^{1/2}$$

is harmonic in the plane except at the point (h_1, h_2). This function could also be written in more concise form as $\log|x+h|^{1/2}$, using vector length notation.

Example 7.2 Under a rotation through angle α, the harmonic functions (7.1.2) become

$$\begin{cases} 1,\ r^n \cos n(\theta + \alpha),\ r^n \sin n(\theta + \alpha), & n = 1, 2, \ldots \\ \log r,\ r^{-n} \cos n(\theta + \alpha),\ r^{-n} \sin n(\theta + \alpha), & n = 1, 2, \ldots. \end{cases}$$

The functions in the first line are harmonic in the plane; those in the second are harmonic in the plane except at the origin.

Example 7.3 In the plane, a simple function that is invariant under scaling is $\frac{x_2}{x_1}$. More generally, if $g(s)$ is a function of one variable, then $u(x_1, x_2) = g(\frac{x_2}{x_1})$ is also invariant under scaling. A natural question to ask is: Are there any harmonic functions of this form? Computing the derivatives, we have

$$\partial_{11} u = \frac{2x_2}{x_1^3} g'(\frac{x_2}{x_1}) + \frac{x_2^2}{x_1^4} g''(\frac{x_2}{x_1}) \quad \text{and} \quad \partial_{22} u = \frac{1}{x_1^2} g''(\frac{x_2}{x_1}).$$

If we set $s = \frac{x_2}{x_1}$, then

$$\partial_{11} u + \partial_{22} u = \frac{1}{x_1^2} \big[(1 + s^2) g''(s) + 2s g'(s)\big].$$

Hence, for u to be harmonic, the function g must satisfy the ODE $[(1+s^2)g']' = 0$. The general solution to this equation is

$$g(s) = c_1 \tan^{-1}(s) + c_2,$$

where c_1 and c_2 are constants.

One conclusion from this computation is that the function $v(x_1, x_2) = \tan^{-1}(\frac{x_2}{x_1})$ is a harmonic function (clearly only when $x_1 \neq 0$). This is a reflection of the fact that the function $u(r, \theta) = \theta$, where θ is polar angle, is harmonic on any domain that does not contain a closed curve encircling the origin. Thus

$u(r,\theta) = v(x_1, x_2)$ in the right half-plane $\{x \mid x_1 > 0\}$. On the other hand, in the upper half-plane $\{x \mid x_2 > 0\}$, $u(r,\theta) = \frac{\pi}{2} - \tan^{-1}(\frac{x_1}{x_2})$. To see the latter, note that a clockwise rotation by $\frac{\pi}{2}$ takes the upper half-plane to the right half-plane with the point (x_1, x_2) being mapped to the point $(x_2, -x_1)$. From the preceding discussion, $\theta - \frac{\pi}{2} = \tan^{-1}\left(\frac{-x_1}{x_2}\right)$, which is equivalent to the claim.

Another useful type of transformation that preserves harmonicity is inversion in a circle (planar case) or sphere (in three dimensions). We begin by defining inversion in a circle. Let $a > 0$ and let $S_a(O)$ denote the circle of radius a centered at the origin O. Using polar coordinates (r,θ) in the plane and (r^*, θ^*) in the transformed plane, inversion in the circle $S_a(O)$ is the transformation defined by the equations

$$rr^* = a^2,\ \theta^* = \theta. \tag{7.1.3}$$

It is easy to solve for (r,θ) in terms of (r^*,θ^*), and vice versa. Notice that under this inversion, the circle $S_a(O)$ remains fixed and its interior and exterior are mapped to its exterior and interior, respectively.

Claim 7.1 If $u = u(r,\theta)$ is harmonic in a domain D, then the function $u(\frac{a^2}{r^*}, \theta^*)$ is harmonic in D^*, the region obtained by inversion of D through the circle $S_a(O)$.

Proof Set $v(r,\theta) = u(\frac{a^2}{r}, \theta)$ (we have dropped the star notation). Then

$$v_r = -\frac{a^2}{r^2} u_r(\frac{a^2}{r}, \theta),\quad v_{rr} = \frac{2a^2}{r^3} u_r(\frac{a^2}{r}, \theta) + \frac{a^4}{r^4} u_{rr}(\frac{a^2}{r}, \theta),\quad v_{\theta\theta} = u_{\theta\theta}(\frac{a^2}{r}, \theta).$$

It follows that

$$\begin{aligned} v_{rr} + \frac{1}{r} v_r + \frac{1}{r^2} v_{\theta\theta} &= \frac{a^4}{r^4} u_{rr}(\frac{a^2}{r}, \theta) + \frac{a^2}{r^3} u_r(\frac{a^2}{r}, \theta) + \frac{1}{r^2} u_{\theta\theta}(\frac{a^2}{r}, \theta) \\ &= \frac{a^4}{r^4}\left[u_{rr}(\frac{a^2}{r}, \theta) + \frac{1}{(\frac{a^2}{r})} u_r(\frac{a^2}{r}, \theta) + \frac{1}{(\frac{a^2}{r})^2} u_{\theta\theta}(\frac{a^2}{r}, \theta) \right]. \end{aligned}$$

The conclusion follows because the term in brackets is zero. □

Inversion in a sphere is defined in a similar fashion. Let $S_a(O)$ be the sphere of radius $a > 0$ centered at the origin. Using the spherical coordinates (ρ, θ, ϕ) and $(\rho^*, \theta^*, \phi^*)$, we obtain the defining relations for the inversion:

$$\rho\rho^* = a^2,\quad \theta^* = \theta,\quad \phi^* = \phi. \tag{7.1.4}$$

Under this map, the interior and exterior of the sphere $S_a(O)$ are mapped to its exterior and interior, respectively. Analogous to Claim 7.1 we have the following.

Claim 7.2 If $u(r,\theta,\phi)$ is harmonic in a domain D, then the function

$$\left(\frac{a}{\rho*}\right)u(\frac{a^2}{\rho*},\theta^*,\phi^*)$$

is harmonic in the region D^* obtained from D by inversion in the sphere $S_a(O)$.

The proof of Claim 7.2 is left as an exercise; it is similar to the previous chain rule computation for inversion in a circle.

The transformation is conveniently expressed in vector form by letting ξ and ξ^* be the position vectors of points with coordinates (ρ,θ,ϕ) and (ρ^*,θ^*,ϕ^*), respectively. Then the vector form for inversion is

$$\xi^* = \frac{a^2}{|\xi|^2}\xi. \tag{7.1.5}$$

This equation holds in both two and three dimensions.

Example 7.4 In the domain $|x| > 1$ in three dimensions, the bounded function $u(x) = 1$ under inversion in the unit sphere becomes $u^*(x) = \frac{1}{|x|}$, which is singular at $x = 0$. On the other hand, in the domain $|x| > 1$ in two dimensions, inversion in the unit circle applied to the function $u(x) = 1$ leads to $u^*(x) = 1$, with no singular points. The reason for the difference between these examples is the presence of the additional factor in Claim 7.2 that produces a harmonic function in three dimensions.

Example 7.5 If $x \in \mathbb{R}^2$ is fixed, then the function $\xi \to \log|x-\xi|$ is harmonic in the plane except at the point x. Under inversion in the circle $S_a(O)$, we obtain the function

$$\log\left|x - \frac{a^2}{|\xi^*|^2}\xi^*\right|.$$

Eliminating the prime notation and simplifying, we get

$$\log\frac{a}{|\xi|} + \log\left|\frac{|\xi|}{a}x - \frac{a}{|\xi|}\xi\right|.$$

The singular points, also called poles, are at the origin and at the point $\xi = \frac{a^2}{|\xi|^2}x$. The pole at the origin comes from the fact that the point at infinity is mapped to the origin under inversion. Also note that the original function $\log|x-\xi|$ is also harmonic as a function of x. Under the above transformation, it becomes the function $\log|x-\xi^*|$ whose singular point is at $x = \xi^*$.

Example 7.6 If we fix $x \in \mathbb{R}^3$, then the function $\xi \to \frac{1}{|x-\xi|}$ is harmonic in $\mathbb{R}^3$ except at x. Inversion in the sphere $S_a(O)$ yields the function

$$\frac{a}{|\xi^*|}\frac{1}{\left|x - \frac{a^2}{|\xi^*|^2}\xi^*\right|}$$

or, dropping the star,

$$\frac{a}{|\xi|}\frac{1}{\left|x - \frac{a^2}{|\xi|^2}\xi\right|}. \tag{7.1.6}$$

The pole of the transformed function is at $\xi = \frac{|\xi|^2}{a^2}x$. As in Example 7.5, the original function $\frac{1}{|x-\xi|}$ is also harmonic in x. Hence, transforming in the ξ variable as before, it becomes the function $\frac{|\xi^*|}{a}\frac{1}{|x-\xi^*|}$, which has pole at $x = \xi^*$.

Remark 7.3 Example 7.6 has a useful physical interpretation. Let ξ be a fixed point inside the sphere $S_a(O)$. Then a unit positive charge placed at ξ produces an electric field with potential given by the harmonic function $x \to \frac{1}{|x-\xi|}$. Likewise, a point charge of magnitude $\frac{-a}{|\xi|}$ placed at $\xi^* = \frac{a^2}{|\xi|^2}\xi$ produces an electric field with potential given by the negative of (7.1.6). The resultant potential of the two charges is the sum of these functions, i.e.,

$$\frac{1}{|x-\xi|} - \frac{a}{|\xi|}\frac{1}{|x-\xi^*|}. \tag{7.1.7}$$

Now for $|x| = a$, we have

$$\begin{aligned} |x-\xi^*|^2 &= a^2 - 2\left(\frac{a^2}{|\xi|^2}\right)x\cdot\xi + \left(\frac{a^2}{|\xi|^2}\right)|\xi|^2 \\ &= \left(\frac{a^2}{|\xi|^2}\right)[|\xi|^2 - 2x\cdot\xi + a^2] = \left(\frac{a^2}{|\xi|^2}\right)|x-\xi|^2. \end{aligned}$$

Hence, for $|x| = a$ the resultant potential (7.1.7) is zero. In other words, the resultant potential of the two charges located at points corresponding to inversion in the sphere vanishes on the sphere. This fact will be useful in constructing solutions to the Dirichlet problem for Laplace's equation in a ball (see Section 3.4).

Likewise, it is left to the reader to show that in the two-dimensional case,

$$\log|x-\xi| - \log\frac{|\xi|}{a}|x-\xi^*| = 0 \tag{7.1.8}$$

for $|x| = a$.

Exercises

7.1 Show that any homogeneous harmonic polynomial of degree 2 in three dimensions can be written as a linear combination of the six functions in (7.1.1).

7.2 Let A be a linear transformation on $\mathbb{R}^2$ that preserves harmonicity. Show that $A = \lambda B$ where λ is a scalar and B is a rotation. HINT: Write $A = \begin{bmatrix} a_{11} & a_{12} \\ a_{21} & a_{22} \end{bmatrix}$. Use the fact that the functions $x_1 x_2$ and $x_1^2 - x_2^2$ are harmonic and the assumption that A preserves harmonicity to determine conditions on the matrix entries. A rotation has orthogonal rows and each row is a unit vector.

7.3 Determine a formula in terms of x_1 and x_2 for the harmonic function $v(r, \theta) = \theta$ in the following half-planes.

a. $\{x \,|\, x_2 < 0\}$

b. $\{x \,|\, x_2 > x_1\}$

7.4 Demonstrate Claim 7.2.

7.5 Apply inversion in the unit circle to the functions in (7.1.2) and show that no new harmonic functions are generated.

7.6 Verify the formula (7.1.8) when $|x| = a$.

7.7 Suppose a positive unit charge is placed at the point $p = (p_1, p_2, p_2) \in \mathbb{R}^3$ with $p_3 > 0$. At what point in space can we place a negative unit charge such that the resultant potential on the plane $x_3 = 0$ is zero?

7.2 Qualitative Properties of Harmonic Functions

The Fundamental Representation Formula

The basic result of this section provides a representation formula for a harmonic function on a domain in terms of its values and those of its normal derivative on the boundary of the domain. Motivation for this formula comes from electrostatics as follows. Let ξ be a fixed point in space where a charge q is placed. The potential generated by this charge at a point $x \in \mathbb{R}^3$ is given by

$$u(x) = \frac{q}{|x - \xi|};$$

the negative gradient of this function gives the force exerted by this charge on a unit charge at x. Instead of being concentrated at the single point, if the charge is distributed through a domain $D \subset \mathbb{R}^3$ with density $\sigma(\xi)$, then the resulting potential is given by

$$u(x) = \int_D \frac{\sigma(\xi)}{|x-\xi|} dV(\xi). \tag{7.2.1}$$

This is called a **volume potential**. Similarly, if charge is distributed over the surface ∂D with density $\sigma(\xi)$, $\xi \in \partial D$, the potential is given by

$$u(x) = \int_{\partial D} \frac{\sigma(\xi)}{|x-\xi|} dS(\xi). \tag{7.2.2}$$

The function (7.2.2) is called a **single-layer potential**. The third type of potential, a **double-layer potential**, takes more effort to describe. Imagine two charges q and $-q$ placed at points ξ and ξ', respectively. Let ξ' approach ξ along the line segment between them and let the charge value q increase according to the rule

$$q|\xi - \xi'| = m,$$

where m is a constant. The potential at a point x of the resulting configuration is given by

$$u(x) = m \frac{d}{dn_\xi}\left(\frac{1}{|x-\xi|}\right),$$

where the directional derivative is in the ξ variable. This is called a **dipole potential** at the point ξ of strength m. If the dipoles are distributed on the surface ∂D with strength density $\sigma(\xi)$, then the **double-layer potential** is given by

$$u(x) = \int_{\partial D} \sigma(\xi) \frac{d}{dn_\xi}\left(\frac{1}{|x-\xi|}\right) dS(\xi). \tag{7.2.3}$$

These potentials are fundamental in the study of electromagnetic fields.[1]

The following result, due to Poisson, is cornerstone in the study of harmonic functions. It states that any function defined and with sufficient smoothness on a domain can be written as a sum of a single-layer, a double-layer, and a volume potential. For a succinct statement we recall the fundamental solutions of the Laplacian

$$K(x) = K_d(x) = \begin{cases} -\frac{1}{4\pi}\frac{1}{|x|}, & d = 3 \\ -\frac{1}{2\pi}\log\frac{1}{|x|}, & d = 2. \end{cases}$$

1. A classic reference to the study of the electromagnetic field is the text by Eygens [8]. Although at the graduate level, the introductory chapters are quite readable by anyone who has taken basic courses in physics and provide a wealth of applications of partial differential equations.

Further, in two dimensions, boundary integrals should be interpreted as oriented path integrals (orientation being counterclockwise about the interior of D), the surface element dS interpreted as the arc element, and the volume element dV as the area element in the plane. Unless otherwise stated, these interpretations need to be made in two dimensions for all integral formulas in this chapter.

Theorem 7.4 (Poisson's Formula) Let $D \subset \mathbb{R}^d$ be a bounded domain and let $u \in C^2(\bar{D})$. Then for any $x \in D$,

$$u(x) = \int_{\partial D} \left[u(\xi) \frac{dK}{dn_\xi}(x-\xi) - K(x-\xi)\frac{du}{dn_\xi}(\xi) \right] dS(\xi) + \int_D \nabla^2 u(\xi) K(x-\xi) dV(\xi). \quad (7.2.4)$$

More generally, the formula holds if $u \in C^2(D) \cap C^1(\bar{D})$ and the volume integral exists.

Proof We give the proof for $d = 3$ and the assumption that $u \in C^2(\bar{D})$; the two-dimensional case is similar and left as an exercise. Take $r > 0$ such that $\overline{B_r(x)} \subset D$ and consider the domain $D_r = D - \overline{B_r(x)}$ (see Figure 7.1). Effectively we have isolated the point $x \in D$ and can apply Green's identity (3.1.6) to u and K on the domain D_r to get

$$\left\{ \int_{\partial D} - \int_{S_r(x)} \right\} \left[u(\xi)\frac{dK}{dn_\xi}(x-\xi) - K(x-\xi)\frac{du}{dn_\xi}(\xi) \right] dS(\xi) = -\int_{D_r} \nabla^2 u(\xi) K(x-\xi) dV(\xi),$$

where we have used $\nabla^2 K = 0$ in D_r. Note: The minus on the left-hand side comes from the fact that the directional derivatives are in the direction of the

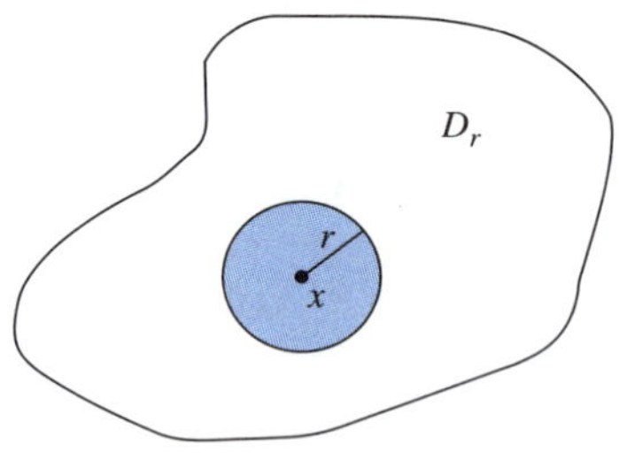

Figure 7.1 The Domain D_r

outward unit normal. The idea is to find the limit as $r \to 0$. It is easy to see that the integral on the right-hand side of this formula converges to

$$-\int_D \nabla^2 u(\xi) K(x-\xi) dV(\xi).$$

Further, the first integral on the left-hand side does not depend on r; call the second integral $I(x,r)$. For $\xi \in S_r(x)$, we can write $K(x,\xi) = -\frac{1}{4\pi r}$ where $r = |x-\xi|$. It follows that

$$\frac{dK}{dn} = \frac{1}{4\pi r^2}, \quad \xi \in S_r(x). \tag{7.2.5}$$

Using this, the integral of interest becomes

$$I(x,r) = \frac{1}{4\pi r^2}\int_{S_r(x)} u(\xi)dS(\xi) + \frac{1}{4\pi r}\int_{S_r(x)} \frac{du}{dn}(\xi)dS(\xi).$$

Since u is continuous, the limit of the first integral is easy:

$$\lim_{r\to 0}\frac{1}{4\pi r^2}\int_{S_r(x)} u(\xi)dS(\xi) = u(x).$$

For the second integral, we use the fact that $|\frac{du}{dn}(\xi)| \le M$ for some constant M and all $\xi \in B_r(x)$. Then it follows that

$$\left|\frac{1}{4\pi r}\int_{S_r(x)} \frac{du}{dn}(\xi)dS(\xi)\right| \le \frac{M}{4\pi r}(4\pi r^2) = Mr \to 0, \quad \text{as } r \to 0.$$

Thus $\lim_{r\to 0} I(x,r) = u(x)$. Putting the pieces together gives the desired representation formula. □

Notice that if the function u is harmonic, the third term on the right-hand side of the representation formulas drops out. Hence any harmonic function on a bounded domain can be expressed as a sum of a single-layer and double-layer potential. More precisely, we have the following.

Corollary 7.5 Let $D \subset \mathbb{R}^d$ be a bounded domain and let $u \in C^2(D) \cap C^1(\bar{D})$ be harmonic in D. Then for any $x \in D$,

$$u(x) = \int_{\partial D}\left[u(\xi)\frac{dK}{dn_\xi}(x-\xi) - K(x-\xi)\frac{du}{dn}(\xi)\right]dS(\xi). \tag{7.2.6}$$

Mean Value Formula and Maximum Principle

Let $A_d(r)$ denote the area of the sphere $S_r(x)$ in $\mathbb{R}^d$. It is useful to note that $A_3(r) = 4\pi r^2$ and $A_2(r) = 2\pi r$ (area in two dimensions is the length of the circle). Given a function $u(x)$ on $\mathbb{R}^d$, its mean value at a point x is defined as the average of u over a sphere centered at x of radius r, i.e.,

$$M^r u(x) = \frac{1}{A_d(r)} \int_{S_r(x)} u(\xi) dS(\xi). \tag{7.2.7}$$

The function $M^t u(x)$ is called the **spherical mean** of u; the map $u \to M^t u(x)$ is called the **spherical mean operator**. Since $dS(\xi) = r^{d-1} d\omega$, where $d\omega$ is the surface element on the unit sphere $S_1(O)$, this can also be written as

$$M^r u(x) = \frac{1}{A_d(1)} \int_{S_1(O)} u(x + r\omega) d\omega,$$

A simple consequence of the representation theorem is that harmonic functions equal their mean values.

Theorem 7.6 (Mean Value Formula) Let $u(x)$ be harmonic in a domain D. Then for every $x \in D$ and $r > 0$ such that $\overline{B_r(x)} \subset D$, $M^r u(x) = u(x)$.

Proof We consider the case of three dimensions. Apply the representation formula to the domain $B_r(x)$ to obtain

$$u(x) = \int_{S_r(x)} \left[u(\xi) \frac{dK}{dn_\xi}(x - \xi) - K(x - \xi) \frac{du}{dn}(\xi) \right] dS(\xi).$$

Using (7.2.5), we can rewrite this in the form

$$u(x) = \frac{1}{4\pi r^2} \int_{S_r(x)} u(\xi) dS(\xi) + \frac{1}{4\pi r} \int_{S_r(x)} \frac{du}{dn}(\xi) dS(\xi).$$

The second integral on the right-hand side is the flux of the harmonic function u through the surface $S_r(x)$ and hence is zero. The first integral on the right-hand side is precisely the mean value of u over the sphere $S_r(x)$, concluding the proof. □

Let $V_r(x)$ denote the volume (area in two dimensions) of the ball $B_r(x)$. Then, using spherical coordinates, the mean value formula can be reformulated as

$$u(x) = \frac{1}{V_r(x)} \int_{B_r(x)} u(\xi) dV(\xi). \tag{7.2.8}$$

A striking converse to the mean value property is valid as follows.

Theorem 7.7 Let $D \subset \mathbb{R}^d$ be a domain and let $u \in C(D)$ such that for every $x \in D$ and $r > 0$ with $\overline{B_r(x)} \subset D$, $M^r u(x) = u(x)$. Then $u \in C^\infty(D)$ and u is harmonic in D.

Proof Let $\phi(x)$ be a radial test function with $\phi(x) = 0$ for $|x| > 1$ and $\int \phi = 1$. For $\varepsilon > 0$, let $\phi_\varepsilon(x) = \varepsilon^{-d}\phi(x/\varepsilon)$; note that $\int \phi_\varepsilon = 1$. Define $u = 0$ outside of D and set $u_\varepsilon(x) = (u * \phi_\varepsilon)(x)$, i.e.,

$$u_\varepsilon(x) = \int_{\mathbb{R}^d} u(x-\xi)\phi_\varepsilon(\xi)dV(\xi) = \int_{\mathbb{R}^d} u(\xi)\phi_\varepsilon(x-\xi)dV(\xi).$$

By differentiating under the integral sign, we find that u_ε is a smooth function. By a change of variable we can express this in a form amenable to using the mean value hypothesis as:

$$u_\varepsilon(x) = \int_{\mathbb{R}^n} u(x - \varepsilon\xi)\phi(\xi)dV(\xi).$$

Expressing the integral in terms of spherical coordinates and setting $\widetilde{\phi}(\rho) = \phi(\xi)$ for $|\xi| = \rho$, we get

$$\begin{aligned} u_\varepsilon(x) &= \int_0^1 \int_{S^{d-1}} u(x - \varepsilon\rho\omega)\tilde{\phi}(\rho)\rho^{n-1}\, d\omega\, d\rho \\ &= A_d(1)\int_0^1 M^{\varepsilon\rho}u(x)\,\tilde{\phi}(\rho)\rho^{n-1}\, d\rho. \end{aligned}$$

For $x \in D$ and ε sufficiently small, $\overline{B_\varepsilon(x)} \subset D$ and $M^{\varepsilon\rho}u(x) = u(x)$. In this case, the formula becomes

$$u_\varepsilon(x) = u(x).$$

Since the left-hand side is a smooth function of x, so is the right-hand side, and hence $u(x)$ is in $C^\infty(D)$. To see that u is harmonic, let $x \in D$ and take $r > 0$ such that $\overline{B_r(x)} \subset D$. By Green's identity,

$$\int_{B_r(x)} \nabla^2 u(\xi)dV(\xi) = \int_{S_r(x)} \frac{du}{dn}(\xi)dS(\xi).$$

Since $dS(\xi) = r^{d-1}d\omega$ (recall $d\omega$ is the surface element on the unit sphere), this can be rewritten as

$$\begin{aligned} \int_{B_r(x)} \nabla^2 u(\xi)dV(\xi) &= r^{d-1}\int_{S_1(x)} u_r(r\omega)d\omega \\ &= r^{d-1}\frac{\partial}{\partial r}\int_{S_1(x)} u(r\omega)d\omega. \end{aligned}$$

Using the formula for the spherical mean we finally arrive at

$$\frac{1}{A_d(1)r^{d-1}}\int_{B_r(x)} \nabla^2 u(y)dV(y) = \frac{\partial}{\partial r}M^r u(x).$$

The right-hand side is zero since $M^r u(x) = u(x)$. It now follows that

$$\int_{B_r(x)} \nabla^2 u(y) dV(y) = 0$$

for any $r > 0$ such that $\overline{B_r(x)} \subset D$. This implies that $\nabla^2 u(x) = 0$ by the mean value theorem for multiple integrals. □

It is an immediate consequence of this result that harmonic functions, which are strong solutions to Laplace's equation, are in fact in class C^∞.

In the first chapter, we gave a heuristic argument for a maximum principle for the one-dimensional heat equation. As Laplace's equation arises from steady-state heat flow, it is reasonable to expect that a maximum principle holds for harmonic functions. Here is the result; it is a second application of the mean value theorem for harmonic functions.

Theorem 7.8 (Maximum [Minimum] Principle for Harmonic Functions) Let $D \subset \mathbb{R}^d$ be a bounded connected domain and let $u(x)$ be harmonic in D and continuous on $\bar{D}$. Let $M = \max_{x \in \bar{D}} u(x)$ (or $m = \min_{x \in \bar{D}} u(x)$). Then either $u(x) < M$ (or $m < u(x)$) for $x \in D$ or u is identically a constant.

Proof Suppose there is a point $p \in D$ such that $u(p) = M$. Let $B_r(p)$ be the largest ball centered at p that is contained in $\overline{D}$. For any $r' < r$ we know that

$$u(p) = \frac{1}{V_d(r')} \int_{B_{r'}(p)} u(\xi) dV(\xi). \tag{7.2.9}$$

The claim is that $u(x) = M$ throughout $B_{r'}(p)$. To see this, suppose $q \in B_{r'}(p)$ is such that $u(q) < M$. Then there is a $\delta > 0$ such that if $\xi \in B_\delta(q)$,

$$u(\xi) \leq \frac{1}{2}(u(q) + M) < M.$$

Let B be the complement of $B_\delta(q)$ in $B_{r'}(p)$ and set $V(B)$ to be the volume of B. Note that $V_n(\delta) + V(B) = V_n(r')$. Then

$$\begin{aligned}
\int_{B_{r'}(p)} u(\xi) dV(\xi) &= \left\{ \int_{B_\delta(q)} + \int_B \right\} u(\xi) dV(\xi) \\
&\leq \frac{1}{2}(u(q) + M) V_d(\delta) + M\, V(B) \\
&< \frac{1}{2}(M + M) V_d(\delta) + M\, V(B) = M\, V_d(r').
\end{aligned}$$

This contradicts (7.2.9) and demonstrates the claim. It follows that $u(x) = M$ throughout $B_r(p)$ since $r' < r$ was arbitrary and throughout $\overline{B_r(p)}$ by

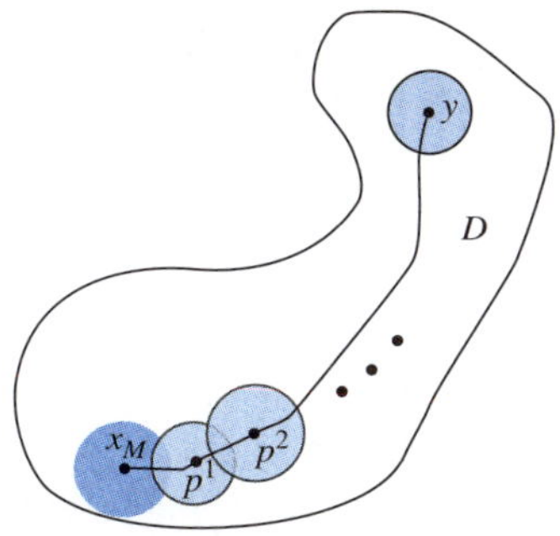

Figure 7.2 Maximum Principle Argument

continuity. In particular, u also achieves its maximum at a point on ∂D. Let y be any point of D. We claim that $u(y) = M$. To see this, we let C be a curve in D that connects p to y. Since the distance from C to the boundary of D is positive, there is a sequence of points $p^0 = p$, p^1, p^2, $\ldots, p^n = y$ and a finite sequence of positive numbers δ_0, $\delta_1, \ldots, \delta_n$ such that

$$(1)\ B_{\delta_k}(p^k) \subset D; \quad (2)\ p^k \in B_{\delta_{k-1}}(p^{k-1}).$$

See Figure 7.2. The second claim is proved by applying the first claim to each ball in succession. In other words, the first claim shows that $u(x) = M$ throughout $B_{\delta_0}(p^0)$ and in particular $u(p^1) = M$. Applying the first claim again gives that $u(x) = M$ for all $x \in B_{\delta_1}(p^1)$. Iterating this argument ultimately gives that $u(p^n) = u(y) = M$, as desired. Since y was any point of D, it follows that $u(x) = M$ throughout D. The minimum principle can be deduced by applying the maximum principle to the function $-u$. □

Exercises

7.8 Write out explicitly the representation formula in two and three dimensions in the case where $D = B_R(O)$ using polar, respectively, spherical coordinates.

7.9 Write out explicit formulas for $M^r f(x)$ for $d = 2$ and $d = 3$ using polar, respectively, spherical coordinates for the integration variables.

7.10 Suppose $u(x)$ is harmonic in $\{|x| < 2\}$ in $\mathbb{R}^3$ and in spherical coordinates has values $u = \sin^2\theta \sin\phi$ on $|x| = 1$. Find the value of u at the origin.

7.11 Demonstrate the volume form of the mean value formula (7.2.8).

7.12 Deduce the minimum principle for harmonic functions.

7.13 Verify the maximum principle for the harmonic function $u(x_1, x_2) = \frac{1-x_1^2-x_2^2}{1-2x_1+x_1^2+x_2^2}$ on the unit disk.

7.3 Discussion of Well-Posedness of BVPs

Recall that for a boundary value problem to be well posed three things must hold: (i) the problem has a solution; (ii) the solution is unique; and (iii) the solution is stable under small changes in the boundary data, i.e., the solution depends continuously in some sense on the boundary data. The problem of existence of a solution for a BVP involving Laplace's equation or Poisson's equation is both conceptually and technically difficult. In this text we will focus on construction of solutions for domains with simple geometry and leave the technical details of existence results to the references ([22, 28]). The purpose of this section is to explore aspects of the uniqueness and stability problems for Laplace's equation. Our principal tools will be the maximum–minimum principle, inversion in a circle/ball, and the energy method.

We begin with an example that indicates that, in general, specifying the unknown function and its normal derivative on the boundary leads to an ill-posed problem.

Example 7.7 (Hadamard's Example) Consider the boundary value problem for Laplace's equation

$$\text{PDE: } u_{xx} + u_{yy} = 0, \quad x \in \mathbb{R}, \quad y > 0$$

$$\text{BC: } \begin{cases} u(x,0) = 0 \\ u_y(x,0) = \frac{1}{n}\sin nx, \end{cases}$$

where n is an integer. An easy calculation shows that $u_n(x,y) = \frac{1}{n^2}\sinh ny \sin nx$ is a solution to this problem for each n. Furthermore, by taking n large enough, the boundary data can be made arbitrarily small whereas the solution can be made large for $y > 0$. Observe that $v(x,y) = y$ is a solution to Laplace's equation with boundary data

$$u(x,0) = 0 \quad \text{and} \quad u_y(x,0) = 1,$$

and $w(x,y) = y + u_n(x,y)$ is also a solution to Laplace's equation with boundary data

$$u(x,0) = 0 \quad \text{and} \quad u_y(x,0) = 1 + \frac{1}{n}\sin nx.$$

Taking the difference between v and w shows that specifying the values of u and the normal derivative u_y on the boundary leads to instability in the solution. In other words, this is an example of a problem that is not well posed.

The Dirichlet Problem on Bounded Domains Consider the Dirichlet problem

$$\begin{aligned} \text{PDE: } \nabla^2 u = 0, &\quad x \in D \\ \text{BC: } u = f, &\quad x \in \partial D, \end{aligned} \tag{7.3.1}$$

where D is a bounded connected domain and f is a continuous function on ∂D. A solution to this problem is a harmonic function u that is continuous on $\bar{D}$. For this problem, the maximum principle is directly applicable to the questions of uniqueness and stability as follows.

Theorem 7.9 (Uniqueness) Let $f \in C(\partial D)$. Then the Dirichlet problem (7.3.1) has at most one solution.

Proof Suppose that $v(x)$ and $w(x)$ are two solutions to this problem. Then the function $u = v - w$ is harmonic in D, continuous on $\bar{D}$, and satisfies the boundary condition $u = 0$ for $x \in \partial D$. By the maximum principle, u achieves its maximum and minimum on the boundary. Since the maximum and minimum values are both zero, it follows that $u = 0$ in D. Thus $v = w$ and uniqueness is proved. □

Example 7.8 Let D be the unit disk in the plane. Suppose $f(\theta) = 2 - 3\sin\theta + 4\cos 3\theta$ is the boundary data for the BVP (7.3.1). On the basis of discussion in Section 3.1 (see equation (7.1.2)), the unique solution of the problem is given by

$$u(r,\theta) = 2 - 3r\sin\theta + 4r^3\cos 3\theta.$$

Theorem 7.10 (Stability) Let f and g be in $C(\partial D)$. Suppose that $v(x)$ and $w(x)$ are the solutions to the problem (7.3.1) corresponding to boundary data f and g, respectively. Given $\varepsilon > 0$, if

$$\max_{x \in \partial D} |f(x) - g(x)| < \varepsilon \tag{7.3.2}$$

then

$$\max_{x \in D} |v(x) - w(x)| < \varepsilon. \tag{7.3.3}$$

Proof From the maximum principle we have for all $x \in D$,

$$\min_{y \in \partial D}(f(y) - g(y)) \le v(x) - w(x) \le \max_{y \in \partial D}(f(y) - g(y)).$$

The right-hand side is clearly less than or equal to $\max_{y\in\partial D}|f(y) - g(y)|$. Furthermore, it is easily seen that

$$\min_{y\in\partial D}(f(y) - g(y)) \geq -\max_{y\in\partial D}|f(y) - g(y)|. \tag{7.3.4}$$

Putting these together we have

$$|v(x) - w(x)| \leq \max_{y\in\partial D}|f(y) - g(y)|$$

from which the conclusion follows. □

Remark 7.11 The notion of stability given in Theorem 7.10 is called **uniform stability**. The hypothesis (7.3.2) implies that $|f(x) - g(x)| < \varepsilon$ for all $x \in \partial D$ and similarly for the conclusion, hence the adjective "uniform." Uniform stability is natural when considering continuous boundary data. Indeed, a continuous function on a closed bounded set achieves its maximum and minimum. In the preceding framework, ∂D is a closed bounded set, hence for any function $f \in C(\partial D)$, the quantity

$$\|f\|_{\partial D,\infty} = \max_{x\in\partial D}|f(x)|$$

is finite. This quantity is a gauge on the size of f and is a *norm* (review the definition in Chapter 4) on the vector space $C(\partial D)$. Notice that from the proof, the conclusion could have been written in stronger form via norms:

$$\|v - w\|_{D,\infty} \leq \|f - g\|_{\partial D,\infty}, \tag{7.3.5}$$

where $\|\cdot\|_{D,\infty}$ is defined analogously.

Dirichlet Problems on Unbounded Domains

In $\mathbb{R}^d$, let $D = \mathbb{R}^d - \overline{B_1(O)}$ be the domain exterior to the unit disk, and consider the following Dirichlet problem.

$$\begin{aligned} \text{PDE: } \nabla^2 u &= 0, \quad x \in D \\ \text{BC: } u(x) &= 1, \quad x \in \partial D. \end{aligned}$$

In two dimensions, two solutions are apparent, i.e.

$$u_1(r,\theta) = 1 \quad \text{and} \quad u_2(r,\theta) = 1 + \log r.$$

Likewise, in three dimensions we have solutions

$$u_3(\rho,\theta,\phi) = 1 \quad \text{and} \quad u_4(\rho,\theta,\phi) = \frac{1}{\rho}.$$

By taking appropriate linear combinations of these, other solutions can be constructed, e.g., $\frac{1}{3} + \frac{2}{3\rho}$ is a solution when $d = 3$. So in either dimension, uniqueness is violated and the problem is not well posed. This difficulty can be circumvented by imposing conditions on the solution at infinity and can be understood by examining the solutions under inversion. Indeed, applying Claims 7.1 and 7.2, respectively, the solutions under inversion in the unit circle (two dimensions) and unit sphere (three dimensions) become

$$\begin{array}{lll} n = 2 & u_1^*(r,\theta) = 1, & u_2^*(r,\theta) = 1 + \log \frac{1}{r}; \\ n = 3 & u_3^*(\rho,\theta,\phi) = \frac{1}{\rho} & u_4^*(\rho,\theta,\phi) = 1. \end{array}$$

Notice that u_1^* in two dimensions and u_4^* in three dimensions are the unique solutions to Laplace's equation in the interior of the unit disk with boundary data $u = 1$ on the boundary. The other two solutions are singular at the origin. From this perspective, a desired condition at infinity that will lead to uniqueness of solutions can be thought of as an additional **boundary condition at infinity**.

Definition 7.12 A domain $D \subset \mathbb{R}^d$ is called an **exterior domain** if D is the complement of the closure of a bounded domain.

The following result formalizes the preceding discussion.

Theorem 7.13 Let $D \subset \mathbb{R}^d$ be an exterior domain and let $f \in C(\partial D)$. Then the Dirichlet problem

$$\begin{aligned} \text{PDE: } \nabla^2 u = 0, \quad & x \in D \\ \text{BC: } u = f, \quad & x \in \partial D \end{aligned} \tag{7.3.6}$$

has at most one solution $u \in C^2(D)$ with $u \in C(\bar{D})$, provided the following boundary condition at infinity is assumed: (i) $u(x)$ is bounded in the case $d = 2$; (ii) $u(x) \to 0$ as $|x| \to \infty$ in the case $d = 3$.

The idea behind the proof is as follows. By using translation (and scaling if necessary), we can assume that the complement of D contains the unit ball in its interior. Suppose that $v(x)$ and $w(x)$ are solutions to the problem (7.3.6) and satisfy the boundary condition at infinity. Then the function $u = v - w$ satisfies the same problem with $f = 0$ along with the boundary condition at infinity. Let D^* be the domain obtained from D via inversion in the unit circle/sphere and let u^* be the function obtained from u via the inversion (Claims 7.1 or 7.2). Then u^* is harmonic in $D^* - \{O\}$ and satisfies the boundary condition $u^* = 0$ on ∂D^*. In other words, u^* could be singular/undefined at the origin. Under inversion, conditions (i) and (ii) become, respectively, (i′) $u^*(x^*)$ is bounded and (ii) $|x^*|\,|u^*(x^*)| = |u(x)| \to 0$ as $x^* \to O$. Conditions (i′) in two dimensions and (ii′) in three dimensions are precisely the conditions needed to prove that

u^* is defined at the origin and is harmonic there.[2] Thus u^* is harmonic in the bounded domain D^* with boundary data zero. By uniqueness of solutions for the latter problem, $u^* = 0$, which in turn implies $u = 0$ as desired.

Example 7.9 Let D be the exterior of the unit ball in $\mathbb{R}^d$. Then $u(r, \theta) = 1$ is the unique solution of the BVP (7.3.6) in two dimensions that remains bounded on D. Likewise, $u(\rho, \theta, \phi) = \frac{1}{\rho}$ is the only solution of this BVP on D that tends to zero as $|\rho| \to \infty$.

Example 7.10 Let D be the exterior of the circle $|x| = a$ in $\mathbb{R}^2$. Let the boundary data for the BVP (7.3.6) be given in polar form as

$$f(\theta) = \cos 2\theta - 3\cos 5\theta.$$

We seek the unique bounded solution to this problem. The form of the solution would be

$$u(r, \theta) = c_1 r^{-2} \cos 2\theta + c_2 r^{-5} \cos 5\theta.$$

Substituting $r = a$ gives $u(a, \theta) = c_1 a^{-2} \cos 2\theta + c_2 a^{-5} \cos 5\theta$. Therefore, $c_1 a^{-2} = 1$ and $c_2 a^{-5} = -3$. The desired solution is then

$$u(r, \theta) = \left(\frac{a}{r}\right)^2 \cos 2\theta - 3\left(\frac{a}{r}\right)^5 \cos 5\theta.$$

Remark 7.14 The methodology employed here can also be applied to more general unbounded domains provided they contain a ball, e.g., a half-plane. Since the boundary of such a domain extends to infinity, the behavior of the boundary data must be taken into account to guarantee a unique solution.

The Neumann Problem

Here we consider the Neumann problem for Laplace's equation:

$$\begin{aligned} \text{PDE: } & \nabla^2 u = 0, \quad x \in D \\ \text{BC: } & \left.\tfrac{du}{dn}\right|_{\partial D} = g. \end{aligned} \tag{7.3.7}$$

To make the proofs conceptually simpler, we assume $g \in C(\partial D)$ and that solutions of this problem are in $C^1(\bar{D})$. It is clear that any constant is always a solution to this problem, therefore uniqueness can only be obtained up to an

2. To be more precise, the conditions imply that u has a removable singularity at the origin; see, e.g., [28].

additive constant. For example, if $g = 0$, then $u(x) = c$ is a solution to this problem for any constant c. The key result for this BVP is as follows.

Theorem 7.15 Let D be a bounded domain. (i) A necessary condition for the Neumann problem (7.3.7) to have a solution is that $\int_{\partial D} g dS = 0$. (ii) Two solutions to the Neumann problem (7.3.7) that are in $C^1(\bar{D})$ differ by a constant.

Proof Green's identity (formula (3.1.5) with $f = 1$) is

$$\int_{\partial D} \nabla u \cdot \mathbf{n}_o \, dS = \int_D \nabla^2 u \, dV.$$

If u is harmonic, this immediately gives part (i) as $\nabla u \cdot n_o = g$. For part (ii), if $v(x)$ and $w(x)$ are solutions to this problem, then the function $u = v - w$ is a solution where the boundary data $g = 0$. We again apply Green's identity (formula (3.1.5)) with $f = g = u$ to obtain

$$\int_{\partial D} u \nabla u \cdot \mathbf{n}_o \, dS = \int_D \left[u \nabla^2 u + |\nabla u|^2 \right] dV \tag{7.3.8}$$

to conclude that $\int_D |\nabla u|^2 dV = 0$. It follows that $\nabla u = 0$ throughout $\bar{D}$ and hence u is a constant. □

One can interpret the BVP (7.3.7) as a steady-state heat flow problem. Physically it follows that the total heat flux through the boundary would be zero, making the condition $\int_{\partial D} g dS = 0$ natural.

It may seem troubling that the solution is only determined up to a constant. However, this constant can usually be determined in physical problems as Example 1.4 indicated and the following example reiterates.

Example 7.11 Consider the diffusion problem in the disk $D = \{|x| < 2\}$:

$$\begin{aligned} \text{PDE: } & c\rho u_t = k \nabla^2 u, \quad x \in D \\ \text{BC: } & \left.\tfrac{du}{dn}\right|_{\partial D} = g \\ \text{IC: } & u(x, 0) = 1, \end{aligned}$$

where g is given in terms of polar angle $g(\theta) = \sin\theta - \sin 2\theta$. Recall that c, ρ, and k are physical constants. The steady-state solution $U = U(r, \theta)$ is a solution to Laplace's equation and must satisfy the boundary condition. Reflecting on the set of solutions (7.1.2) and the fact that $\frac{d}{dn} = \frac{\partial}{\partial r}$ (the domain is a disk), a natural guess for the steady-state solution is

$$U(r, \theta) = C + c_1 r \sin\theta + c_2 r^2 \sin 2\theta,$$

where c_1 and c_2 are constants to be determined and C is the arbitrary constant resulting from Theorem 7.15 (ii). Differentiating with respect to r, we have $\frac{\partial U}{\partial r} = c_1 \sin\theta + 2c_2 r \sin 2\theta$. Substituting $r = 2$ and comparing with the boundary condition yields the equations $1 = c_1$ and $-1 = 4c_2$. Therefore, the steady-state has form

$$U(r,\theta) = C + r\sin\theta - \frac{1}{4}r^2 \sin 2\theta.$$

To determine the constant C, we compute the energy of the initial data and compare it to the energy of the steady state. The result is that $C = 1$. Therefore, the desired solution to the steady-state Neumann problem of importance to this diffusion problem is

$$U(r,\theta) = 1 + r\sin\theta - \frac{1}{4}r^2 \sin 2\theta.$$

Remark 7.16 The considerations given for Dirichlet problems on unbounded domains are also valid for Neumann problems and use the same ideas. In particular, Theorem 7.15 is valid on unbounded domains with the assumption that $u(x)$ is bounded in two dimensions and tends to zero in three dimensions.

The Dirichlet Principle

Here we return to the Dirichlet problem for Laplace's equation on a bounded domain D. It will be assumed that a unique solution u exists for this problem and that $u \in C^2(\bar{D})$. This latter assumption can be weakened using approximation arguments, the technical details of which will be omitted for simplicity. The identity (7.3.8) used in the study of the Neumann problem involves the integral of $|\nabla u|^2$ over the domain D and from a physical perspective relates to potential energy. In our framework, we define the energy to be

$$E(u) = \frac{1}{2}\int_D |\nabla u|^2 \, dV.$$

The Dirichlet principle states that among all functions $w(x)$ that satisfy the boundary condition

$$w(x) = f(x), \quad x \in \partial D, \tag{7.3.9}$$

the lowest energy occurs for the harmonic function that satisfies this boundary condition. Mathematically, the Dirichlet principle is given in the following result.

Theorem 7.17 Let $u(x)$ be the unique harmonic function that satisfies (7.3.9) and is in $C^1(\bar{D})$. If $w \in C^1(D)$ satisfies (7.3.9), then $E(w) \geq E(u)$.

Proof Write $w = u + v$ so that $v \in C^1(\bar{D})$ and $v(x) = 0$ for $x \in \partial D$. The energy of w is given by

$$E(w) = \frac{1}{2}\int_D |\nabla(u+v)|^2 dV = E(u) + \int_D \nabla u \cdot \nabla v \, dV + E(v).$$

Green's first identity written for the functions v and u is

$$\int_{\partial D} v\nabla u \cdot \mathbf{n}_o \, dS = \int_D \nabla v \cdot \nabla u \, dV + \int_D v\nabla^2 u \, dV.$$

It follows that $E(w) = E(u) + E(v)$ and since $E(v) \geq 0$, the result follows. □

Theorem 7.17 suggests that an alternative method for solving the Dirichlet problem would be to find the function u that minimizes the energy functional $E(w)$ over all functions that satisfy the boundary condition. This converse to Dirichlet's principle is also valid.

Proposition 7.18 Let $u \in C^2(D) \cap C^1(\bar{D})$ satisfy (7.3.9) and minimize the energy functional $E(w)$. Then u is harmonic in D.

An outline of the proof is as follows. Let ε be a constant and take $v \in C^1(\bar{D})$ such that $v(x) = 0$ for $x \in \partial D$. Then

$$\begin{aligned} E(u) \leq E(u+\varepsilon v) &= E(u) + \varepsilon \int_D \nabla u \cdot \nabla v \, dV + \varepsilon^2 E(v) \\ &= E(u) - \varepsilon \int_D v\nabla^2 u \, dV + \varepsilon^2 E(v). \end{aligned}$$

It follows that $\int_D v\nabla^2 u \, dV = \varepsilon E(v)$ for any function v satisfying the stated conditions and any constant ε. Hence, for the same functions v,

$$\int_D v\nabla^2 u = 0. \qquad (7.3.10)$$

The difficulty here is that v must be C^1. Let $x^0 \in D$ and take $r > 0$ such that $B_r(x^0) \subset D$. The claim is that $\int_{B_r(x^0)} \nabla^2 u \, dV = 0$. Once this is shown, since it is true for all smaller balls centered at x^0, it follows by continuity that $\nabla^2 u(x^0) = 0$ as desired. The claim would follow if (7.3.10) held for the function $v(x)$, which was 1 on $B_r(x^0)$ and zero outside this ball. However, such function does not satisfy the C^1 condition. Nonetheless, this choice of v can be approximated suitably by one that does; the details are omitted.

Exercises

7.14 Let D be a bounded connected domain and consider the Dirichlet problem for Poisson's equation:

$$\begin{aligned} \text{PDE: } \nabla^2 u = f, \quad & x \in D \\ \text{BC: } u = g, \quad & x \in \partial D. \end{aligned} \qquad (7.3.11)$$

Show that this problem has at most one solution in the class $C^2(D) \cap C(\bar{D})$.

7.15 Let $D = \{a < \rho < b\} \subset \mathbb{R}^3$ be the region between concentric spheres. Assume $a > 0$. Find the unique solution to the BVP (7.3.11) where $g = 0$ and $f = 1$.

7.16

a. Find the steady-state temperature distribution in the region outside a sphere of radius $R > 0$ if the surface of the sphere is held at constant temperature T under the assumption that the temperature tends to zero at infinity.

b. Find the steady-state temperature distribution in the region exterior to an infinite cylinder of radius $R > 0$ if the surface of the cylinder is held at constant temperature T under the assumption that the temperature is bounded.

7.17 Use inversion in a circle to demonstrate a stability result for the Dirichlet problem for Laplace's equation in an exterior domain under the assumption that the solution is bounded. Specifically, obtain the norm inequality (7.3.5).

7.18 Let $D = \{(x, y)\,|\, x > 1,\, y > 1\} \subset \mathbb{R}^2$ be a quarter-plane. Find the unique harmonic function $u(x, y)$ in D that satisfies the boundary conditions $u(1, y) = 1$ and $u(x, 1) = 0$. HINT: Consider the polar angle relative to a coordinate system with the origin at the point $(1, 1)$.

7.19 Let $D = \{(x, y)\,|\, x \in \mathbb{R},\, y > 0\}$ be the upper half-plane in $\mathbb{R}^2$. Show that the Dirichlet problem for Laplace's equation with boundary condition $u(x, 0) = 1$ has infinitely many solutions without the assumption that u is bounded.

7.20 Let $D \subset \mathbb{R}^3$ be a bound domain and $h = h(x) \in C(D)$ be a positive function. Show that the BVP

$$\begin{aligned} \text{PDE: } & \nabla^2 u = f, \quad x \in D \\ \text{BC: } & \tfrac{du}{dn} + hu = g, \quad x \in \partial D \end{aligned}$$

has at most one solution in the class $C^2(D) \cap C^1(\bar{D})$.

7.21 Let $h > 0$ be a constant and $D = \{|x| < a\}$ in the plane. Find the harmonic function in D that satisfies the Robin condition

$$\frac{du}{dn} + hu = \sin 2\theta$$

(polar coordinates) on the boundary of D.

7.22 Let D be a bounded domain and consider the Neumann problem

$$\begin{aligned} \text{PDE: } \nabla^2 u = f, \quad & x \in D \\ \text{BC: } \tfrac{du}{dn} = g, \quad & x \in \partial D. \end{aligned}$$

a. Show that a necessary condition for this problem to have a solution is $\int_D f\, dV = \int_{\partial D} g\, dS$.

b. Under the condition of part (a), show that any two solutions differ by a constant.

7.23 Let $D = \{a < \rho < b\} \subset \mathbb{R}^3$ be the region between concentric spheres. Find the solution of Poisson's equation $\nabla^2 u = 1$, $x \in D$ with boundary conditions $u = 0$ on $\rho = a$ and $\frac{du}{dn} = 0$ on $\rho = b$.

7.4 Green's Functions

Recall the fundamental representation formula for a function on a bounded domain, restated here for convenience:

$$u(x) = \int_{\partial D} \left[u(\xi) \frac{dK}{dn_x}(x - \xi) - K(x - \xi) \frac{du}{dn_x}(\xi) \right] dS(\xi) \\ + \int_D \nabla^2 u(\xi) K(x - \xi) dV(\xi). \quad (7.4.1)$$

This formula gives $u(x)$ as a sum of a double-layer, single-layer, and volume potential. As such, it does not provide a solution formula for Poisson's equation because the specification of the function and its normal derivative on the boundary does not provide a well-posed problem—as Hadamard's example in the previous section showed.

However, formula (7.4.1) does suggest that a Dirichlet problem could be solved using a double-layer potential and a Neumann problem by a single-layer potential. The idea here is to introduce a modification of the function K that will eliminate the first term in the formula and hence effectively provide a solution formula for the Dirichlet problem. Alternatively, we could modify K so as to eliminate the second term and obtain a solution formula for a Neumann problem.[3]

Let $v(x)$ be any harmonic function in D and set $G(x, \xi) = K(x - \xi) + v(x)$. Using Green's second identity, observe that

$$\int_D v \nabla^2 u\, dV + \int_{\partial D} \left[u \frac{dv}{dn} - v \frac{du}{dn} \right] dS = 0.$$

3. The resulting function is often called the Neumann function, or the Green's function for the Neumann problem. For the details, see [28].

Combining this with the representation formula provides a more general form of the latter:

$$u(\xi) = \int_{\partial D} \left[u(\xi) \frac{dG}{dn_x}(x,\xi) - G(x,\xi) \frac{du}{dn}(\xi) \right] dS(\xi) + \int_D \nabla^2 u(\xi) G(x,\xi) dV(\xi). \quad (7.4.2)$$

Definition 7.19 The **Dirichlet–Green's function** for the Laplace operator on a bounded domain D is a function $G(x,\xi)$ that satisfies the following three properties:

1. $x \to G(x,\xi)$ is harmonic in D and of class $C^2(\bar{D})$ except at the point $x = \xi$.
2. $G(x,\xi) = 0$ for $x \in \partial D$.
3. $G(x,\xi) = K(x,\xi) + v(x,\xi)$ where for each $\xi \in D$, $x \to v(x,\xi)$ is defined and harmonic in D and of class $C^1(\bar{D})$.

Remark 7.20 Two remarks are pertinent at this point.

1. The Green's function is symmetric:

   $$G(x,\xi) = G(\xi,x), \text{ for } \quad x \neq \xi. \quad (7.4.3)$$

   From a physical perspective this is known as the reciprocity principle: a point charge placed at x will have the same effect at the point ξ as a point charge at ξ has at x. The proof of the symmetry property is left to the reader. As a consequence of this symmetry, x and ξ can be interchanged in properties (1–3) in the definition. In particular, $\xi \to G(x,\xi)$ is harmonic in D.

2. An alternative formulation of the definition can be given based on the fact that K is a fundamental solution to the Laplace operator: $\nabla^2 K = \delta$. Indeed, any fundamental solution will have the form $G(x,\xi) = K(x-\xi) + v(x,\xi)$ where $x \to v(x,\xi)$ is harmonic in D. Then a function $G(x,\xi)$ is the Dirichlet–Green's function if and only if $x \to G(x,\xi)$ is $C^2(\bar{D})$ except at $x = \xi$, is a fundamental solution for the Laplacian ($\nabla^2_{(x)} G = \delta(x-\xi)$), and condition (2) of the definition holds.

To effectively find the Dirichlet–Green's function, we must solve the following Dirichlet problem for each $x \in D$:

$$\text{PDE: } \nabla^2 v = 0, \quad x \in D$$
$$\text{BC: } v(x) = -K(x-\xi), \quad x \in \partial D.$$

On the surface it seems we are trying to find the Green's function to solve Dirichlet problems and to do so, we need to solve a Dirichlet problem. However, the above Dirichlet problem for v has a boundary data of simple form and can be solved easily in certain domains. The physical interpretation of the Green's function is that $x \to G(x, \xi)$ represents the potential in the domain D of a unit charge placed at $\xi \in D$ where the potential is held to zero at the boundary.

Historical Comment The Dirichlet–Green's function was first studied by mathematical physicist George Green in his monograph *An Essay on the Application of Mathematical Analysis to the Theories of Electricity and Magnetism* [13], a fascinating historical document combining physical insights with mathematics. The physical interpretation of the Green's function along with his knowledge of various electrostatic experiments led Green to the belief that such a function would exist for any bounded region. It is now known that the geometry of the domain is the key factor for being able to prove that the Green's function exists. For example, it is known (see [22]) that if D satisfies the exterior sphere condition (for any point on the boundary of D there is a sphere in the complement of D that intersects $\bar{D}$ in precisely this point), then the Green's function exists.

Assuming that Green's function for a domain D exists, formula (7.4.2) simplifies to

$$u(x) = \int_{\partial D} u(\xi) P(x, \xi)\, dS(\xi) + \int_D \nabla^2 u(\xi) G(x, \xi) dV(\xi),$$

where

$$P(x, \xi) = \tfrac{dG}{dn_\xi}(x, \xi)$$

is called the **Poisson kernel**. Assuming we can find the Green's function for the domain D, the preceding formula for u suggests that the solution to the BVP

$$\begin{aligned} \text{PDE: } \nabla^2 u = F, \quad x \in D \\ \text{BC: } u = f, \quad x \in \partial D \end{aligned} \tag{7.4.4}$$

is given by

$$u(\xi) = \int_{\partial D} f(\xi) P(x, \xi) dS(\xi) + \int_D F(\xi) G(x, \xi) dV(\xi). \tag{7.4.5}$$

To explore this suggestion, consider the volume potential term

$$v(x) = \int_D F(\xi) G(x, \xi) dV(\xi).$$

Under the assumption that $F \in C(\bar{D})$, $v(\xi)$ is continuous on D and defines a weak solution of Poisson's equation $\nabla^2 u = F$. To see this, let $\phi(x)$ be a test function on D, so that $\phi(x) = 0$ near the boundary of D. We need to show that $\langle \nabla^2 u, \phi \rangle = \langle F, \phi \rangle$. First, using the definition and interchanging the order of integration (justified as the integrand is absolutely convergent), we have

$$\begin{aligned} \langle \nabla^2 u, \phi \rangle = \langle u, \nabla^2 \phi \rangle &= \int_D \left[\int_D F(\xi) G(x, \xi) dV(\xi) \right] \nabla^2 \phi(x) dV(x) \\ &= \int_D F(\xi) \left[\int_D G(x, \xi) \nabla^2 \phi(x) dV(x) \right] dV(\xi). \end{aligned}$$

Since $\nabla^2_{(x)} G(x, \xi) = \delta(x - \xi)$, the inner integral becomes

$$\int_D G(x, \xi) \nabla^2 \phi(x) dV(x) = \langle G(\cdot, \xi), \nabla^2 \phi(\cdot) \rangle = \langle \delta(x - \xi), \phi(x) \rangle = \phi(\xi).$$

Hence, $\langle \nabla^2 u, \phi \rangle = \int_D F(\xi)\phi(\xi) dV(\xi) = \langle F, \phi \rangle$, as desired. Does $v(x)$ define a strong solution to Poisson's equation? No, not without further smoothness assumption on F. An explicit example is given in Exercise 7.27.

The double-layer potential term in (7.4.5) turns out to be better behaved. The formula

$$u(x) = \int_{\partial D} f(\xi) P(x, \xi) dS(\xi) \tag{7.4.6}$$

is known as **Poisson's integral formula** and defines a harmonic function in D provided f is continuous. This is an application of the converse to the mean value property (Theorem 7.7); see Exercise 7.28.

The special case of $F = 0$ and $f = 1$ in BVP (7.4.4) has solution $u(x) = 1$. Hence in this case our solution formula provides a basic property of the Poisson kernel (provided it exists):

$$\int_{\partial D} P(x, \xi) dS(\xi) = 1.$$

The Green's Function on a Ball

Let $D_a = B_a(O)$ be the ball centered at the origin of radius $a > 0$ in $\mathbb{R}^d$. Based on the discussion regarding inversion in a circle and sphere, in particular Remark 7.3, we can write the Green's function for the Laplacian on D_a:

$$G(x, \xi) = \begin{cases} K_2(x - \xi) - \frac{1}{2\pi} \log \left[\frac{|\xi|}{a} |x - \xi^*| \right], & d = 2 \\ K_3(x - \xi) - \left(\frac{a}{|\xi|} \right) K_3(x - \xi^*), & d = 3, \end{cases} \tag{7.4.7}$$

where $\xi^* = \frac{a^2}{|\xi|^2} \xi$ is the transformed point under inversion. Since ξ^* is not in D_a, the function $x \to G(x, \xi)$ is harmonic in D_a when $x \neq \xi$ for each $\xi \in D_a$.

Furthermore, by Remark 7.3, $G(x,\xi) = 0$ for $x \in S_a(O) = \partial D_a$. This verifies the properties, so $G(x,\xi)$ is indeed the Green's function.

The Poisson kernel can be computed by first finding the gradient of G for $|\xi| = a$. The first claim is that for $|\xi| = a$,

$$\frac{\partial G}{\partial \xi_i} = \frac{\xi_i}{A_d(1)|x-\xi|^d}\left(\frac{a^2-|x|^2}{a^2}\right) \tag{7.4.8}$$

in both two and three dimensions. (Recall that $A_d(1)$ is the area of the unit sphere in d dimensions.) Here are the details for $d = 3$. Using standard rules,

$$\frac{\partial G}{\partial x_i} = \frac{1}{4\pi}\left[\frac{x_i-\xi_i}{|x-\xi|^3} - \frac{a}{|\xi|}\frac{x_i-\xi_i^*}{|x-\xi^*|^3}\right].$$

Since $G(x,\xi) = G(\xi,x)$, we have

$$\frac{\partial G}{\partial \xi_i} = \frac{1}{4\pi}\left[\frac{\xi_i-x_i}{|x-\xi|^3} - \frac{a}{|x|}\frac{\xi_i-x_i^*}{|\xi-x^*|^3}\right]$$

by interchanging x and ξ in the previous formula. If we let $|\xi| = a$, then $|\xi - x^*| = \frac{a}{|x|}|\xi - x|$ and the partial derivative becomes

$$\frac{\partial G}{\partial \xi_i} = \frac{1}{4\pi}\left[\frac{\xi_i-x_i}{|x-\xi|^3} - \frac{|x|^2}{a^2}\frac{\xi_i-x_i^*}{|x-\xi|^3}\right].$$

Now using $x_i^* = \frac{a^2}{|x|^2}x_i$ and simplifying proves (7.4.8). The computations for $d = 2$ are similar. For $|\xi| = a$, the outward unit normal is given by $\frac{\xi}{a}$, so the Poisson kernel is given by

$$P(x,\xi) = \boldsymbol{\nabla} G \cdot \frac{\xi}{a} = \frac{a^2-|x|^2}{A_d(1)\,a|x-\xi|^d}, \qquad x \in \partial D_a. \tag{7.4.9}$$

At this point we can write the Poisson integral formula in explicit form as

$$u(x) = \frac{a^2-|x|^2}{A_d(1)\,a}\int_{|x|=a}\frac{f(\xi)}{|x-\xi|^d}\,dS(\xi). \tag{7.4.10}$$

Writing $\xi = a\omega'$ and $x = \rho\omega$, where ω and $\omega' \in S_1(O)$, the formula has a polar/spherical form as

$$u(\rho\omega) = \frac{a^{d-2}(a^2-\rho^2)}{A_d(1)}\int_{S_1(O)}\frac{f(a\omega')}{[a^2-2a\rho(\omega\cdot\omega')+\rho^2]^{d/2}}\,d\omega', \tag{7.4.11}$$

where $d\omega'$ is the surface element on the unit sphere. (Recall that the surface element on $S_a(O)$ is $dS(x) = a^d d\omega'$.) The following theorem provides the solution to the Dirichlet problem for Laplace's equation on the ball.

Theorem 7.21 Let $f \in C(S_a(O)$. Define $u(x)$ on $\overline{B_a(O)}$ by (7.4.10) for $|x| < a$ and $u(x) = f(x)$ for $|x| = a$. Then u is harmonic on $B_a(O)$ and continuous on $\overline{D_a}$.

The fact that u is harmonic and hence continuous on $B_a(O)$ follows from previous discussions. In order to show that $u \in C(\overline{D_a})$, we need to show that

$$\lim_{x \to x^0, |x|<a} u(x) = f(x^0) \tag{7.4.12}$$

for every boundary point x^0, $|x^0| = a$. The proof of this limit relation is technical, and the details are discussed at the end of this section. An intuitive grasp of this relation can be obtained by explicitly writing out the formulas for the Poisson integral for $d = 2$ and $d = 3$, as is done in the following example.

Example 7.12 The preceding formulas for the solution of the Dirichlet problem are written in terms of d and, in fact, hold in any dimension. Here they are rewritten explicitly for dimensions $d = 2$ and $d = 3$. In the former case, write $\omega = (\cos\theta, \sin\theta)$, $\omega' = (\cos\theta', \sin\theta')$, and the boundary data as $f(a\omega') = f(\theta')$. Then (7.4.11) becomes

$$u(r, \theta) = \frac{1}{2\pi}\int_{-\pi}^{\pi} \frac{(a^2 - r^2)f(\theta')}{[a^2 - 2ar\cos(\theta - \theta') + r^2]}\, d\theta'. \tag{7.4.13}$$

An intuitive understanding of the limit relation (7.4.12) can be seen from this expression as follows. Take the case where $a = 1$ and $\theta = 0$. Set

$$P_r(\theta') = \frac{1}{2\pi}\frac{1 - r^2}{1 - 2r\cos\theta' + r^2},$$

which is also called the Poisson kernel in two dimensions. Then $\int_{-\pi}^{\pi} P_r(\theta')d\theta' = 1$, since the integral of the Poisson kernel over the boundary sphere is 1. Thus, the integral (7.4.13) can be thought of as averaging the values of f using weights given by the Poisson kernel. Plots of the P_r are given in Figure 7.3. The important observation is the localization of this kernel for small r, and hence, the Poisson integral formula provides local averages of f, which intuitively should approximate the function f.

In three dimensions, we can also explicitly write the Poisson integral formula. Write $\omega = (\sin\phi\cos\theta, \sin\phi\sin\theta, \cos\phi)$, and similarly for ω'. The boundary data becomes $f(a\omega') = f(\theta', \phi')$. In this case the Poisson integral formula becomes

$$u(\rho, \theta, \phi) = \frac{a}{4\pi}\int_{-\pi}^{\pi}\int_{0}^{\pi} \frac{(a^2 - \rho^2)f(\theta', \phi')\sin\phi'}{[a^2 - 2a\rho\cos\Theta + \rho^2]^{3/2}}\, d\phi'\, d\theta', \tag{7.4.14}$$

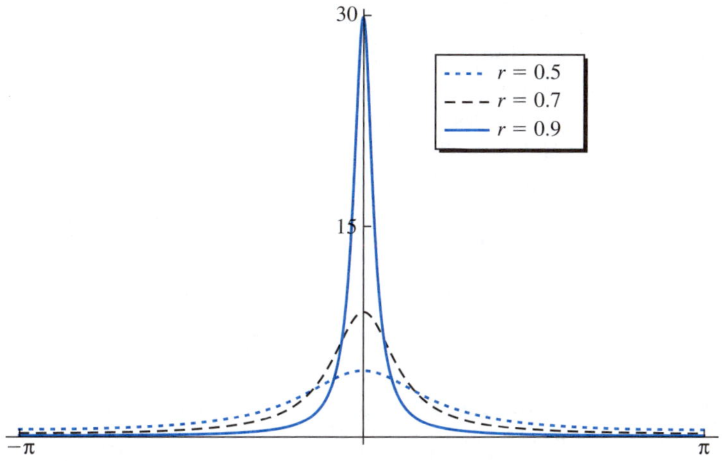

Figure 7.3 The Poisson Kernel for the Disk

where Θ is the angle between ω and ω'. Note that $\cos\Theta = \cos(\theta-\theta')\sin\phi\sin\phi' + \cos\phi\cos\phi'$. Details are left to the reader.

The plot of the Poisson kernel in two dimensions is somewhat reminiscent of the same for the Dirichlet kernel (Figure 5.5). A key difference is the lack of oscillation away from the origin in the Poisson kernel. The following extends Theorem 7.21 in two dimensions to the case where the boundary data is piecewise continuous. The proof is analogous to that given for pointwise convergence of Fourier series (in fact it is easier) and omitted.

Theorem 7.22 Let $f \in \mathrm{PC}[-\pi,\pi]$ and let

$$f^*(\theta) = \begin{cases} f(\theta), & \text{if } f \text{ is continous at } \theta \\ \frac{1}{2}[\tilde{f}(\theta+0) + \tilde{f}(\theta-0)], & \text{otherwise,} \end{cases} \tag{7.4.15}$$

where $\tilde{f}$ is the $2\pi-$periodic extension of f. Define $u(r,\theta)$ by (7.4.13) for $r < a$ and $u(a,\theta) = f^*(\theta)$. Then u is harmonic for $r < a$ and

$$\lim_{r\to a^-} u(r,\theta) = f^*(\theta).$$

Effectively this result solves the Dirichlet problem on the disk where the boundary data is piecewise continuous.

As another application of the ideas presented in this chapter, the following example provides the solution to the exterior problem for a ball in three dimensions; the two-dimensional case is left as an exercise.

Example 7.13 (The Exterior Problem in Three Dimensions) Let D_a^* be the exterior domain in three dimensions: $D_a^* = \mathbb{R}^3 - \overline{B_a(O)}$. The exterior Dirichlet problem with boundary data f and the condition that $\lim_{|x|\to\infty} u(x) = 0$ can be solved by applying inversion in the sphere $|x| = a$ to the formula (7.4.10), or what is effectively the same thing, applying inversion in the x-variable for the Poisson kernel. Since $x = \frac{a^2}{|x^*|^2}x^*$, applying inversion to the Poisson kernel produces

$$\frac{a}{|x^*|}\frac{a^2 - \frac{a^4}{|x^*|^4}|x^*|^2}{4\pi a\left|\frac{a^2}{|x^*|^2}x^* - \xi\right|^3} = \frac{a^2(|x^*|^2 - a^2)|x^*|^3}{4\pi\left|a^2x^* - |x^*|^2\xi\right|^3}.$$

Dropping the stars and writing $P^*(x,\xi)$ for the Poisson kernel on this exterior domain, after simplifying this becomes

$$P^*(x,\xi) = \frac{(|x|^2 - a^2)}{4\pi a\left[a^2 - 2x\cdot\xi + |x|^2\right]^{3/2}}.$$

Keep in mind that $|x| > a$ and $|\xi| = a$ in this formula. Hence, the solution to the exterior Dirichlet problem is given by

$$u(x) = \frac{|x|^2 - a^2}{4\pi a}\int_{|\xi|=a}\frac{f(\xi)}{\left[a^2 - 2x\cdot\xi + |x|^2\right]^{3/2}}\,dS(\xi).$$

The Poisson integral formula is not particularly useful for computing solutions of a Dirichlet problem even for boundary data of very simple form. This difficulty will be rectified in the next section, where Fourier methods are applied. Nonetheless, the Poisson formula can be used to obtain other properties of harmonic functions. Here is one example.

Theorem 7.23 (Liouville's Theorem) If $u(x)$ is a harmonic function in $\mathbb{R}^d$ and is bounded below or above, then $u(x)$ is a constant.

Proof Suppose that $u(x)$ is harmonic in $\mathbb{R}^d$ and is bounded below, say $u(x) \geq m$ for all $x \in \mathbb{R}^d$. We may suppose that $m \geq 0$ (if not, simply add a constant function to u so that it is). We will show that $u(x) = u(O)$ for all $x \in \mathbb{R}^d$. Take $x \in \mathbb{R}^d$ and let a be a number greater than $|x|$. Then (7.4.11) can be used in the form

$$u(x) = \frac{a^{d-2}(a^2 - \rho^2)}{A_d(1)}\int_{S_1(O)}\frac{u(a\omega')}{\left[a^2 - 2a\rho(\omega\cdot\omega') + \rho^2\right]^{d/2}}\,d\omega',\ x = \rho\omega.$$

Since $-1 \leq \omega\cdot\omega' \leq 1$, it follows that $(a+\rho)^2 \leq a^2 - 2a\rho(\omega\cdot\omega') + \rho^2 \leq (a-\rho)^2$.

Hence, since $u \geq 0$,

$$\frac{a^{d-2}(a^2-\rho^2)}{(a-\rho)^d}\frac{1}{A_d(1)}\int_{S_1(O)} u(a\omega')\,d\omega' \leq u(x)$$
$$\leq \frac{a^{d-2}(a^2-\rho^2)}{(a+\rho)^d}\frac{1}{A_d(1)}\int_{S_1(O)} u(a\omega')\,d\omega'.$$

Both the integrals in this inequality equal $A_d(1)$ times the mean value of u on the unit sphere. Applying the mean value theorem for harmonic functions, we have

$$\frac{a^{d-2}(a^2-\rho^2)}{(a-\rho)^d}u(O) \leq u(\xi) \leq \frac{a^{d-2}(a^2-\rho^2)}{(a+\rho)^d}u(O).$$

Thus, letting $a \to \infty$ we obtain $u(O) \leq u(x) \leq u(O)$ and the conclusion follows. The proof of the result in the case where u is bounded above is left to the reader.

□

The Green's Function on a Half-Space

In the discussion of the Green's functions, we assumed that the underlying domain was bounded, although in the last example, we constructed the Poisson kernel on an exterior domain via inversion transformations (see also Exercise 7.31). In unbounded domains, the boundary can also be unbounded, which may lead to convergence difficulties. Nonetheless, simple reflection in the boundary of a half-space or half-plane leads quickly to the Green's function and a situation where convergence difficulties are not present due to the decay of the fundamental solution K.

Let $H^d = \{x \in \mathbb{R}^d | \, x_d > 0\}$. For $\xi \in H^d$, write $\xi = (\xi', \xi_d)$ where $\xi' \in \mathbb{R}^{d-1}$. The reflection of this point through the boundary $x_d = 0$ is the point $\xi^* = (\xi', -\xi_d)$. Furthermore, the function $x \to K(x - \xi^*)$ is harmonic in H^d as ξ^* is not in the upper half-space. We define

$$G(x, \xi) = K(x - \xi) - K(x - \xi^*).$$

All the properties of the Green's function are now immediate from this formula. In three dimensions, there is an easy electrostatic interpretation. A unit positive charge placed at ξ produces a free-space potential given by $-4\pi K(x-\xi)$. Placing a unit negative charge at ξ^* produces a potential of $4\pi K(x-\xi^*)$. The resultant potential of the two charges is the sum of these two, and along the plane $x_3 = 0$, the potential cancels out. We used a similar principle and interpretation in constructing the Green's function for a ball. Within the discipline of physics, this viewpoint is called the electrostatic method of images; the charge at ξ^* is a fictitious image charge (it lies outside the domain of the problem) with position chosen to create a resultant potential of zero on the boundary of the domain.

Example 7.14 To further illustrate the method of images, we construct the Green's function on the quarter-space $D = \{x \in \mathbb{R}^3 | x_2 > 0, x_3 > 0\}$. Let a positive unit charge be placed at $\xi \in D$, producing a potential of $-4\pi K(x-\xi)$. Reflecting this point through the x_1x_3−plane gives the point $\xi^1 = (\xi_1, -\xi_2, \xi_3)$. If a unit negative charge is placed here, the resulting free-space potential is $4\pi K(x-\xi^1)$, which is harmonic in D. The resultant potential of the two particles is $-4\pi K(x-\xi) + 4\pi K(x-\xi^1)$, which vanishes on the plane $x_2 = 0$. To this function we add the terms

$$4\pi K(x-\xi^*) - 4\pi K(x-\xi^{1*}),$$

corresponding to a unit negative charge at ξ^* and a unit positive charge at ξ^{1*}. We define $G(x,\xi)$ to be the negative of this function divided by 4π, i.e.,

$$G(x,\xi) = K(x-\xi) - K(x-\xi^1) + K(x-\xi^{1*}) - K(x-\xi^*).$$

It is easily verified that G satisfies the properties of a Green's function for the quarter-space D. Similar ideas will work in two dimensions, although the electrostatic interpretation disappears and the fundamental solution used is logarithmic.

Returning to the half-space domain, the Poisson kernel is given by

$$P(x,\xi) = -\frac{\partial G}{\partial \xi_d}(x,\xi), \quad \xi \in \partial H^d.$$

Computing the derivative (keep in mind $x_d = 0$) we obtain

$$P(x,\xi) = \frac{2x_d}{A_d(1)\,[|x' - \xi'|^2 + x_d^2]^{d/2}}, \quad \xi \in \partial H^d. \tag{7.4.16}$$

Given a function f defined on ∂H^d, the Poisson integral takes the form

$$u(x) = \frac{2x_d}{A_d(1)} \int_{\mathbb{R}^{d-1}} \frac{f(\xi')}{[|x' - \xi'|^2 + x_d^2]^{n/2}}\, dV(\xi'), \tag{7.4.17}$$

and it is clear that convergence of the integral is in question. However, for fixed $x \in H^d$, $P(x,\xi) \le C\,|\xi'|^{-d}$, so if f is a bounded function, sufficient decay of the integrand for large x' is present to guarantee convergence of the integral.

Example 7.15 Take $d = 2$. Then the Poisson integral is

$$u(x) = \frac{x_2}{\pi} \int_{\mathbb{R}} \frac{f(\xi_1)}{[(x_1 - \xi_1)^2 + x_2^2]} \, d\xi_1. \tag{7.4.18}$$

In this case, if f is a bounded function, say$|f(\xi_1)| \leq M$ for all $\xi_1 \in \mathbb{R}$, then

$$|u(x)| \leq \frac{M}{\pi} \int_{\mathbb{R}} \frac{x_2}{[\xi_1^2 + x_2^2]} \, d\xi_1 = M.$$

This illustrates the convergence of the integral defining the Poisson integral. In three dimensions, the Poisson integral takes the form

$$\begin{aligned} u(x) &= \frac{x_3}{2\pi} \int_{\mathbb{R}^2} \frac{f(\xi')}{[|x' - \xi'|^2 + x_3^2]^{3/2}} \, dV(\xi') \\ &= \frac{x_3}{2\pi} \int_{\mathbb{R}} \int_{\mathbb{R}} \frac{f(\xi_1, \xi_2) \, d\xi_1 \, d\xi_2}{[(x_1 - \xi_1)^2 + (x_2 - \xi_2)^2 + x_3^2]^{3/2}}. \end{aligned}$$

The analog of Theorem 7.21 for a half-space is the following result.

Theorem 7.24 Let $f \in C(\mathbb{R}^{d-1})$ be bounded. Define $u(\xi)$ on $\overline{H^d}$ by (7.4.17) for $x \in H^d$ and $u(x) = f(x)$ for $x \in \partial H^d$. Then u is harmonic in H^d and is continuous on $\overline{H^d}$.

Theorem 7.24 effectively provides the solution to the Dirichlet problem for a half-space through the Poisson integral. We just needed the additional hypothesis that the boundary data is bounded to ensure convergence of the integral. The proof is discussed at the end of this section.

Example 7.16 Here we take $d = 2$ and consider the boundary data

$$f(x_1, 0) = \begin{cases} -1, & -\infty < x_1 \leq -\pi \\ \frac{x_1}{\pi}, & -\pi < x_1 \leq \pi \\ 1, & \pi < x_1 < \infty. \end{cases}$$

It is left as an exercise in integration to compute the solution of the Dirichlet problem in the upper half-plane as

$$\begin{aligned} u(x) = \frac{1}{\pi}\Big(1 + \frac{x_1}{\pi}\Big)\tan^{-1}\Big(\frac{\pi + x_1}{x_2}\Big) - \frac{1}{\pi}\Big(1 - \frac{x_1}{\pi}\Big)\tan^{-1}\Big(\frac{\pi - x_1}{x_2}\Big) \\ + \frac{x_2}{2\pi^2} \log\left[\frac{(\pi - x_1)^2 + x_2^2}{(\pi + x_1)^2 + x_2^2}\right]. \end{aligned}$$

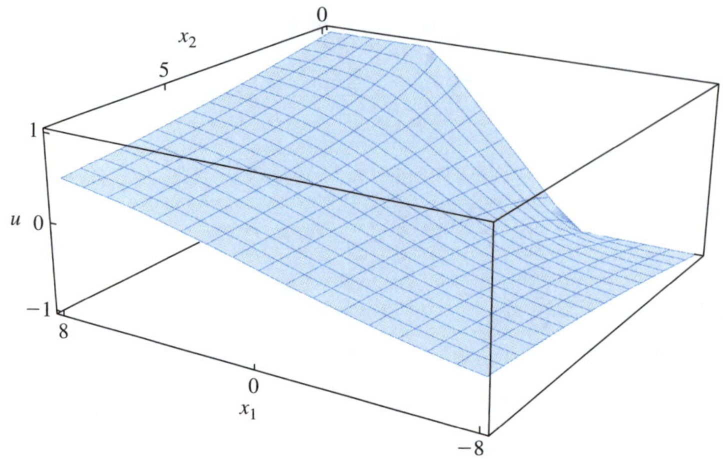

Figure 7.4 Harmonic Function on a Half-Plane

It is easily checked that $\lim_{x_2 \to 0^+} u(x) = f(x_1, 0)$. For example, if $x_1 > \pi$, then

$$\lim_{x_2 \to 0^+} u(x) = \frac{1}{\pi}\left(1 + \frac{x_1}{\pi}\right)\frac{\pi}{2} - \frac{1}{\pi}\left(1 - \frac{x_1}{\pi}\right)\left(-\frac{\pi}{2}\right) = 1.$$

Similarly for the other two intervals. A plot of this solution is given in Figure 7.4, illustrating Theorem 7.2.4.

Analogous to the case of the disk, an extension of Theorem 7.24 in two dimensions to boundary data that are piecewise continuous is sometimes useful.

Theorem 7.25 Let $f \in \mathrm{PC}(\mathbb{R})$ be bounded and let

$$f^*(x) = \begin{cases} f(x), & \text{if } f \text{ is continous at } x \\ \frac{1}{2}[f(x+0) + f(x-0)], & \text{otherwise.} \end{cases}$$

Define $u(x)$ on H^2 by the Poisson integral formula (7.4.18) and set $u(x) = f^*(x')$ if $x \in \partial H^2$. Then u is harmonic in the upper half-plane and the following limit relation holds:

$$\lim_{x_2 \to 0^+} u(x) = f^*(x').$$

An example illustrating Theorem 7.25 is found in the exercises.

Discussion of the Proofs of Theorems 7.21 and 7.24

To conclude the proof of Theorem 7.21, we must demonstrate the limit relation (7.4.12). Let x^0 be a point on the boundary sphere. Since $\int_{|x|=a} P(x,\xi)dS(\xi) = 1$, then

$$u(x) - f(x^0) = \int_{|\xi|=a} P(x,\xi)[f(\xi) - f(x^0)]dS(\xi).$$

The idea is to split the integral into two pieces, the first integrating over the part of the sphere sufficiently near x^0 where continuity of f will be applied and the second on the rest of the sphere where the function f is bounded and properties of the Poisson kernel will apply. Specifically, let $\varepsilon > 0$ and choose $\delta > 0$ such that if $\xi \in B_\delta(x^0) \cap S_a(O)$, then $|f(\xi) - f(x^0)| < \varepsilon/2$. Let $C_\delta = B_\delta(x^0) \cap S_a(O)$ and let C_δ^* be the complement in the sphere, i.e., $C_\delta^* = \{\xi : |\xi - x^0| \geq \delta\} \cap S_a(O)$. We split the integral as follows:

$$u(x) - u(x^0) = \left\{\int_{C_\delta} + \int_{C_\delta^*}\right\} P(x,\xi)[f(\xi) - f(x^0)]dS(\xi) = I_\delta(x) + I_\delta^*(x).$$

For the first piece,

$$|I_\delta| \leq \int_{C_\delta} P(x,\xi)|f(\xi) - f(x^0)|dS(\xi) < \frac{\varepsilon}{2}\int_{C_\delta} P(x,\xi)dS(\xi)$$
$$\leq \frac{\varepsilon}{2}\int_{|x|=a} P(x,\xi)dS(\xi) = \frac{\varepsilon}{2},$$

where the fact that $\int_{|x|=a} P(x,\xi)dS(\xi) = 1$ has been used in the last step. Since f is continuous on the boundary sphere, there is a constant M such that $|f(\xi)| \leq M$ for all $|\xi| = a$. Let x be such that $|x| < a$ and $|x - x^0| \leq \delta/2$. Then if $\xi \in C_\delta^*$, $|x - \xi| \geq \delta/2$ and hence

$$P(x,\xi) \leq \left(\frac{2}{\delta}\right)^d \frac{1}{A_d(1)a}(a^2 - |x|^2).$$

We now have the estimate

$$|I_\delta^*| \leq 2M\left(\frac{2}{\delta}\right)^d \frac{1}{A_d(1)a}(a^2 - |x|^2),$$

and hence choosing $|x|$ sufficiently close to a will force $|I_\delta^*| < \varepsilon/2$. Putting the estimates together, it follows that $|u(x) - f(x^0)| < \varepsilon$, which is the desired limit relation.

The proof of Theorem 7.24 is strikingly similar to that of Theorem 1.2 given at the end of Chapter 1. The heuristic reason for this similarity can be understood by comparing the plots of the Gaussian heat kernel, Figure 1.5, with that of the Poisson kernel, Figure 7.3. Details are left to the interested reader.

Exercises

7.24 Let $\xi^1$and ξ^2 be distinct points of a bounded domain D and let $\varepsilon > 0$ be chosen so that $\overline{B_\varepsilon(\xi^k)} \subset D$, $k = 1, 2$. Let D_ε be the domain obtained by deleting both of these balls from D. Let $G(x, \xi)$ be the Green's function on D and set $u_1(x) = G(x, \xi^1)$ and $u_2(x) = G(x, \xi^2)$ (both are harmonic in D). Write Green's second identity on the domain D_ε for the functions u_1 and u_2 and let $\varepsilon \to 0$. Conclude the symmetry property for the Green's function: $G(\xi^2, \xi^1) = G(\xi^1, \xi^2)$.

7.25 Let D be a domain and $G(x, \xi)$ be the Green's function on D. Show that $G(x, \xi) \geq 0$ for $x \neq \xi$. (Hint: Apply the maximum principle to the function $u(x) = G(x, \xi)$ on $D - \overline{B_\varepsilon(\xi)}$.

7.26 Let $G(x, \xi)$ be the Green's function on a domain D. Show that $\frac{dG}{dn_x} \leq 0$ on the boundary of D.

7.27

a. On the interval $[0, 1]$, solve the ODE $u'' + \frac{2}{\rho}u' = 1$ subject to the conditions that $u(0)$ is finite and $u(1) = 0$. Interpreting ρ as radial distance in three dimensions, $\rho = |x|$, show that $u(x) = u(\rho)$ gives a strong solution to the problem $\nabla^2 u = 1$ in $B_1(O)$ with $u = 0$ on the boundary.

b. On the interval $[0, 1]$, solve the ODE $u'' + \frac{2}{\rho}u' = \rho$ subject to the same conditions as in part (a). With the same interpretation as in part (a), show that $u(x) = u(\rho)$ is not a C^2 function on $B_1(O)$. ($u(x)$ in this case is a weak solution of Poisson's equation, but not a strong solution!)

7.28 Let D be a domain and let $f \in C(\partial D)$. Show that the function given by (7.4.6) defines a harmonic function in D. HINT: Show that u is continuous on D and compute the spherical mean of u over spheres centered at ξ contained in D. Recall Theorem 7.7.

7.29 Derive the formula for the Poisson kernel (7.4.9) in two dimensions.

7.30 Derive (7.4.14).

7.31 Show that the Poisson kernel for the exterior Dirichlet problem for a disk of radius a in two dimensions is given by

$$P^*(x, \xi) = \frac{|x|^2 - a^2}{2\pi a\left[a^2 - 2x \cdot \xi + |x|^2\right]}.$$

7.32 In $\mathbb{R}^2$, let $D_{\pi/k} = \{(r,\theta)\,|\,r > 0,\, 0 < \theta < \pi/k\}$ (a wedge) where k is a positive integer. Compute the Green's function on the following.

a. $D_{\pi/2}$

b. $D_{\pi/4}$

7.33 Find the Green's function for the upper hemisphere $\{x \in \mathbb{R}^3 |\, x \in B_a(O),\ x_3 > 0\}$.

7.34 Find the Green's function for the domain $D = \{x \in \mathbb{R}^3 |\, 0 < x_3 < 1\}$ (the region between two parallel planes). HINT: An infinite sequence of charges need to be chosen so that at the resultant potential on each plane $x_3 = k$, $k \in \mathbb{Z}$ is zero.

7.35 Show that if $u(x)$ is harmonic in $\mathbb{R}^d$ and bounded above, then u is constant. HINT: The function $-u(\xi)$ is harmonic and bounded below.

7.36 Let $f(x_1) = 1$ for $|x_1| < \pi$ and zero elsewhere. Find the solution to the Dirichlet problem in the upper half-plane with f as boundary data. Produce a surface plot of the solution and use this example to illustrate Theorem 7.25.

7.5 Fourier Methods and Laplace's Equation

In this section we apply separation of variables and Fourier methods to Laplace's equation. Along the way we will unfold the connection with solution formulas obtained via the Green's functions.

Problems with Circular Symmetry in the Plane

The Disk We begin with a problem we have already solved—the Dirichlet problem for Laplace's equation in a disk:

$$\begin{aligned} \text{PDE: } \nabla^2 u &= 0, \quad x \in D_a \\ \text{BC: } u|_{\partial D_a} &= f, \end{aligned}$$

where $D_a = \{x \in \mathbb{R}^2 \,|\, |x| < a\}$ and f is a function defined on the boundary of this disk. From the geometric perspective in Example 3.4 and from the perspective of separation of variables in Example 5.1, we know that the separated solutions of Laplace's equation in the plane have the form $r^{|n|}e^{in\theta}$ (now is a good time to revisit these examples). Therefore, a solution to the BVP should take the form

$$u(r,\theta) = \sum_{n=-\infty}^{\infty} c_n r^{|n|} e^{in\theta},$$

for appropriate choice of coefficients $\{c_n\}$. The ideas of Chapter 5 tell us the natural choice for these: $c_n a^{|n|} = \widehat{f}(n)$, where $\widehat{f}(n)$ is the nth complex Fourier coefficient of the boundary data f. Thus the conjectured solution to the BVP is

$$u(r,\theta) = \sum_{-\infty}^{\infty} \widehat{f}(n)\Big(\frac{r}{a}\Big)^{|n|} e^{in\theta}. \tag{7.5.1}$$

This formula was also found in Example 5.1 where $a = 1$ and is called the Fourier series representation of the solution.

To justify that this is indeed the solution to the problem, we need to show two things: (1) the series defines a solution to Laplace's equation, i.e., can be differentiated term by term; (2) the boundary condition is satisfied, i.e., $u(a,\theta) = f(\theta)$. For both, the results on Fourier series in Chapter 5 come into play. We begin with (2) and make the assumption that $f \in C(\mathbf{T})$ with a piecewise continuous derivative. Under these conditions the Fourier series of f converges uniformly to f (Theorem 5.21), and this just tells us that $u(a,\theta) = f(\theta)$ as desired. The assumption on the boundary data can be weakened to $f \in \mathrm{PC}[-\pi,\pi]$ with a piecewise continuous derivative, in which case we get the conclusion that $u(a,\theta) = f^*(\theta)$, where f^* is given by (7.4.15). This follows from the pointwise convergence theorem for Fourier series, Theorem 5.19.

To verify that $u(r,\theta)$ as defined is a solution to Laplace's equation in the disk, we must show that the series can be differentiated term by term twice in each variable. The conclusion (1) then follows as the general term of the series defining u is a solution to Laplace's equation. The details are carried out as an application of results in Appendix A. We need to show that the series defining u and the series of term-by-term derivatives are uniformly convergent in any subdisk D_{r_0}, where $r_0 < a$. Here is a sketch of the details. Under either of the preceding hypotheses on the boundary data f, we know that the sequence of Fourier coefficients is bounded, say $|\widehat{f}(n)| \le M$ for all n. Then the general term of the series for u is bounded by $M\left(\frac{r_0}{a}\right)^{|n|}$. Since $\sum\left(\frac{r_0}{a}\right)^{|n|}$ converges, it follows by the Weierstrass M-test that the series defining u is uniformly convergent in D_{r_0}. We now turn to the series of term-by-term derivatives in the θ-variable. Differentiating the series for u m times with respect to θ yields the series

$$\sum \widehat{f}(n)\left(\frac{r}{a}\right)^{|n|}(in)^m e^{in\theta}.$$

The general term of this series is bounded by $M\left(\frac{r_0}{a}\right)^{|n|}|n|^m$. Since $\sum\left(\frac{r_0}{a}\right)^{|n|}|n|^m$ converges, the Weierstrass M-test again applies with conclusion that the series of term-by-term derivatives is uniformly convergent on D_{r_0}. By Theorem A.2, $u(r,\theta)$ as defined by the series (7.5.1) is differentiable in the θ variable, in fact infinitely differentiable since m could be any positive integer in this argument. Furthermore, the mth derivative is given by differentiating the series m times with respect to θ. By a similar argument, the function $u(r,\theta)$ is infinitely differentiable in the r-variable, where the differentiation is carried out term by term. We summarize our conclusions as follows.

Proposition 7.26 Under the hypothesis that the boundary data $f \in \mathrm{PC}[-\pi, \pi]$ with a piecewise continuous derivative, the series (7.5.1) gives the solution to the Dirichlet problem for Laplace's equation in the disk D_a where the boundary condition is satisfied in the sense that $u(a, \theta) = f^*(\theta)$.

The Fourier series solution to the Dirichlet problem can also be derived from formula (7.4.13), rewritten here for convenience

$$u(r, \theta) = \int_{-\pi}^{\pi} f(\theta') P_r(\theta - \theta') d\theta', \tag{7.5.2}$$

where the Poisson kernel is given by

$$P_r(\theta) = \frac{1}{2\pi} \frac{a^2 - r^2}{[a^2 - 2ar\cos\theta + r^2]}.$$

Indeed, writing $\cos\theta = \frac{1}{2}[e^{i\theta} + e^{-i\theta}]$, factoring the denominator in the Poisson kernel, and using partial fractions gives

$$\begin{aligned} P_r(\theta) &= \frac{1}{2\pi} \frac{a^2 - r^2}{(a - re^{i\theta})(a - re^{-i\theta})} \\ &= \frac{1}{2\pi} \left(\frac{a}{a - re^{i\theta}} + \frac{re^{-i\theta}}{a - re^{-i\theta}} \right) \\ &= \frac{1}{2\pi} \left(\frac{1}{1 - \frac{r}{a} e^{i\theta}} + \frac{\frac{r}{a} e^{-i\theta}}{1 - \frac{r}{a} e^{-i\theta}} \right). \end{aligned}$$

Both terms on the right-hand side can be rewritten using the geometric sum formula as follows:

$$\begin{aligned} P_r(\theta) &= \frac{1}{2\pi} \left[\sum_{n=0}^{\infty} \left(\frac{r}{a}\right)^n e^{in\theta} + \sum_{n=1}^{\infty} \left(\frac{r}{a}\right)^n e^{-in\theta} \right] \\ &= \frac{1}{2\pi} \sum_{n=-\infty}^{\infty} \left(\frac{r}{a}\right)^{|n|} e^{in\theta}. \end{aligned}$$

Since $r < a$, this series is uniformly convergent for $\theta \in [-\pi, \pi]$. Substituting this formula for the Poisson kernel into (7.5.2) and integrating term by term (justified by uniform convergence) leads to the Fourier series representation:

$$u(r, \theta) = \frac{1}{2\pi} \int_{-\pi}^{\pi} f(\theta') \sum_{n=-\infty}^{\infty} \left(\frac{r}{a}\right)^{|n|} e^{in(\theta - \theta')}\, d\theta' = \sum_{-\infty}^{\infty} \widehat{f}(n) \left(\frac{r}{a}\right)^{|n|} e^{in\theta}.$$

It is useful to compare and contrast the two forms of the solution to the Dirichlet problem. Formula (7.5.2) provides a continuous superposition (through integration) of the boundary data with the Poisson kernel (which is a solution to Laplace's equation in the disk). On the other hand, (7.5.1) is a discrete superposition (through the sum) of the solutions to Laplace's equation

$$(r,\theta) \to \left(\frac{r}{a}\right)^{|m|} e^{im\theta}$$

with coefficients given by the Fourier coefficients of the boundary data f. As mentioned, the former can be difficult to compute even for functions of simple type, whereas Fourier coefficients are relatively easy to compute and/or easily approximated numerically. The following examples illustrate this.

Example 7.17 Let $a = 1$ and assume boundary data of the form $f(\theta) = 1 - \frac{|\theta|}{\pi}$. This function is continuous on $[-\pi, \pi]$ and since $f(-\pi) = f(\pi)$, it is continuous when thought of as a function on the boundary of D_1. We compute the Fourier coefficients as

$$\begin{aligned} \widehat{f}(m) &= \frac{1}{2\pi}\int_{-\pi}^{\pi}(1 - \frac{|\theta|}{\pi})e^{-im\theta}\,d\theta \\ &= \frac{1}{\pi}\int_{0}^{\pi}(1 - \frac{\theta}{\pi})\cos m\theta\,d\theta, \end{aligned}$$

where we used Euler's formula ($e^{is} = \cos s + i\sin s$) and the fact that f is even. The last integral may be computed for $m \neq 0$ using integration by parts:

$$\begin{aligned} \widehat{f}(m) &= \frac{1}{\pi}\left[\left(1 - \frac{\theta}{\pi}\right)\frac{\sin m\theta}{m}\bigg|_0^{\pi} + \frac{1}{\pi m}\int_0^{\pi}\sin m\theta\,d\theta\right] \\ &= \frac{1-(-1)^m}{\pi^2 m^2}, \end{aligned}$$

where we used the fact that $\cos m\pi = (-1)^m$. By an easy integration $\widehat{f}(0) = \frac{1}{2}$. Thus the Fourier series of this function is given by

$$1 - \frac{|\theta|}{\pi} \sim \frac{1}{2} + \sum_{m\neq 0}\frac{1-(-1)^m}{\pi^2 m^2}e^{im\theta} = \frac{1}{2} + \sum_{m=1}^{\infty}2\frac{1-(-1)^m}{\pi^2 m^2}\cos m\theta,$$

and the Fourier series solution to the Dirichlet problem is

$$u(r,\theta) = \frac{1}{2} + \sum_{m\neq 0}\frac{1-(-1)^m}{\pi^2 m^2}r^{|m|}e^{im\theta} = \frac{1}{2} + \sum_{m=1}^{\infty}2\frac{1-(-1)^m}{\pi^2 m^2}r^m\cos m\theta.$$

To approximate the solution to the Dirichlet problem, we form the partial sums of the series for u:

$$u_N(r,\theta) = \frac{1}{2} + \sum_{m=1}^{N}2\frac{1-(-1)^m}{\pi^2 m^2}r^m\cos m\theta.$$

Figure 7.5 gives a plot of the first eight nonzero terms, indicating the rapid convergence of the series for u. We can estimate the error in the approximation as follows:

$$\begin{aligned} |u(r,\theta) - u_N(r,\theta)| &\leq \sum_{m=N+1}^{\infty} \left| 2\frac{1-(-1)^m}{\pi^2 m^2} r^m \cos m\theta \right| \\ &\leq \sum_{m=N+1}^{\infty} r^m \\ &= \frac{r^{N+1}}{1-r}. \end{aligned}$$

Here we used the fact that $2\frac{1-(-1)^m}{\pi^2 m^2} \leq 1$ for all m in the second step and applied the geometric sum formula in the last. Let $r_0 < 1$, then $\frac{r^{N+1}}{1-r} \leq \frac{r_0^{N+1}}{1-r_0}$ for $r \leq r_0$. Putting this together with the above, we have the estimate

$$|u(r,\theta) - u_N(r,\theta)| \leq \frac{r_0^{N+1}}{1-r_0}, \text{ provided } r \leq r_0.$$

In other words, $u(r,\theta)$ is very well approximated by $u_N(r,\theta)$ inside the unit disk. Just as an example, take $r_0 = 0.6$ and $N = 8$; the right-hand side is 0.0419.

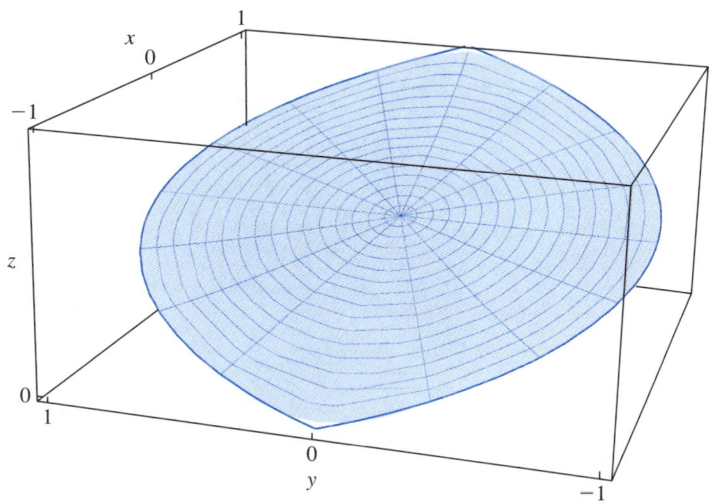

Figure 7.5 Solution Approximation for Example 7.17

Example 7.18 In this example we again take $a = 1$ and piecewise continuous boundary data of form

$$f(\theta) = \begin{cases} -1, & -\pi < \theta < 0 \\ 1, & 0 < \theta < \pi. \end{cases}$$

The Fourier coefficients take the form

$$\widehat{f}(n) = \frac{1}{2\pi}\int_{-\pi}^{\pi} f(\theta)e^{-in\theta}\,d\theta = \frac{-i}{\pi}\int_{0}^{\pi} \sin n\theta\,d\theta = -i\frac{1-(-1)^n}{n\pi}, \quad |n| \geq 1,$$

and because f is an odd function, $\widehat{f}(0) = 0$. The solution formula takes the form

$$\begin{aligned} u(r,\theta) &= -i\sum_{|n|=0}^{\infty} \frac{1-(-1)^n}{n\pi} r^{|n|} e^{in\theta} \\ &= \sum_{n=0}^{\infty} \frac{4}{(2n+1)\pi}\, r^{2n+1} \sin(2n+1)\theta. \end{aligned}$$

We obtain the second line by applying Euler's identity to the first (or by computing the real form of the solution). The solution is plotted in Figure 7.6 using the first eight nonzero terms.

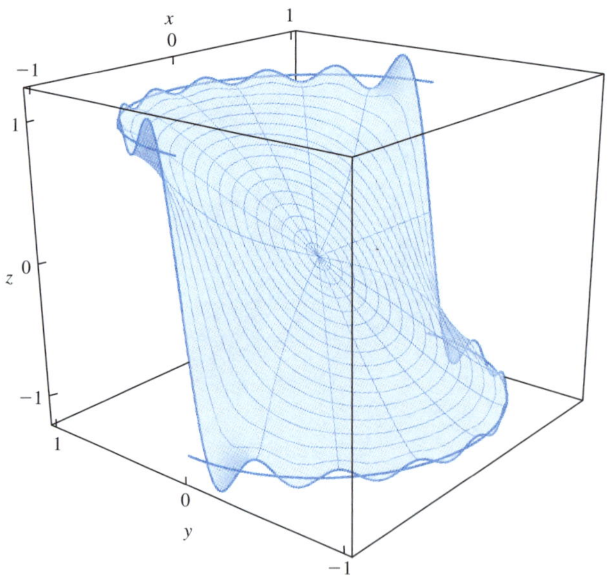

Figure 7.6 Plot for Example 7.18

Remark 7.27 The exterior problem for Laplace's equation can also be solved using the method of separation of variables. The resulting Fourier series representation of the solution is

$$u(r,\theta) = \sum_{n=-\infty}^{\infty} \widehat{f}(n) \left(\frac{a}{r}\right)^{|n|} e^{in\theta}. \tag{7.5.3}$$

The assumption on the boundary data is the same for the disk problem, i.e., $f \in \mathrm{PC}[-\pi,\pi]$ with a piecewise continuous derivative. (We leave the details of the derivation and justification of the solution to the reader.) Further, the Dirichlet boundary condition can be replaced by Neumann or Robin conditions for either the interior or exterior disk problems; these considerations are left as exercises.

The Annulus Here we consider Laplace's equation in the annular region $D_{a,b} = \{a < |x| < b\}$ with Dirichlet boundary conditions written in polar form:

$$u|_{r=a} = g(\theta) \text{ and } u|_{r=b} = f(\theta).$$

Since the origin is not in the annulus, we need to consider the full set of separated solutions for Laplace's equation. The solution would take the form

$$u(r,\theta) = (c_0 + d_0 \log r) + \sum_{|n|\neq 0} \left(c_n r^{|n|} + d_n r^{-|n|}\right) e^{in\theta}.$$

The boundary condition $u(a,\theta) = g(\theta)$ implies the equations

$$c_0 + d_0 \log a = \widehat{g}(0)$$
$$c_n a^{|n|} + d_n a^{-|n|} = \widehat{g}(n).$$

Similarly, the boundary condition $u(b,\theta) = f(\theta)$ implies

$$c_0 + d_0 \log b = \widehat{f}(0)$$
$$c_n b^{|n|} + d_n b^{-|n|} = \widehat{f}(n).$$

This system of linear equations can be solved for the unknown constants $\{c_n\}$ and $\{d_n\}$. The end result leads to the following formula for the Fourier series representation of the solution to the Dirichlet problem on the annulus:

$$\begin{aligned} u(r,\theta) = \frac{\widehat{g}(0)\log b - \widehat{f}(0)\log a}{\log(b/a)} &+ \frac{\widehat{f}(0) - \widehat{g}(0)}{\log(b/a)} \log r \\ &+ \sum_{n\neq 0} \left(\frac{b^{|n|}\widehat{f}(n) - a^{|n|}\widehat{g}(n)}{b^{2|n|} - a^{2|n|}} \right) r^{|n|} e^{in\theta} \\ &+ \sum_{n\neq 0} (ab)^{|n|} \left(\frac{b^{|n|}\widehat{g}(n) - a^{|n|}\widehat{f}(n)}{b^{2|n|} - a^{|n|}} \right) r^{-|n|} e^{in\theta}. \end{aligned}$$

Details are left as an exercise. The solution can be justified under the assumption that the boundary data f and g are piecewise continuous with a piecewise continuous derivative.

The Wedge Laplace's equation on a wedge domain often occurs in applications. The wedge domain in polar form is described as $D_{a,\alpha} = \{0 < r < a,\ 0 < \theta < \alpha\}$, where $0 < \alpha < 2\pi$. We assume Dirichlet boundary conditions of the form

$$\begin{cases} u(r,0) = 0 \\ u(r,\alpha) = 0 \end{cases}, \quad 0 \le r < a \quad \text{and} \quad u(a,\theta) = f(\theta).$$

Separation of variables for Laplace's equation can be applied to this problem as we did in detail in Example 5.1. On the other hand, simple geometry will lead to the correct form for the separated solutions. To handle the circular boundary, we will use Fourier series on the interval $[0,\alpha]$. We build these from the functions $\cos\frac{n\pi}{\alpha}\theta$ and $\sin\frac{n\pi}{\alpha}\theta$ (use of the sine and cosine functions rather than complex exponentials is convenient for the boundary condition at $\theta = 0$). Because of the homogeneous Dirichlet condition at $\theta = 0$, the cosine terms will not appear in the solution. Hence, we need separated solutions of Laplace's equation of the form $u(r,\theta) = R(r)\sin\frac{n\pi}{\alpha}\theta$. At this point, the computations are the same as for the disk: $R(r) = r^{\frac{n\pi}{\alpha}}$ or $R(r) = r^{-\frac{n\pi}{\alpha}}$. The negative powers are ignored as $u = 0$ at the origin. Following the pattern for the disk and the annulus, the solution to the wedge problem is

$$\begin{aligned} u(r,\theta) &= \sum_{n=1}^{\infty} \widehat{f}_s(n)\left(\frac{r}{a}\right)^n \sin\frac{n\pi}{\alpha}\theta \\ \widehat{f}_s(n) &= \frac{2}{\alpha}\int_0^{\alpha} f(\theta)\sin\frac{n\pi}{\alpha}\theta\, d\theta. \end{aligned}$$

We justify this solution formula for the wedge problem following the pattern used for the disk.

Remark 7.28 The Dirichlet problem on a wedge domain can be solved using Green's function methods provided the angle α has form $\frac{\pi}{m}$ for some integer m (see, e.g., Exercise 7.32). However, the solution via Fourier methods is far simpler as well as easily computed for explicit boundary data.

Rectangular Domains

Separation of variables is highly useful in rectangular domains where the method of images fails, as we illustrate in this subsection. We also demonstrate the use of the Fourier method in computing the Green's function on rectangular regions.

Let $R_{a,b} = \{(x,y) : 0 < x < a,\ 0 < y < b\}$ be a rectangle in the plane. The problem of interest here concerns solving Laplace's equation in $R_{a,b}$ with Dirichlet boundary conditions. Modifications of the method to include Neumann, Robin, or mixtures of boundary conditions follow the same pattern.

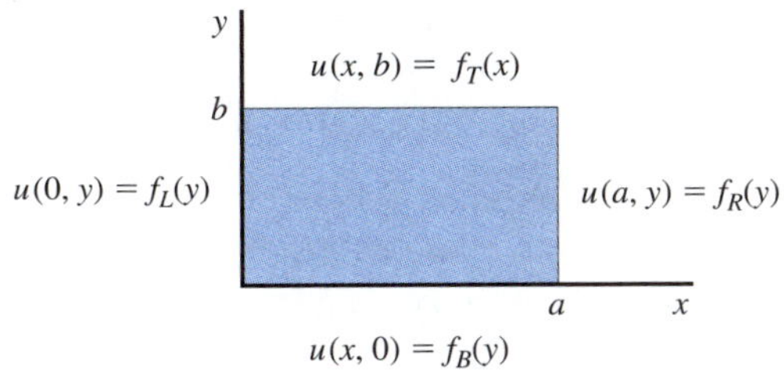

Figure 7.7 Rectangular Region $R_{a,b}$

The formulation of the boundary conditions is easiest with the aid of a picture; see Figure 7.7. On each side of the rectangle, the unknown function is specified as labeled. The natural approach to finding the harmonic function in $R_{a,b}$ satisfying these boundary conditions is to break the problem into four subproblems in which the boundary conditions on three of the sides are replaced by homogeneous boundary conditions. Separation of variables applies to each of the subproblems. After solving each of the subproblems, we add their solutions to obtain the solution to the full problem, as depicted in the figure. More explicitly, the subproblems have boundary conditions of the form

$$\text{B:}\begin{cases} u(0,y) = u(a,y) = 0 \\ u(x,0) = f_B(x),\ u(x,b) = 0, \end{cases} \qquad \text{R:}\begin{cases} u(x,0) = u(x,b) = 0 \\ u(0,y) = 0,\ u(a,y) = f_R(y), \end{cases}$$

and similarly for the other two sides. Subscripting the solution to each subproblem in the same fashion as the boundary data, the solution to the Dirichlet problem corresponding to Figure 7.7 would then be

$$u(x,y) = u_B(x,y) + u_R(x,y) + u_T(x,y) + u_L(x,y).$$

To illustrate the application of separation of variables to the subproblems, consider the following problem corresponding to the subproblem of replacing f_B by f and zeroing out the other boundary conditions.

$$\begin{aligned} &\text{PDE: } \nabla^2 u = 0, \quad (x,y) \in R_{a,b} \\ &\text{BC: } \begin{cases} u(0,y) = u(a,y) = 0, & 0 < y < b \\ u(x,0) = f(x),\ u(x,b) = 0, & 0 < x < a \end{cases} \end{aligned} \tag{7.5.4}$$

Assume that the boundary function $f \in \text{PC}[0,a]$ with a piecewise continuous derivative. We seek separated solutions $u(x,y) = X(x)Y(y)$ that satisfy the three homogeneous boundary conditions. Substituting into Laplace's equation, introducing a separation constant $-\mu$, and invoking the boundary conditions leads to the following ODE boundary value problems.

$$\begin{gathered} X'' + \mu X = 0, \quad 0 < x < a \\ X(0) = X(a) = 0 \\ Y'' - \mu Y = 0, \quad 0 < y < b \\ Y(b) = 0 \end{gathered}$$

The X BVP has eigenvalues $\mu = \left(\frac{n\pi}{a}\right)^2$, $n = 1, 2, \ldots$ with corresponding eigenfunctions $X_n(x) = \sin \frac{n\pi}{a}x$. With these values of μ, the general solution of the ODE for Y can be expressed in terms of hyperbolic sine and cosine. To simplify the application of the boundary condition at $y = b$, we express this general solution in the form

$$Y(y) = c_1 \cosh \frac{n\pi}{a}(b - y) + c_2 \sinh \frac{n\pi}{a}(b - y).$$

The boundary condition then implies $c_1 = 0$ and hence the separated solutions we need take the form

$$u_n(x, y) = \sinh \frac{n\pi}{a}(b - y) \sin \frac{n\pi}{a}x, \quad n = 1, 2, \ldots.$$

Applying superposition, we seek a solution to the BVP (7.5.4) of the form

$$u(x, y) = \sum_{n=1}^{\infty} c_n \sinh \frac{n\pi}{a}(b - y) \sin nx.$$

We conclude that $c_n \sinh \frac{n\pi b}{a} = \widehat{f}_s(n)$, the latter being the Fourier sine coefficients of f on the interval $[0, a]$, and the conjectured solution to the boundary value problem is

$$u(x, y) = \sum_{n=1}^{\infty} \widehat{f}_s(n) \frac{\sinh \frac{n\pi}{a}(b - y)}{\sinh \frac{n\pi b}{a}} \sin \frac{n\pi}{a}x. \tag{7.5.5}$$

Setting $y = b$, the right-hand side of this series becomes the Fourier sine series of the boundary data f and hence converges by the pointwise convergence theorem for Fourier series. As the separated solutions satisfy the other three boundary conditions, we conclude that our conjectured solution indeed satisfies the boundary conditions (7.5.4).

The second thing that we must show is that the series defining u can be differentiated term by term twice in each variable. As in the case of the disk, this is an application of the Weierstrass M-test in showing that the series for u and those for the term-by-term derivatives are uniformly convergent. Here we will be content with showing that the series for u is uniformly convergent on any subrectangle of form $R_{a,\delta} = \{(x, y) : 0 \le x \le a,\ \delta \le y \le b\}$ where δ is any number satisfying $0 < \delta < b$. Similar details demonstrate that the series of term-by-term derivatives in either the x- or y-variable is also uniformly convergent on such rectangles.

As in the case of the disk, $|\widehat{f}_s(n)| \leq M$, for all n and some constant M. Set $\eta = b - y$, then

$$\frac{\sinh \frac{n\pi}{a}\eta}{\sinh \frac{n\pi b}{a}} = e^{\frac{n\pi}{a}(\eta - b)} \frac{1 - e^{-\frac{2n\pi}{a}\eta}}{1 - e^{-\frac{2n\pi}{a}b}} \leq Ce^{\frac{n\pi}{a}(\eta - b)},$$

where $C = \frac{1}{1-e^{-\frac{2\pi b}{a}}}$. Thus the ratio of hyperbolic sine functions in (7.5.5) is bounded by $Ce^{-\frac{n\pi}{a}y} \leq Ce^{-\frac{n\pi}{a}\delta}$, the latter since $\delta \leq y \leq b$. It follows that the general term of the series defining u is bounded by a constant times $e^{-\frac{n\pi}{a}\delta}$ for all $(x, y) \in R_{a,\delta}$. Since $\sum e^{-\frac{n\pi}{a}\delta}$ converges, the Weierstrass M-test applies and the series for u is uniformly convergent on this subrectangle.

Example 7.19 As an explicit example, take $a = b = \pi$ and $f(x) = 1$. Then, computing the sine coefficients of f, the series (7.5.5) becomes

$$u(x, y) = \sum_{n=1}^{\infty} \frac{2(1 - (-1)^n)}{\pi n} \frac{\sinh n(\pi - y)}{\sinh n\pi} \sin nx.$$

Truncating this sum to the first ten nonzero terms, an approximation to the solution is given in Figure 7.8(a).

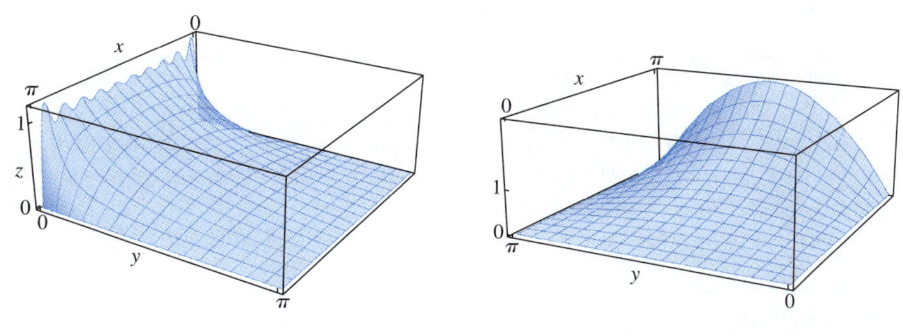

(a) Example 7.19 (b) Example 7.20

Figure 7.8 Solutions to Laplace's Equation on a Rectangle

Example 7.20 Again take $a = b = \pi$ and let $f_R(y) = g(y) = y(\pi - y)$ and $f_B = f_T = f_L = 0$. The separated solutions for this problem take the form

$$u_n(x, y) = \sinh nx \sin ny,\ n = 1, 2, \ldots.$$

Following the last example, the solution to this problem is given by the series

$$u(x, y) = \sum_{n=1}^{\infty} \widehat{g}_s(n) \frac{\sinh nx}{\sinh n\pi} \sin ny.$$

The Fourier sine coefficients of g are easily computed as $\widehat{g}_s(n) = \frac{4}{\pi n^3}(1 - (-1)^n)$. A plot of this solution using the first three nonzero terms is given in Figure 7.8(b).

Remark 7.29 It is useful to contrast the plots of these two examples. In the first case, the Gibb's effect is clearly visible. It is due to the fact that the odd extension of the boundary data f has jump discontinuities. In the second example, the function g has continuous odd extension and the Fourier sine coefficients of g have sufficient decay that the series for the corresponding solution is uniformly convergent on the entire rectangle.

In the previous section, the Green's function for the Laplacian was defined and strictly associated with Dirichlet boundary conditions. Here we are going to take a broader view of Green's functions by allowing other types of boundary conditions. Multiple Fourier expansions as discussed in Chapter 6 provide a complete set of orthogonal eigenfunctions of the Laplacian on rectangles satisfying a variety of different possible boundary conditions on the sides. As such, they are a natural candidate to build series expansions of Green's functions. The following example is illustrative and typical.

Example 7.21 In this example, a double Fourier series expansion of the Green's function for the Laplacian on $R_{a,b}$ is given, assuming homogeneous Dirichlet boundary conditions on all sides. The eigenfunctions of the Laplacian that satisfy the Dirichlet boundary conditions are the functions $\sin \frac{n\pi}{a} x \sin \frac{m\pi}{b} y$, where $n, m = 1, 2, \ldots$. Let $(\xi, \eta) \in R_{a,b}$. Then the Green's function can be expanded in a double Fourier sine series of form

$$G(x, y, \xi, \eta) = \sum_{n,m=1}^{\infty} a_{nm}(\xi, \eta) \sin \frac{n\pi}{a} x \sin \frac{n\pi}{b} y.$$

The delta functional $\delta_{(\xi,\eta)}(x, y) = \delta_\xi(x)\delta_\eta(y)$ has expansion of the form

$$\delta_{(\xi,\eta)}(x, y) = \sum_{n,m=1}^{\infty} \sin \frac{n\pi}{a} \xi \sin \frac{m\pi}{b} \eta \sin \frac{n\pi}{a} x \sin \frac{m\pi}{b} y.$$

Since $\nabla^2_{(x,y)} G = \delta_{(\xi,\eta)}$, we can compare the two expansions and see that

$$a_{nm}(\xi,\eta) = -\frac{\sin \frac{n\pi}{a}\xi \sin \frac{m\pi}{b}\eta}{\left[\left(\frac{n\pi}{a}\right)^2 + \left(\frac{m\pi}{b}\right)^2\right]}.$$

The convergence of the series for G is in the weak sense. In terms of pointwise convergence it can be shown that the series for G is only conditionally convergent.[4] This is in contrast to the situation on one dimension; recall Example 5.21.

Example 7.21 is typical of multiple Fourier expansions of the Green's functions for the Laplacian corresponding to any given set of boundary conditions on the sides. Unfortunately, the resulting series is not of much value for numerical computation because of the conditional convergence. A more effective series representation for the Green's function can be obtained by using the Fourier expansion in one variable and employing the Green's function for a related ordinary differential operator. The following example determines this type of expansion for Example 7.21.

Example 7.22 It is convenient to work with Poisson's equation $\nabla^2 u = F(x,y)$ in the rectangle $R_{a,b}$ with Dirichlet boundary conditions on the sides. We proceed with a formal computation; justification is discussed afterward. Expand u and F in a Fourier sine series on the interval $[0,a]$:

$$u(x,y) = \sum_{n=1}^{\infty} u_n(y) \sin \frac{n\pi}{a}x, \quad u_n(y) = \frac{2}{a}\int_0^a u(x,y) \sin \frac{n\pi}{a}x\, dx$$

$$F(x,y) = \sum_{n=1}^{\infty} F_n(y) \sin \frac{n\pi}{a}x, \quad F_n(y) = \frac{2}{a}\int_0^a F(x,y) \sin \frac{n\pi}{a}x\, dx.$$

Apply the Laplace operator to u term by term and compare with the expansion for F to obtain the conclusion that $u_n(y)$ must satisfy the differential equation

$$u_n'' - \left(\frac{n\pi}{a}\right)^2 u_n = F_n.$$

4. Conditional convergence means that the series is convergent pointwise, but not absolutely convergent.

Further, u_n must satisfy the Dirichlet boundary conditions $u_n(0) = u_n(b) = 0$. From the methods of Chapter 4, we can express the solution for this BVP in the form

$$u_n(y) = \int_0^b G_n(y,\eta)F_n(\eta)d\eta,$$

where $G_n(y,\eta)$ is the Green's function for the ODE boundary value problem. For this problem, the latter has the form

$$G_n(y,\eta) = \begin{cases} -\dfrac{a\sinh\frac{n\pi y}{a}\sinh\frac{n\pi(b-\eta)}{a}}{n\pi\sinh\frac{n\pi b}{a}}, & 0 < y \le \eta \\ -\dfrac{a\sinh\frac{n\pi\eta}{a}\sinh\frac{n\pi(b-y)}{a}}{n\pi\sinh\frac{n\pi b}{a}}, & \eta \le y < b. \end{cases}$$

Comparing the formulas, the Green's function has expansion

$$G(x,y,\xi,\eta) = \frac{2}{a}\sum_{n=1}^{\infty} G_n(y,\eta)\sin\frac{n\pi}{a}\xi\sin\frac{n\pi}{a}x.$$

This expansion and the series of term-by-term derivatives are absolutely convergent provided $y \neq \eta$. This can be seen by approximating the hyperbolic sine functions for large n, $\sinh nz \approx \frac{1}{2}e^{nz}$. Using this it follows that for $y < \eta$,

$$\frac{\sinh\frac{n\pi y}{a}\sinh\frac{n\pi(b-\eta)}{a}}{\sinh\frac{n\pi b}{a}} \approx \frac{e^{n\pi y/a}e^{n\pi(b-\eta)/a}}{2e^{n\pi b/a}} = \frac{1}{2}e^{n\pi(y-\eta)/a}.$$

This shows that the coefficients $G_n(y,\eta)$ provide exponential decay for $y < \eta$. A similar computation holds for $y > \eta$ as well as for the series of term-by-term derivatives in any variable.

Remark 7.30 The methodology given in Example 7.22 can be used directly to compute solutions of Poisson's equation on rectangles with homogeneous boundary conditions.

Exercises

7.37 Determine the solution of Laplace's equation in the unit disk where $f(\theta) = 1$ for $\pi/2 \le \theta < \pi$, $f(\theta) = -1$ for $-\pi \le \theta \le -\pi/2$ and $f(\theta) = 0$ elsewhere. Plot an approximate solution in three dimensions.

7.38 Find the Fourier solution of Laplace's equation in the disk D_a with Neumann boundary conditions of the form $\frac{du}{dn} = g$ where $g = g(\theta)$. Assume $\int_{-\pi}^{\pi} g(\theta)d\theta = 0$, which is needed for existence.

7.39 Find the Fourier series solution of Laplace's equation in the disk D_a with Robin boundary condition of the form $\frac{du}{dn}+hu = g$ where h is a positive constant and $g = g(\theta)$.

7.40 Work out the details for the solution formula for the Dirichlet problem on an annulus.

7.41 Find the steady-state temperature distribution inside the annulus $A_{1,2}$ where the outer circle is insulated and the inner circle is held at temperature $\sin^2\theta$. (The solution is $u = u(r,\theta)$, so the boundary conditions can be written $\frac{\partial u}{\partial r}(2,\theta) = 0,\ u(1,\theta) = \sin^2\theta$.) Plot the solution.

7.42 Solve Laplace's equation in the exterior disk ($r > a$) with the boundary condition $u(a,\theta) = f(\theta)$ and the assumption that u is bounded as $r \to \infty$.

7.43 Find the explicit solution for Exercise 7.42 when $f(\theta) = 1 - 2\sin 2\theta$.

7.44 Solve the Dirichlet problem on the wedge where the boundary condition $u(b,\theta) = f(\theta)$ is replaced by $u_r(b,\theta) = f(\theta)$.

7.45 For each of the following boundary conditions, solve Laplace's equation in the square $R_{\pi,\pi}$.

a. $u_x(0,y) = u_x(\pi,y) = 0, \quad u(x,0) = 0, \quad u(x,\pi) = x$

b. $u(0,y) = u_x(\pi,y) = 0, \quad u(x,0) = 0, \quad u(x,\pi) = x(\pi - x)$

7.46 Consider the Laplace operator on the square $R_{\pi,\pi}$ with boundary conditions $u(0,y) = u_x(\pi,y) = 0$ and $u(x,0) = u(x,\pi) = 0$.

a. Find the double Fourier expansion for the Green's function.

b. Find an expansion for this Green's function of the form

$$\sum_{n=0}^{\infty} G_n(y,\eta)\sin\left(n+\frac{1}{2}\right)x,$$

where $G_n(y,\eta)$ is the Green's function for a particular ODE boundary value problem. HINT: For part (b), see Exercise 4.22.

7.47 Find the solution to Laplace's equation on the cube $R_{\pi,\pi,\pi}$ that satisfies the following boundary conditions: $u(0,y,z) = u_x(\pi,y,z) = 0,\ u(x,0,z) = u_y(x,\pi,z) = 0,\ u(x,y,0) = 0,\ u(x,y,\pi) = 1$.

7.48 Find a solution formula for the following problem.

$$\text{PDE: } u_{xx} + u_{yy} = 0,\ 0 < x < \pi, \quad 0 < y < \pi$$

$$\text{BC: } \begin{cases} u(x,0) = u(x,\pi) = 0, & 0 < x < \pi \\ \begin{cases} u(0,y) = 1 \\ u(\pi,y) = -1 \end{cases}, & 0 < y < \pi \end{cases}$$

Chapter 8

Fourier Expansions on a Sphere

In this chapter, we construct functions defined on the unit sphere—the analog of Fourier series expansions of functions defined on the unit circle. Our approach is to determine the eigenfunctions of the spherical Laplace operator, ∇^2_{sp}, using spherical coordinates. These eigenfunctions are called spherical harmonics and are the analog of complex exponentials $e^{in\theta}$, thought of as functions on the circle. We describe the general idea in the first section. Separation of variables is the key tool leading to the associated Legendre equation. In Section 8.2, we develop the solutions of this equation: Legendre polynomials, associated Legendre functions, and their properties. In Section 8.3, Fourier expansions on the sphere are defined and we put them to use for potential problems in the final section.

8.1 Eigenfunctions on the Sphere: General Ideas

On the interval $(-\pi, \pi)$ with periodic boundary conditions, the orthogonal eigenfunctions of the operator $-\frac{d^2}{d\theta^2}$ are $\{e^{in\theta}\}_{n=-\infty}^{\infty}$. This can also be thought of as the collection of eigenfunctions of this operator on the unit circle $\mathbf{T}$. Our objective in this section is to construct the analog of this framework on the unit sphere in three dimensions. Henceforth, the unit sphere will be denoted S^2; the superscript 2 indicates this is a two-dimensional embedded surface in three-dimensional space. (Likewise, we could denote the unit circle $\mathbf{T}$ by S^1.) Points $\omega \in S^2$ are written $\omega = (\cos\theta \sin\phi, \sin\theta \sin\phi, \cos\phi)$ where $\phi \in [0, \pi]$ and $\theta \in [-\pi, \pi]$. The natural surface element on S^2 will be denoted $d\omega$; in terms of spherical coordinates it is given by $d\omega = \sin\phi \, d\phi \, d\theta$. We need to define what a smooth function is on the sphere. Although this is usually done using notions from differential geometry, here we are simply going to discuss smoothness relative to the spherical coordinate system. A function on the sphere would be written $u(\omega) = u(\theta, \phi)$. We assume u is smooth in both variables for $\theta \in (-\pi, \pi)$ and $\phi \in (0, \pi)$. As θ is a polar angle, clearly we need to impose periodic boundary conditions in the θ-variable, i.e., $u(-\pi, \phi) = u(\pi, \phi)$ for all $\phi \in (0, \pi)$ and similarly for all partial derivatives. The coordinates $\phi = 0$ and $\phi = \pi$ correspond to the north and south poles of the sphere, respectively. At these points, the θ-variable becomes ambiguous. It is natural to assume that $u_\theta(\theta, 0) = 0 = u_\theta(\theta, \pi)$ and similarly for all other partial derivatives.

We are interested in finding the eigenvalues and eigenfunctions of the operator $-\nabla^2_{\text{sp}}$, where in terms of spherical coordinates we have

$$\nabla^2_{\text{sp}} = \frac{\partial^2}{\partial \phi^2} + \cot\phi \frac{\partial}{\partial \phi} + \frac{1}{\sin^2\phi}\frac{\partial}{\partial \theta}. \tag{8.1.1}$$

A straightforward computation using integration by parts will show that for smooth functions $u(\omega)$ and $v(\omega)$ on the sphere

$$\int_{S^2} (\nabla^2_{\text{sp}} u)\bar{v}\, d\omega = \int_{S^2} u(\overline{\nabla^2_{\text{sp}} v}) d\omega. \tag{8.1.2}$$

In other words, $-\nabla^2_{\text{sp}}$ is a symmetric operator with respect to the inner product

$$\langle u, v \rangle = \int_{S^2} u\bar{v}\, d\omega.$$

We carry out this computation by writing $u(\omega) = u(\theta, \phi)$ and similarly for v and expressing the integrals in terms of spherical coordinates. As consequence, eigenfunctions of $-\nabla^2_{\text{sp}}$ corresponding to distinct eigenvalues are orthogonal with respect to this inner product. Symbolically, let $\Omega_1(\omega)$ and $\Omega_2(\omega)$ be eigenfunctions of $-\nabla^2_{\text{sp}}$ with corresponding distinct eigenvalues μ_1 and μ_2, respectively. In other words, they are solutions of the equation

$$\nabla^2_{\text{sp}} u + \mu u = 0 \tag{8.1.3}$$

on the sphere. Then by (8.1.2) we have

$$(\mu_1 - \mu_2) \int_{S^2} \Omega_1 \overline{\Omega_2}\, d\omega = 0,$$

which gives the orthogonality immediately since $\mu_1 \neq \mu_2$.

With $\omega = (\cos\theta \sin\phi, \sin\theta \sin\phi, \cos\phi)$, the natural coordinate curves given by

$$\phi \to (\cos\theta \sin\phi, \sin\theta \sin\phi, \cos\phi)$$

and

$$\theta \to (\cos\theta \sin\phi, \sin\theta \sin\phi, \cos\phi)$$

(latitude and longitude, respectively) have unit tangent vectors denoted $\mathbf{e}_\phi$ and $\mathbf{e}_\theta$. Notice that $\mathbf{e}_\phi \cdot \mathbf{e}_\theta = 0$. The geometry of the sphere, the coordinate curves,

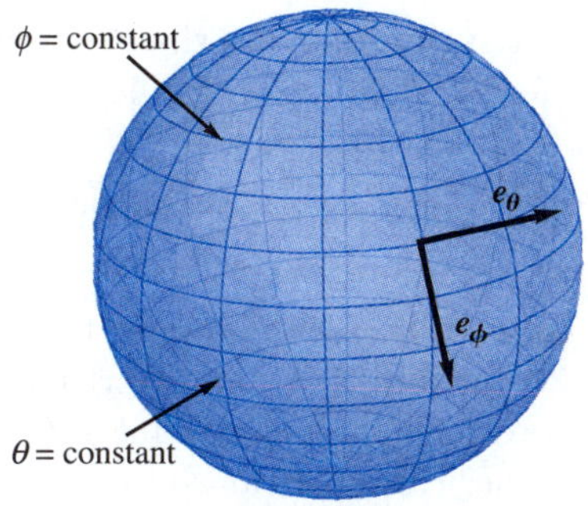

Figure 8.1 Sphere Geometry

and these tangent vectors are seen in Figure 8.1. Consider the vector differential operator defined by

$$\boldsymbol{\nabla}_{\text{sp}} u = \frac{1}{\sin\phi}\frac{\partial u}{\partial\theta}\mathbf{e}_\theta + \frac{\partial u}{\partial\phi}\mathbf{e}_\phi.$$

This is called the surface gradient for the sphere. Using a similar integration by parts computation it follows that

$$\int_{S^2} \boldsymbol{\nabla}_{\text{sp}} u \cdot \overline{\boldsymbol{\nabla}_{\text{sp}} v}\, d\omega = -\int_{S^2} (\nabla^2_{\text{sp}} u)\bar{v}\, d\omega. \tag{8.1.4}$$

Now, if Ω is an eigenfunction of $-\nabla^2_{\text{sp}}$ with eigenvalue μ, then using this formula with $u = v = \Omega$ we get

$$\mu \int_{S^2} |\Omega(\omega)|^2 d\omega = \int_{S^2} |\boldsymbol{\nabla}_{\text{sp}} \Omega(\omega)|^2 d\omega.$$

It then follows that all eigenvalues of $-\nabla^2_{\text{sp}}$ are nonnegative.

This discussion provides the framework in which to construct the eigenvalues and eigenfunctions. It is completely analogous to the corresponding problem on the unit circle. We now turn our attention to the actual determination of the eigenfunctions and eigenvalues. For this, we write $\Omega(\omega) = \Omega(\theta, \phi)$ and consider the equation

$$\frac{\partial^2\Omega}{\partial\phi^2} + \cot\phi\frac{\partial\Omega}{\partial\phi} + \frac{1}{\sin^2\phi}\frac{\partial^2\Omega}{\partial\theta^2} + \mu\Omega = 0. \tag{8.1.5}$$

Solutions of this equation will be preserved under translation in the θ variable. Further, this variable is the longitudinal variable and ranges in the interval $[-\pi, \pi]$. Hence any solution must satisfy periodic boundary conditions in θ. This motivates seeking separated solutions of the equation of the form

$$\Omega(\theta, \phi) = \Phi(\phi)e^{im\theta}, \tag{8.1.6}$$

where m is an integer. Substituting this into the equation and simplifying yields the following ordinary differential equation for Φ:

$$\Phi'' + \cot\phi\Phi' + \left(\mu - \frac{m^2}{\sin^2\phi}\right)\Phi = 0.$$

As $\phi \in [0, \pi]$, this equation has singular points at the endpoints. The equation becomes more tractable by introducing the invertible change of variable $s = \cos\phi$. This change of coordinates maps the interval $[0, \pi]$ onto the interval $[-1, 1]$. Using the change rule, this equation becomes

$$(1 - s^2)\Phi'' - 2s\Phi' + \left(\mu - \frac{m^2}{1 - s^2}\right)\Phi = 0, \tag{8.1.7}$$

known as the *associated Legendre equation of order* m. In the case where $m = 0$, it is called **Legendre's equation.**

8.2 The Legendre and Associated Legendre Equations

The previous section suggests a path to find the eigenfunctions of $-\nabla^2_{\text{sp}}$. The focus of this section is the construction of all smooth solutions of the associated Legendre equation (8.1.7). We begin with some observations that reduce the problem to the case $m = 0$, i.e., Legendre's equation. Write the associated Legendre equation in the form

$$[(1 - s^2)\Phi']' + \left(\mu - \frac{m^2}{1 - s^2}\right)\Phi = 0.$$

If we apply the usual product rule $(m + 1)$ times, we obtain

$$(fg)^{(m+1)} = \sum_{k=0}^{m+1} \frac{(m+1)!}{k!(m+1-k)!} f^{(k)} g^{(m+1-k)}.$$

We then apply this to the choice $f = 1 - s^2$ and $g = \Phi'$:

$$[(1 - s^2)\Phi']^{(m+1)} = (1 - s^2)\Phi^{(m+2)} - 2(m + 1)s\Phi^{(m+1)} - m(m + 1)\Phi^{(m)}.$$

Differentiating Legendre's equation m times and using this identity gives

$$(1 - s^2)\Phi^{(m+2)} - 2(m + 1)s\Phi^{(m+1)} - m(m + 1)\Phi^{(m)} + \mu\Phi^{(m)} = 0. \tag{8.2.1}$$

At this point we let

$$\Psi = (1 - s^2)^{m/2}\Phi^{(m)},$$

then

$$(1-s^2)\Psi' = -ms(1-s^2)\Phi^{(m)} + (1-s^2)^{(m/2)+1}\Phi^{(m+1)},$$

and differentiating again plus simplification leads to

$$[(1-s^2)\Psi']' = (1-s^2)^{m/2} \left[(1-s^2)\Phi^{(m+2)} - 2(m+1)s\Phi^{(m+1)} + \frac{m^2}{1-s^2}\Phi^{(m)} - m(m+1)\Phi^{(m)}\right].$$

Substituting from (8.2.1), this equation simplifies to

$$[(1-s^2)\Psi']' = -\mu\Psi + \frac{m^2}{1-s^2}\Psi.$$

In other words, if Φ satisfies Legendre's equation, then $\Psi = (1-s^2)^{m/2}\frac{d^m\Phi}{ds^m}$ satisfies the associated Legendre equation of order m. This differentiation trick is a significant reduction in the problem of finding all solutions of the associated Legendre equation.

Legendre Polynomials

We now consider Legendre's equation

$$(1-s^2)\Phi'' - 2s\Phi' + \mu\Phi = 0 \tag{8.2.2}$$

on the interval $[-1,1]$ with the conditions that $\Phi(\pm 1)$ is finite. On the sphere S^2, solutions of Legendre's equation correspond to functions $\Omega(\theta,\phi) = \Phi(\phi)$ that are constant on latitudes $\phi =$ constant; functions on the sphere with this property are called **zonal**. Geometrically, a function $\Omega(\omega)$ is zonal if and only if for every rotation A that leaves the north pole fixed, $\Omega(A\omega) = \Omega(\omega)$. The points $s = \pm 1$ correspond with the north and south poles of the sphere, respectively, $\phi = 0$ and $\phi = \pi$. The point $s = 0$ is a regular point for Legendre's equation, hence we seek solutions in the form of a power series

$$\Phi(s) = \sum_{k=0}^{\infty} a_k s^k.$$

Computing the derivatives, substituting into Legendre's equation, and combining like powers of s gives the formal relation:

$$\sum_{k=2}^{\infty} [(k+1)(k+2)a_{k+2} + (\mu - k(k+1))a_k]\, s^k + [6a_3 + (\mu-2)a_2]s + (2a_2 + \mu a_0) = 0.$$

Hence, for a solution to exist in the form of a power series, the coefficients must satisfy the recursion formula

$$a_{k+2} = -\frac{\mu - k(k+1)}{(k+1)(k+2)} a_k, \quad k = 0, 1, 2, \ldots. \tag{8.2.3}$$

This formula says that from a known value of a_0, all coefficients for even values of k can be computed and, similarly, from a value for a_1, all coefficients for odd values of k can be found. Hence, the recursion formula should provide two linearly independent solutions of Legendre's equation, one involving even powers of s, the other odd powers.

The key observation to make is that if $\mu = l(l+1)$ for some nonnegative integer l, then the sequence of coefficients will terminate and Legendre's equation will have a polynomial solution. For example, if $l = 2n$, we take $a_0 \neq 0$ and $a_1 = 0$, then use the recursion formula to compute

$$a_0 \to a_2 \to a_4 \to \cdots \to a_{2n} \quad \text{and} \quad 0 = a_{2n+2} = a_{2n+4} = \cdots.$$

Likewise, assuming $l = 2n+1$ and taking $a_0 = 0$ and $a_1 \neq 0$, a polynomial solution of degree $2n+1$ is generated. Thus, for each $\mu = l(l+1)$ where l is a nonnegative integer, there is a polynomial solution of Legendre's equation of precise degree l. Normalizing each polynomial so that its leading coefficient is $\frac{(2l)!}{2^l (l!)^2}$, the resulting polynomial is called **Legendre's polynomial of degree l,** and is denoted $P_l(s)$. The choice of normalization may seem arbitrary; however, as will be seen shortly, it leads to the property $P_l(1) = 1$ for all l. For each l, Legendre's equation has a second linearly independent solution, $Q_l(s)$, which satisfies the asymptotic (see [20])

$$Q_l(s) \approx c_l \log \frac{1+s}{1-s} \quad \text{as} \quad s \to \pm 1,$$

where c_l is a nonzero constant. In particular, $Q_l(s)$ is unbounded at the endpoints and is not square integrable on $[-1, 1]$.

We will now list a few of the properties of Legendre polynomials.

Orthogonality Legendre polynomials form an orthogonal set with respect to the inner product $\langle f, g \rangle = \int_{-1}^{1} f(s)g(s)ds$. This follows from the fact that $P_l(\cos\phi)$ is an eigenfunction of $-\nabla^2_{\text{sp}}$ with eigenvalue $l(l+1)$.

Indeed, abusing notation and writing $P_l(\omega) = P_l(\cos\phi)$, we have for $k \neq l$,

$$\begin{aligned} 0 = \int_{S^2} P_l(\omega)P_k(\omega)d\omega &= 2\pi \int_{-\pi}^{\pi} P_l(\cos\phi)P_k(\cos\phi)\sin\phi d\phi \\ &= \int_{-1}^{1} P_l(s)P_k(s)ds, \end{aligned}$$

where the substitution $s = \cos\phi$ has been used in the last step. The orthogonality can also be shown from the fact that Legendre polynomials are eigenfunctions of the singular Sturm–Liouville problem

$$[(1-s^2)\Phi']' + \mu\Phi = 0,$$
$$\Phi(\pm 1) \text{ is finite.}$$

Rodrigues Formula Legendre polynomials are given by the formula

$$P_l(s) = \frac{(-1)^l}{2^l l!}\left(\frac{d}{ds}\right)^l (1-s^2)^l. \tag{8.2.4}$$

To see this, let $p(s) = \left(\frac{d}{ds}\right)^l (1-s^2)^l$, then $p(s)$ is a polynomial of degree l. As such, it can be written as

$$p(s) = \sum_{k=0}^{l} c_k P_k(s)$$

for appropriate choice of constants c_k. By the orthogonality of the Legendre polynomials, $c_k = 0$ for $k = 0, 1, \ldots, l-1$. We conclude that

$$p(s) = c_l P_l(s).$$

The constant c_l can be found by computing the leading coefficients of $p(s)$ using differentiation formulas and comparing it with that of $P_l(s)$. Details are left to the reader.

Explicit Formula Left as an exercise, the Legendre polynomials are given by

$$P_l(s) = \frac{1}{2^l}\sum_{k\le \frac{l}{2}} \frac{(-1)^k(2l-2k)!}{k!(l-k)!(l-2k)!} s^{l-2k}.$$

From this follows the formulas $P_{2l-1}(0) = 0$ and $P_{2l}(0) = \frac{(-1)^l(2l)!}{2^{2l}(l!)^2}$.

L^2–Norm This norm is given by the formula

$$\|P_l\|_2^2 = \int_{-1}^{1} [P_l(s)]^2 ds = \frac{2}{2l+1}. \tag{8.2.5}$$

The proof of this is based on the Rodrigues formula and integration by parts l times as follows:

$$\begin{aligned}\|P_l\|_2^2 &= \frac{(-1)^l}{2^l l!}\int_{-1}^{1} P_l(s)\left(\frac{d}{ds}\right)^l (1-s^2)^l ds \\ &= \frac{1}{2^l l!}\int_{-1}^{1}\left[\left(\frac{d}{ds}\right)^l P_l(s)\right](1-s^2)^l ds.\end{aligned}$$

The boundary terms that come up in the integration by parts vanish because the $(1-s^2)^l$ is a polynomial that has $s=\pm 1$ as zeros each of order l. Hence,

$$\left(\frac{d}{ds}\right)^{l-k}(1-s^2)^l = 0$$

at $s=\pm 1$ for each $k=0,1,\ldots,l-1$. Now,

$$\left(\frac{d}{ds}\right)^{l} P_l(s) = l! \times \text{the leading coefficient of } P_l,$$

hence

$$\|P_l\|_2^2 = \frac{1}{2^l}\frac{(2l)!}{2^l(l!)^2}\int_{-1}^{1}(1-s^2)^l ds.$$

The latter integral has known value given by $\frac{2^{l+1}l!}{1\cdot 3\cdot 5\cdots(2l+1)}$. (This can be derived in a number of ways; see Exercise 8.2.) Notice that $\frac{(2l)!}{2^l l!} = \frac{1\cdot 2\cdot 3\cdots(2l)}{2\cdot 4\cdots(2l)} = 1\cdot 3\cdot 5\cdots(2l-1)$. Using this and the value for the integral gives the result.

Recursion Formulas The Legendre polynomials satisfy the following formulas.

$$(l+1)P_{l+1}(s) - (2l+1)sP_l(s) + lP_{l-1}(s) = 0 \tag{8.2.6}$$

$$P'_{l+1}(s) - P'_{l-1}(s) = (2l+1)P_l(s) \tag{8.2.7}$$

Since $P_0(s)=1$ and $P_1(s)=s$, formula (8.2.6) implies $P_l(1)=1$ for all $l=2,3,\ldots$.

We will demonstrate the first identity and leave the second one as an exercise. The polynomial $sP_l(s)$ has degree $l+1$ and hence can be written

$$sP_l(s) = \sum_{k=0}^{l+1} c_k^l P_k(s)$$

for suitable constants c_k^l. The constants can be determined from the orthogonality conditions:

$$c_k^l = \frac{2k+1}{2}\int_{-1}^{1} sP_l(s)P_k(s)ds. \tag{8.2.8}$$

Substituting the Rodrigues formula for P_l this can be written

$$c_k^l = \frac{(-1)^l(2k+1)}{2^{l+1}l!}\int_{-1}^{1} sP_k(s)\left(\frac{d}{ds}\right)^l(1-s^2)^l ds.$$

Since $sP_k(s)$ is a polynomial of degree $k+1$, using integration by parts, the integral is zero for $k+1 < l$. Thus

$$sP_l(s) = c_{l+1}^l P_{l+1}(s) + c_l^l P_l(s) + c_{l-1}^l P_{l-1}(s).$$

We determine $c_{l+1}^l = \frac{l+1}{2l+1}$ by comparing the leading coefficients of $sP_l(s)$ and $P_{l+1}(s)$. Furthermore, $sP_l(s)$ has no term of degree l, hence $c_l^l = 0$. Now since $sP_{l-1}(s) = \frac{l}{2l-1}P_l(s) + c_{l-2}^{l-1}P_{l-2}(s)$, using (8.2.8) with $k = l-1$ we have

$$c_{l-1}^l = \frac{2l-1}{2}\frac{l}{2l-1}\int_{-1}^{1}[P_l(s)]^2 ds = \frac{l}{2l+1}.$$

This completes the proof. □

Generating Function The following formula holds:

$$\frac{1}{\sqrt{1-2\rho x + x^2}} = \sum_{l=0}^{\infty} \rho^l P_l(x). \tag{8.2.9}$$

The series is uniformly convergent for $|\rho| \le \rho_0 < 1$ and $x \in [-1,1]$. The function on the left-hand side is called the generating function for the Legendre polynomials.

To see this, let $F(\rho, x)$ denote the function on the left-hand side. By direct calculation, F satisfies the equation

$$(1-2\rho x + x^2)\frac{\partial F}{\partial \rho} = (x-\rho)F. \tag{8.2.10}$$

We expand F in a power series about $\rho = 0$ to obtain

$$F(\rho, x) = \sum_{l=0}^{\infty} a_l(x)\rho^l.$$

It turns out that the radius of convergence is 1. Setting $\rho = 0$ it follows that $a_0(x) = 1 = P_0(x)$. Further, differentiating both sides with respect to ρ and setting $\rho = 0$ we get that $a_1(x) = x = P_1(x)$. Substituting the series into the differential identity (8.2.10) and rearranging terms, we can show that the functions $a_l(x)$ satisfy the same recursion relation as the Legendre polynomials (8.2.6). This immediately implies that $a_l(x) = P_l(x)$ as desired.

Historical Comment The Legendre polynomials (and associated Legendre functions defined shortly) belong to a broad class of functions called the special functions of mathematical physics. Many of the latter arise from concrete boundary value problems associated with physical laws. We have arrived at the

Legendre polynomials via separation of variables applied to Laplace's equation, which is quite natural from the perspective of partial differential equations. Their properties were developed from the structure of Legendre's equation. In the late 1700s, Legendre was investigating the gravitational potential (related to F above), explored the series expansion for F, and proved that the coefficients $a_l(x)$ were polynomials. From there, he derived the various other properties of the polynomials that now bear his name. One can also start with the Rodrigues formula as the definition and develop the properties from there (see, e.g., [20, 16]).

Formulas for Legendre polynomials can be determined using either the Rodrigues formula or, starting from $P_0(s) = 1$ and $P_1(s) = s$, using the recursion formula. Figure 8.2 gives the plot of the first few Legendre polynomials. These plots hint at two things:

- Each $P_l(s)$ has precisely l zeros in the interval $[-1, 1]$. Further, these zeros are interlaced with those of $P_{l+1}(s)$.
- $|P_l(s)| \leq 1$.

For a proof of the first item, see for example [16]. The second item will be seen as a consequence of the addition theorem for spherical harmonics given in Section 8.3. Substituting $s = \cos\phi$, a zonal function on the sphere that is an eigenfunction of $-\nabla^2_{\text{sp}}$ with eigenvalue $l(l+1)$ is given by $P_l(\cos\phi)$. A few of these are plotted on the surface of the sphere in Figure 8.3.

The Fourier–Legendre series and coefficients of a function $f \in L^2[-1, 1]$ are given by the following formulas (following the algorithm of Section 3.3).

$$f \sim \sum_{l=0}^{\infty} \widehat{f}(l) P_l(s)$$

$$\widehat{f}(l) = \frac{2l+1}{2} \int_{-1}^{1} f(s) P_l(s) ds$$

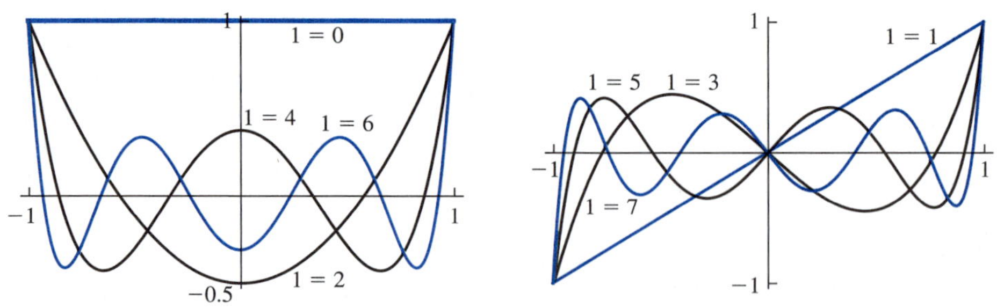

Figure 8.2 Examples of Legendre Polynomials

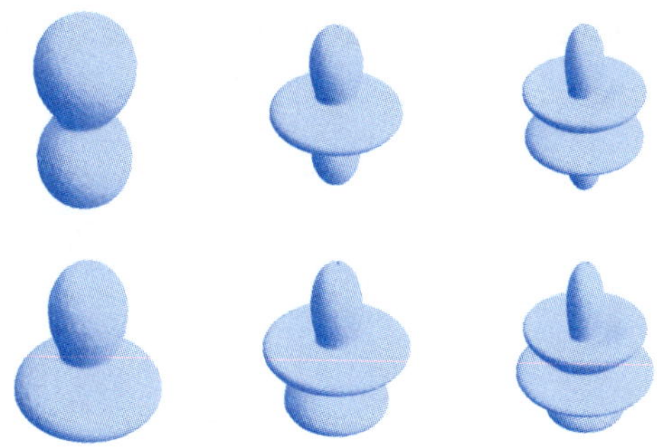

Figure 8.3 Three-Dimensional Plots of $P_l(\cos\phi)$, $l = 2, 4, 6, 3, 5, 7$

Example 8.1 For the function $f(s) = \begin{cases} -1, & -1 < x < 0 \\ 1, & 0 < x < 1 \end{cases}$, the Fourier–Legendre coefficients can be computed explicitly as follows. Since f is an odd function, $\widehat{f}(2l) = 0$ and by a simple substitution

$$\widehat{f}(2l+1) = \frac{4l+3}{2} \int_0^1 P_{2l+1}(s)ds = P_{2l}(0) - P_{2l+2}(0),$$

where (8.2.7) was used in the last step. The Fourier–Legendre series of f is

$$f \sim \sum_{l=0}^{\infty} [P_{2l}(0) - P_{2l+2}(0)] P_{2l+1}(s).$$

Some typical partial sums are plotted in Figure 8.4, which indicates pointwise convergence on the interval $(-1, 1)$; note also the apparent Gibb's effect near the jump discontinuity $x = 0$.

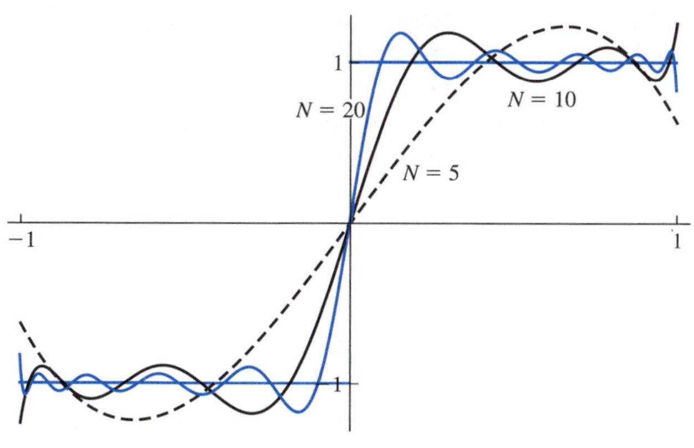

Figure 8.4 Legendre Series Plot

The following result gives the L^2–convergence and pointwise convergence results together.

Theorem 8.1 (1) The orthogonal set $\{P_l(s)\}_{l=0}^{\infty}$ is complete in $L^2[-1,1]$. Parseval's equality takes the form

$$\sum_{l=0}^{\infty} \frac{2}{2l+1}|\widehat{f}(l)|^2 = \|f\|_2^2.$$

(2) If $f \in \mathrm{PC}[-1,1]$ with f' also piecewise continuous, then for all $x \in (-1,1)$,

$$\lim_{N\to\infty} S_N f(x) = \frac{1}{2}[f(x+0) + f(x-0)],$$

where $S_N f(x)$ is the Nth partial sum of the Fourier–Legendre series of f.

A sketch of the key ideas underlying the proof is as follows. We will demonstrate Parseval's equality for continuous functions by showing that for $f \in C[-1,1]$, $\|S_N f - f\|_2 \to 0$ as $N \to \infty$. (The general case can be deduced from this using approximation arguments.) The Weierstrass approximation theorem (see Theorem A.8) states that any continuous function on a closed interval can be uniformly approximated by a polynomial. Hence, given $\varepsilon > 0$, there is a polynomial $p(s)$ such that $\|f-p\|_\infty < \varepsilon/4$. It follows that $\|f-p\|_2 < \varepsilon/2$. Now, if $N > \mathrm{degree}(p)$, then the Fourier–Legendre series of p satisfies $S_N p(s) = p(s)$. Add and subtract appropriate terms and use the triangle inequality to see that

$$\|S_N f - f\|_2 \le \|S_N f - S_N p\|_2 + \|S_N p - p\|_2 + \|p - f\|_2.$$

The middle term is zero by the above observation, and Bessel's inequality applies to the first term to give

$$\|S_N f - S_N p\|_2 = \|S_N(f-p)\|_2 \le \|f-p\|_2 < \frac{\varepsilon}{2}.$$

Thus, if $N > \mathrm{degree}(p)$, then $\|S_N f - f\|_2 < \varepsilon$, which is what we wanted to prove. The proof of part (2) is strikingly similar to that given for Fourier series and is left for the reader to pursue in the literature (e.g. [20]). A heuristic perspective can be given by writing the formula for the partial sums as an integral. Substituting the formula for the Fourier–Legendre coefficients into the partial sums and interchanging the integral and sum gives

$$S_N f(s) = \int_{-1}^{1} f(x) D_N(x,s)\,dx,$$

where

$$D_N(x,s) = \sum_{l=0}^{N} \frac{2l+1}{2} P_l(x) P_l(s).$$

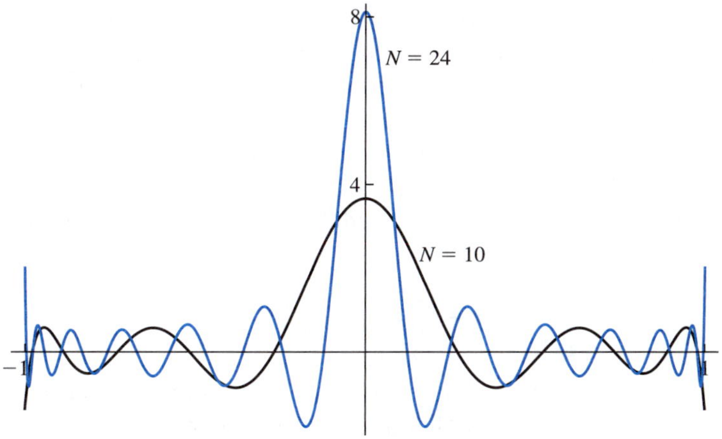

Figure 8.5 Legendre Dirichlet Kernel

A closed form for the kernel D_N is known (Exercise 8.4). Plots of this kernel are given in Figure 8.5 for $s = 0$, bearing striking resemblance to the Dirichlet kernel for Fourier series. Heuristically, the partial sums $S_N f(0)$ are averages of the values of f near zero just like in the case of Fourier series (the same holds at other points $s \in (-1, 1)$).

Theorem 8.1 partially fulfills the main objective of this section. If $f \in L^2(S^2)$ and is invariant under all rotations that leave the north pole fixed, i.e., $f(A\omega) = f(\omega)$ for such rotations, then f depends only on the angle ϕ and we write $f(\omega) = f(\cos\phi) = f(\omega \cdot \mathbf{e}_3)$. By the above result, f can be expanded in terms of the eigenfunctions $P_l(\cos\phi)$ of the operator $-\nabla^2_{\text{sp}}$. We summarize these observations as follows.

Corollary 8.2 Let $f \in L^2(S^2)$ be zonal. Then f can be expanded in L^2−norm as

$$f(\omega) \sim \sum_{l=1}^{\infty} \widehat{f}(l) P_l(\omega \cdot \mathbf{e}_3),$$

where $\widehat{f}(l) = \frac{2l+1}{4\pi} \int_{S^2} f(\omega) P_l(\omega \cdot \mathbf{e}_3) d\omega$.

Legendre's equation (8.2.2) with boundary conditions $X(\pm 1)$ is finite is an example of a singular Sturm–Liouville problem. We have shown that Legendre polynomials are eigenfunctions of this problem. The following result tells us that they are the only ones.

Proposition 8.3 Let $Y(s)$ be an eigenfunction of the singular problem consisting of Legendre's equation with boundary conditions $X(\pm 1)$ is finite. Then $Y(s) = cP_l(s)$ for some nonnegative integer l and some constant c.

Proof If the eigenvalues corresponding to Y is $\mu \neq l(l+1)$ for all l, then $\langle Y, P_l \rangle = \int_{-1}^{1} Y(s)P_l(s)ds = 0$ for all l. By Theorem 8.1, $Y = 0$. If the corresponding eigenvalue $\mu = l(l+1)$ for some nonnegative integer l, consider the function

$$g(s) = Y(s) - \frac{\langle Y, P_l \rangle}{\|P_l\|_2^2} P_l(s).$$

The second term on the right-hand side is the orthogonal projection of Y on P_l and hence $\langle g, P_l \rangle = 0$. It follows that g and P_l are linearly independent on $[-1, 1]$ and hence g must be a multiple of the second linearly independent solution of Legendre's equation. As mentioned previously, the latter is unbounded as $s \to \pm 1$. Hence, $g = 0$ from which it follows that Y is a multiple of P_l. This concludes the proof. □

The Associated Legendre Functions

We now return to the ideas at the beginning of this section and determine solutions to the associated Legendre equation and discuss their properties. Proofs will not be given; they are based on the ideas and discussion of Legendre polynomials. The key observation previously made was that if $\Phi(s)$ is a solution to Legendre's equation, then $\Psi(s) = (1-s^2)^{m/2} \frac{d^m \Phi}{ds^m}$ is a solution of the associated Legendre equation (8.1.7). From the discussion of Legendre's equation, we will choose the separation constant $\mu = l(l+1)$ where l is a nonnegative integer and $\Phi(s) = P_l(s)$. Then the **associated Legendre functions** are introduced as

$$P_l^m(s) = (1-s^2)^{m/2} \frac{d^m P_l}{ds^m}. \tag{8.2.11}$$

Since P_l is a polynomial of degree l, these are only of interest if $l \geq m$. Nonetheless, for $m = 0, 1, 2, \ldots,$ and $l \geq m$, P_l^m is a solution to the singular Sturm–Liouville problem

$$[(1-s^2)\Psi']' + [l(l+1) - \frac{m^2}{1-s^2}]\Psi = 0$$
$$\Psi(-1) = \Psi(1) = 0.$$

Modifying the techniques used for regular Sturm–Liouville problems and the ideas used in treating Legendre polynomials, the following result gives an analog of part (1) of Theorem 8.1 for the associated Legendre functions.

Proposition 8.4 For each positive integer m, the collection $\{P_l^m\}_{l\geq m}$ forms a complete orthogonal set in $L^2[-1,1]$. Moreover,

$$\|P_l^m\|_2^2 = \frac{(l+m)!}{(l-m)!}\frac{2}{2l+1}.$$

Although we have no direct need of them, it should be remarked that associated Fourier–Legendre series can be defined. However, remember that the associated Legendre functions are a stepping stone to producing a complete set of orthogonal eigenfunctions of $-\nabla^2_{\text{sp}}$, a topic we return to and complete in the next section.

Exercises

8.1 Verify the orthogonality of the Legendre polynomials by using the fact that they are eigenfunctions of a singular Sturm–Liouville problem.

8.2 Let $I_l = \int_{-1}^{1}(1-s^2)^l ds$. Using integration by parts, show that $(2l-1)I_l = 2lI_{l-1}$. Use this to show that $I_l = \frac{2^{l+1}l!}{1\cdot 3\cdots(2l+1)}$.

8.3 Derive formula (8.2.7).

8.4 Show that $\sum_{l=0}^{N}\frac{2l+1}{2}P_l(s)P_l(x) = \frac{N+1}{2}\left[\frac{P_{N+1}(s)P_N(x)-P_N(s)P_{N+1}(x)}{s-x}\right]$. HINT: Multiply the recursion formula (8.2.3) by $P_l(x)$, subtract from the same formula with s and x interchanged, and sum for $l = 1,\ldots,N$.

8.5

a. In the derivation of the generating function formula (8.2.9), provide the details that the functions $a_l(x)$ satisfy the same recursion formula as the Legendre polynomials.

b. Use the generating function to show that $P_l(1) = 1$ and $P_l(-1) = (-1)^l$.

8.6 Let $f(s) = \begin{cases} s, & 0 < s < 1 \\ 0, & -1 < s < 0 \end{cases}$. Find the Fourier–Legendre series of f and plot some typical partial sums.

8.7 Recall the formula (7.4.7) for the Dirichlet–Green's function on the ball $B_a(O)\subset\mathbb{R}^3$. Find an expansion of G in terms of $P_l(\cos\Theta)$, Θ being the angle between x and ξ in the region $|x|<|\xi|$ and in the region $|\xi|<|x|<a$. HINT: Use the generating function for Legendre polynomials.

8.3 Spherical Harmonics and Their Properties

From the discussion of Legendre polynomials, the eigenvalues of $-\nabla^2_{\text{sp}}$ are precisely of the form $l(l+1)$ where l is a nonnegative integer. Furthermore, for each m with $|m| \le l$, the function

$$Y_{lm}(\omega) = P_l^{|m|}(\cos\phi)e^{im\theta}, \quad \omega = (\cos\theta\sin\phi, \sin\theta\sin\phi, \cos\phi)$$

is an eigenfunction with eigenvalue $l(l+1)$. Notice that for each l, there are precisely $2l+1$ linearly independent complex-valued eigenfunctions. In other words, the dimension of the eigenspace is $2l+1$, a dimension that grows with l. Functions in this eigenspace are called **spherical harmonics of order l.** This is in sharp contrast to the case of the circle where the eigenvalues were written as n^2 and the corresponding eigenspace is spanned by the two functions $e^{in\theta}$ and $e^{-in\theta}$, for each integer n.

By way of their construction, the collection of functions $\{Y_{lm}(\omega) : |m| \le l,\, l = 0, 1, 2, \dots\}$ is an orthogonal set in $L^2(S^2)$, i.e.,

$$\langle Y_{lm}, Y_{l'm'}\rangle = \int_{S^2} Y_{lm}(\omega)\overline{Y_{l',m'}(\omega)}d\omega = 0$$

for $l \ne l'$ or $m \ne m'$. Furthermore, their L^2—norms are given by

$$\|Y_{lm}\|_2^2 = \frac{4\pi}{2l+1}\frac{(l+|m|)!}{(l-|m|)!}.$$

Given a function $f \in L^2(S^2)$, its **spherical harmonic expansion** is given by

$$f \sim \sum_{l=0}^{\infty}\sum_{|m|\le l} \widehat{f}_{lm} Y_{lm}(\omega),$$

where the coefficients are given by

$$\begin{aligned}\widehat{f}_{lm} &= \frac{\langle f, Y_{lm}\rangle}{\|Y_{lm}\|_2^2} = \frac{1}{\|Y_{lm}\|_2^2}\int_{S^2} f(\omega)\overline{Y_{lm}(\omega)}d\omega \\ &= \frac{2l+1}{4\pi}\frac{(l-|m|)!}{(l+|m|)!}\int_{-\pi}^{\pi}\int_0^{\pi} f(\theta,\phi)P_l^{|m|}(\cos\phi)e^{-im\theta}\sin\phi\, d\phi\, d\theta.\end{aligned}$$

Combining the results on Legendre polynomials and associated Legendre functions, we obtain the following result—one that is central to the application of spherical harmonic expansions.

Theorem 8.5 The collection of spherical harmonics $\{Y_{lm} : |m| \le l, l = 0, 1, 2, \dots\}$ is a complete orthogonal set in $L^2(S^2)$.

This result tells us that spherical harmonic expansions of functions $f \in L^2(S^2)$ converge in L^2–norm and Parseval's equality holds. The latter takes the following explicit form:

$$\sum_{l=0}^{\infty} \sum_{|m| \leq l} |\widehat{f}_{lm}|^2 \frac{4\pi}{2l+1} \frac{(l+|m|)!}{(l-|m|)!} = \int_{S^2} |f(\omega)|^2 d\omega. \tag{8.3.1}$$

Computation of spherical harmonic coefficients can be a labor-intensive process. However, when the function f can be recognized as a sum of spherical harmonics, computation of the integrals can be avoided.

Example 8.2 Consider the function $f(\theta, \phi) = 1 + 3\cos\phi + 3\cos^2\phi$. As a function on the sphere, f is zonal and hence has a Legendre expansion. We can find this by writing

$$\begin{aligned} f(\theta, \phi) &= 2 + 3\cos\phi + (3\cos^2\phi - 1) \\ &= 2P_0(\cos\phi) + 3P_1(\cos\phi) + 2P_2(\cos\phi). \end{aligned}$$

Example 8.3 Consider the function $g(\theta, \phi) = \sin\phi\cos\theta + \cos\phi\sin\phi\sin\theta$ on the sphere. To find the spherical harmonic expansion we need the following formulas for associated Legendre functions: $P_1^1(\cos\phi) = \sin\phi$ and $P_2^1(\cos\phi) = 3\sin\phi\cos\phi$. Thus

$$g(\theta, \phi) = P_1^1(\cos\phi)\cos\theta + \frac{1}{3}P_2^1(\cos\phi)\sin\theta.$$

Applying Euler's formula the desired expansion is

$$g(\theta, \phi) = \frac{1}{2}Y_{1,1}(\theta, \phi) + \frac{1}{2}Y_{1,-1}(\theta, \phi) - \frac{i}{6}Y_{2,1}(\theta, \phi) + \frac{i}{6}Y_{2,-1}(\theta, \phi).$$

The explicit connection between spherical harmonics and separated solutions of Laplace's equation is brought out in the following result.

Proposition 8.6 Let $x \in \mathbb{R}^3$ be written in spherical form as $x = \rho\omega$ where $\rho > 0$ and $\omega \in S^2$. For each $l = 0, 1, \ldots$ and $|m| \leq l$, the function $\rho^l Y_{lm}(\omega)$ is harmonic in $\mathbb{R}^3$ and the function $\frac{Y_{lm}(\omega)}{\rho^{l+1}}$ is harmonic in $\mathbb{R}^3 - \{O\}$.

Proof Laplace's equation in spherical coordinates is written

$$\frac{\partial^2 u}{\partial \rho^2} + \frac{2}{\rho}\frac{\partial u}{\partial \rho} + \frac{1}{\rho^2}\nabla_{\text{sp}}^2 u = 0$$

and it is completely natural to seek separated solutions of the form

$$u(x) = R(\rho)Y(\omega)$$

(effectively separating the radial and angular variables). Substituting this into the equation and rearranging terms leads to

$$\frac{\rho^2 R'' + 2\rho R'}{R} = -\frac{\nabla^2_{\text{sp}} Y}{Y}.$$

Calling the separation constant μ, we then obtain the equations

$$\rho^2 R'' + 2\rho R' - \mu R = 0 \quad \text{and} \quad \nabla^2_{\text{sp}} Y + \mu Y = 0.$$

We know that the second equation has eigenvalues $\mu = l(l+1)$ with corresponding eigenfunctions given by spherical harmonics Y_{lm}. Thus, we are left with solving the Euler equation

$$\rho^2 R'' + 2\rho R' - l(l+1)R = 0.$$

The solutions of this equation are precisely ρ^l and $\rho^{-(l+1)}$. We now need show that $\rho^l Y_{lm}(\omega)$ is, in fact, a C^2 function on $\mathbb{R}^3$. In fact, we claim that $\rho^l Y_{lm}(\omega)$ is a homogeneous polynomial of degree l. To see this, suppose $m \geq 0$. Then

$$P_l^m(s) = (1 - s^2)^{m/2} q_{l-m}(s),$$

where $q_{l-m}(s)$ is a polynomial of degree $l - m$. We can write

$$q_{l-m}(s) = \sum_{j \leq \frac{l-m}{2}} a_j s^{l-m-2j},$$

and hence

$$P_l^m(\cos\phi) = \sin^m \phi \sum_{j \leq \frac{l-m}{2}} a_j \cos^{l-m-2j} \phi.$$

It follows that

$$\begin{aligned} \rho^l P_l^m(\cos\phi) e^{im\theta} &= \left[\rho e^{i\theta} \sin\phi\right]^m \sum_{j \leq \frac{l-m}{2}} a_j \rho^{2j} \left[\rho \cos\phi\right]^{l-m-2j} \\ &= (x + iy)^m \sum_{j \leq \frac{l-m}{2}} a_j (x^2 + y^2 + z^2)^j z^{l-m-2j}, \end{aligned} \tag{8.3.2}$$

where we have used $x = \rho \cos\theta \sin\phi$, $y = \rho \sin\theta \sin\phi$, and $z = \rho \cos\phi$. Notice that the right-hand side is a homogeneous polynomial of degree l. By a similar calculation for $m < 0$,

$$\rho^l P_l^{-m}(\cos\phi) e^{im\theta} = (x - iy)^m \sum_{j \leq \frac{l-m}{2}} a_j (x^2 + y^2 + z^2)^j z^{l+m-2j}. \tag{8.3.3}$$

This proves the claim. The fact that $\frac{Y_{lm}(\omega)}{\rho^{l+1}}$ is C^2 away from the origin follows from this. □

As a corollary of the preceding proof, any spherical harmonic is the restriction to the sphere of a homogeneous harmonic polynomial. The analog of this in two dimensions is contained in the formulas

$$r^{|n|}e^{in\theta} = (x+iy)^{|n|}, \quad r^{|n|}e^{-in\theta} = (x-iy)^{|n|}.$$

These are an immediate consequence of Euler's formula.

Another useful result concerning spherical harmonics is the addition theorem. This is an analog of the addition theorem for sine and cosine. For simplicity in statement, we will use the orthonormal spherical harmonics defined by

$$\mathcal{Y}_{lm}(\omega) = \frac{Y_{lm}(\omega)}{\|Y_{lm}\|_2}.$$

Proposition 8.7 The orthonormal spherical harmonics satisfy the formula

$$\sum_{m=-l}^{l} \mathcal{Y}_{lm}(\omega)\overline{\mathcal{Y}_{lm}(\omega')} = \frac{2l+1}{4\pi}P_l(\omega\cdot\omega'). \tag{8.3.4}$$

Proof Let A be a rotation. Then the function $\mathcal{Y}_{lm}(A\omega)$ is again a spherical harmonic of order l because the sphere, all homogeneous polynomials, and the Laplace operator are invariant under rotations. It follows that $\mathcal{Y}_{lm}(A\omega)$ can be written as a linear combination of $\mathcal{Y}_{lm}(\omega)$, i.e.,

$$\mathcal{Y}_{lm}(A\omega) = \sum_{j=-l}^{l} a_{jm}\mathcal{Y}_{lj}(\omega). \tag{8.3.5}$$

We know that

$$\int_{S^2} \mathcal{Y}_{lm}(A\omega)\overline{\mathcal{Y}_{lk}(A\omega)}d\omega = \int_{S^2} \mathcal{Y}_{lm}(\omega)\overline{\mathcal{Y}_{lk}(\omega)}d\omega = \delta^k_m.$$

(δ^k_m is the Kronecker delta with value 1 if $k=m$ and zero otherwise.) From the two previous formulas it follows that

$$\sum_{k=-l}^{l} a_{km}\overline{a_{kj}} = \delta^j_k.$$

Now let $F(\omega,\omega')$ denote the left-hand side of (8.3.4). The function $F(\omega,\omega')$ is a spherical harmonic in each variable separately. Using (8.3.5) and this fact, a computation gives

$$F(A\omega, A\omega') = F(\omega,\omega').$$

Now let A' be the rotation that takes ω' to e_3 (the north pole) and let A'' be the rotation that fixes the north pole and takes $A'\omega$ to a point η in the plane determined by e_1 and e_3. For $A = A''A'$, the last formula shows that $F(\eta, e_3) = F(\omega, \omega')$. The left-hand side only depends on the position of η relative to the north pole, i.e., on the dot product $\eta \cdot e_3$. As rotations preserve the dot product, this is the same as $\omega \cdot \omega'$ and hence the function F only depends on $\omega \cdot \omega'$. We write $F(\omega, \omega') = P(\omega \cdot \omega')$. Since F is a spherical harmonic of order l in each variable separately and is zonal, it must be a multiple of $P_l(\omega \cdot \omega')$. We can now write

$$\sum_{m=-l}^{l} \mathcal{Y}_{lm}(\omega)\overline{\mathcal{Y}_{lm}(\omega')} = c\, P_l(\omega \cdot \omega').$$

To determine the constant, set $\omega' = \omega$ and integrate over the sphere to get $2l + 1 = c(4\pi)$, from which the result follows. □

If we set $\omega' = \omega$ in the addition formula, then

$$\sum_{m=-l}^{l} |\mathcal{Y}_{lm}(\omega)|^2 = \frac{2l+1}{4\pi}.$$

It follows that

$$|\mathcal{Y}_{lm}(\omega)| \le \sqrt{\frac{2l+1}{4\pi}}. \tag{8.3.6}$$

As a special case, $\mathcal{Y}_{l0}(\omega) = \sqrt{\frac{2l+1}{4\pi}}\, P_l(\omega \cdot e_3)$ (the zonal spherical harmonic in orthonormal form), and hence $|P_l(s)| \le 1$, a claim made earlier.

We have been using complex-valued functions throughout the preceding discussion, which is convenient from a computational viewpoint. These spherical harmonics can also be written using real-valued functions by applying Euler's formula to the complex exponential in Y_{lm}. For l a nonnegative integer and $0 \le m \le l$, the functions

$$Y_{lm}^{\cos}(\omega) = P_l^m(\cos\phi)\cos m\theta$$
$$Y_{lm}^{\sin}(\omega) = P_l^m(\cos\phi)\sin m\theta$$

are called real spherical harmonics. These are convenient for graphical purposes; Figure 8.6 gives plots of $Y_{5m}^{\cos}$ for $m = 0, 1, \ldots, 5$. The leftmost subfigure in the first row corresponds to $m = 0$ and is a zonal harmonic. The rightmost subfigure in the second row is the case $m = 5$ where there is only longitudinal dependence in the function. Such a spherical harmonic is sometimes called a tesseral harmonic.

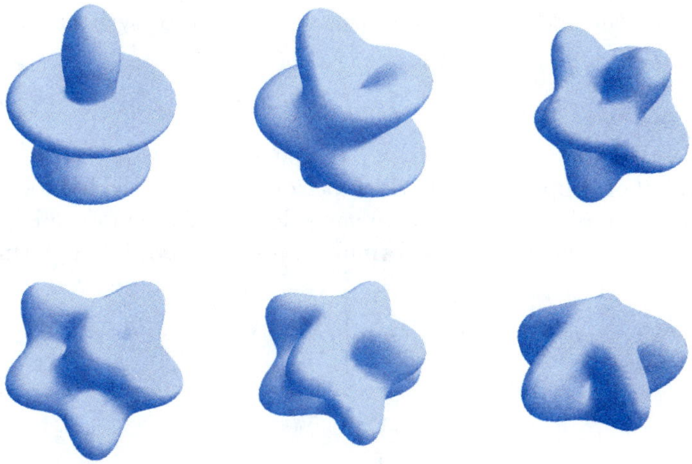

Figure 8.6 $Y_{5m}^{\cos}(\omega)$ for $m = 0, 1, \ldots, 5$

Application to Electrostatics

Suppose charge has distribution in space given by a density function $\sigma(x)$ where $\sigma(x) = 0$ outside a given bounded set $\Omega \subset B_a(O)$. Then the electrostatic potential of this charge distribution is given by the formula

$$u(\xi) = \int_{\mathbb{R}^3} \frac{\sigma(x)}{|x-\xi|}\, dV(x).$$

Writing $\xi = \rho\omega$ and $x = r\omega'$ (spherical coordinates), this formula can be rewritten as

$$u(\rho\omega) = \int_0^a \int_{S^2} \frac{\sigma(r\omega')}{\left[r^2 - 2r\rho(\omega\cdot\omega') + \rho^2\right]^{1/2}} r^2\, d\omega'\, dr.$$

We now assume $\rho > a$ and make use of the generating function for Legendre polynomials:

$$\begin{aligned}\left[r^2 - 2r\rho(\omega\cdot\omega') + \rho^2\right]^{-1/2} &= \rho^{-1}\left[1 - 2\frac{r}{\rho}(\omega\cdot\omega') + \left(\frac{r}{\rho}\right)^2\right]^{-1/2} \\ &= \rho^{-1}\sum_{l=1}^{\infty} P_l(\omega\cdot\omega')\left(\frac{r}{\rho}\right)^l.\end{aligned}$$

Substituting into the formula for u and integrating term by term we obtain

$$u(\rho\omega) = \sum_{l=0}^{\infty} \frac{1}{\rho^{l+1}} \int_0^a \int_{S^2} \sigma(r\omega') r^l P_l(\omega\cdot\omega') r^2\, d\omega'\, dr.$$

Upon applying the addition theorem this further can be written in the form

$$u(\rho\omega) = \sum_{l=0}^{\infty} \frac{1}{\rho^{l+1}} \frac{4\pi}{2l+1} \left[\sum_{m=-l}^{l} \int_0^a \int_{S^2} \sigma(r\omega') r^l \mathcal{Y}_{lm}(\omega') r^2 \, d\omega' \, dr \right] \mathcal{Y}_{lm}(\omega).$$

From the proof of Proposition 8.6, the function $r^l \mathcal{Y}_{lm}(\omega')$ is a homogeneous harmonic polynomial of degree l in three variables; we will denote this polynomial $\mathcal{Y}_{lm}(x)$. It follows that the above formula can be written in the form

$$u(\rho\omega) = \sum_{l=0}^{\infty} \frac{\sqrt{4\pi}}{2l+1} \left[\sum_{m=-l}^{l} M_{lm} \right] \frac{\mathcal{Y}_{lm}(\omega)}{\rho^{l+1}}, \tag{8.3.7}$$

where

$$M_{lm} = \sqrt{4\pi} \int_{\mathbb{R}^3} \sigma(x) \mathcal{Y}_{lm}(x) dV(x)$$

are called the electrostatic moments of the charge density. In the case where $l = m = 0$, M_{00} is simply the total charge. The physical conclusion is that for large distances, i.e., $\rho >> a$, the charge distribution has potential approximately equal to that of a single particle placed appropriately, with charge M_{00}.

Corresponding to $l = 1$ there are three moments, $M_{1,1}$, $M_{1,0}$, and $M_{1,-1}$, the so-called dipole moments. Using formulas (8.3.2) and (8.3.3) in subscripted notation with $l = 1$ these can be written

$$\begin{aligned} M_{1,1} &= 4\pi c_{1,1} \int_{\mathbb{R}^3} \sigma(x)(x_1 + ix_2) dV(x) \\ M_{1,0} &= 4\pi c_{1,0} \int_{\mathbb{R}^3} \sigma(z) x_3 \, dV(x) \\ M_{1,-1} &= 4\pi c_{1,-1} \int_{\mathbb{R}^3} \sigma(x)(x_1 - ix_2) dV(x), \end{aligned}$$

where the $c_{i,j}$ are normalization constants (recall that the $\mathcal{Y}_{lm}$ are orthonormal spherical harmonics). Notice that these can be expressed in terms of the usual moments of the charge density about each of the coordinate planes. For further electrostatic interpretations and variations, see [8].

Exercises

8.8 For degrees $l = 0, 1, 2$, write out a table of the spherical harmonics Y_{lm} and the corresponding homogeneous harmonic polynomials.

8.9 Determine the spherical harmonic expansions for the following functions.

a. $1 + 2\cos\phi + \cos^4\phi$

b. $\sin\phi\cos\theta + \sin^2\phi\sin 2\theta$

8.10 The discussion about the electrostatic moments corresponding to a charge distribution also applies to the gravitational potential of a mass distribution. Suppose a mass of a solid body is distributed with constant density σ in the region defined in spherical coordinates by the inequalities $0 \le \rho \le a + b\sin\phi$, $0 \le \phi \le \pi$, and $-\pi \le \theta \le \pi$. Find all the moments M_{lm} for $l = 0, 1, 2$. Show that M_{00} is the total mass of the solid. Where is the center of mass of this solid located?

8.4 Potential Problems and Spherical Harmonics

The functions $\rho^l \mathcal{Y}_{lm}(\omega)$, where $l = 0, 1, \ldots,$ $|m| \le l$, are all separated solutions to Laplace's equation in $\mathbb{R}^3$ (Proposition 8.6) with the property that

$$\{\mathcal{Y}_{lm}(\omega) \,:\, |m| \le l, l = 0, 1, \ldots, \}$$

forms a complete orthonormal set in $L^2(S^2)$ (Theorem 8.5). This suggests that the series

$$u(\rho\omega) = \sum_{l=0}^{\infty} \left(\frac{\rho}{a}\right)^l \sum_{|m|\le l} \widehat{f}(l,m)\mathcal{Y}_{lm}(\omega), \tag{8.4.1}$$

where

$$\widehat{f}(l,m) = \int_{S^2} f(\omega')\overline{\mathcal{Y}_{lm}(\omega')}d\omega'$$

provides the solution to the Dirichlet problem for Laplace's equation on $B_a(O)$ with boundary data f. This is the analog of the Fourier series representation of the same problem in two dimensions. That the series defines a harmonic function in $B_a(O)$ can be demonstrated using the estimate (8.3.6) for the spherical harmonics and the geometric decay of the term $\left(\frac{\rho}{a}\right)^l$; the proof is a modification of that given for the disk.

As in the case of the disk, the series solution to the Dirichlet problem for the ball can also be derived from the Poisson integral representation. To see this, start with the generating function for Legendre polynomials written in the form

$$\left[1 - 2\rho(\omega\cdot\omega') + \rho^2\right]^{-1/2} = \sum_{l=0}^{\infty} \rho^l P_l(\omega\cdot\omega').$$

The series is a power series in ρ and so can be differentiated term by term inside its circle of convergence ($\rho < 1$). Applying the differential operator $2\frac{\partial}{\partial\rho} + 1$ to both sides and using simple algebra, we have

$$\frac{1-\rho^2}{[1 - 2\rho(\omega\cdot\omega') + \rho^2]^{3/2}} = \sum_{l=0}^{\infty}(2l+1)\rho^l P_l(\omega\cdot\omega').$$

Substitute $\frac{\rho}{a}$ for ρ in the above formula and recall the Poisson integral representation in the form (7.4.11) with $d = 3$. It follows that the solution to the Dirichlet problem can be written in the form

$$u(\rho\omega) = \frac{1}{4\pi}\int_{S^2} f(\omega')\sum_{l=0}^{\infty}(2l+1)\left(\frac{\rho}{a}\right)^l P_l(\omega\cdot\omega')d\omega'.$$

Applying the addition theorem for spherical harmonics (8.3.4) and interchanging the integral and sum (justified by uniform convergence) completes the alternative derivation of the series representation for u.

Recalling Theorem 7.21, it follows that the boundary condition for the Dirichlet problem is satisfied by the series representation in the following limit sense

$$\lim_{\rho\to a^-}\sum_{l=0}^{\infty}\left(\frac{\rho}{a}\right)^l\sum_{|m|\le l}\widehat{f}(l,m)\mathcal{Y}_{lm}(\omega).$$

Similar ideas can be applied to solve the exterior problem for the ball with Dirichlet boundary condition f. This can also be obtained by applying inversion to (8.4.1). Our observations are summarized in the following result.

Theorem 8.8 Let $f \in C(S^2)$. (i) The solution to Laplace's equation on $B_a(O)$ with Dirichlet boundary data given by f is given by the series (8.4.1), the boundary condition satisfied in the sense that $\lim_{\rho\to a^-} u(\rho\omega) = f(\omega)$. (ii) The solution to the corresponding exterior problem for Laplace's equation with boundary data given by f is given by the series

$$u(\rho\omega) = \sum_{l=0}^{\infty}\left(\frac{a}{\rho}\right)^{l+1}\sum_{|m|\le l}\widehat{f}(l,m)\mathcal{Y}_{lm}(\omega)$$

with the boundary condition satisfied in the sense that $\lim_{\rho\to a^+} u(\rho\omega) = f(\omega)$.

Like the case of the disk, the series solution formulas provide an effective means for computing the solution. Of course, computing the spherical harmonic expansion of the boundary data on a sphere is computationally more complex than computing the Fourier series of a function on the circle. Here are some examples.

Example 8.4 In the case where the boundary data can easily be expressed in terms of spherical harmonics, the series solution is immediate. For example, consider the interior problem for the ball $B_a(O)$ where the boundary data is given by $f(\theta,\phi) = \sin\phi\cos\theta + \cos\phi\sin\phi\sin\theta$. In Example 8.3, the spherical harmonic expansion of f was determined as

$$f(\omega) = \frac{1}{2}Y_{1,1}(\omega) + \frac{1}{2}Y_{1,-1}(\omega) - \frac{i}{6}Y_{2,1}(\theta,\phi) + \frac{i}{6}Y_{2,-1}(\omega).$$

Hence, the solution to the Dirichlet problem is given by

$$u(\rho\omega) = \frac{1}{2}\frac{\rho}{a}[Y_{1,1}(\omega) + Y_{1,-1}(\omega)] - \frac{i}{6}\left(\frac{\rho}{a}\right)^2 [Y_{2,1}(\omega) - Y_{2,-1}(\omega)].$$

Example 8.5 (Point Charge Inside a Sphere Held at Constant Potential) A charge q is placed inside a conducting sphere of radius a, which is held at constant potential u_0. The problem is to determine the potential inside the sphere. Assuming the point charge is at the point $\xi = c\omega'$ where $c < a$ and $\omega' \in S^2$, the potential is governed by Poisson's equation in the form (Coulomb's law)

$$\nabla^2 u = -4\pi q\delta_\xi(x)$$

with the boundary condition $u|_{S_a(O)} = u_0$. Now the free-space potential of the point charge is given by $\frac{q}{|x-\xi|}$. Hence, the solution to our BVP will have the form

$$u(x) = \frac{q}{|x-\xi|} + v(x),$$

where v is a harmonic function and must satisfy the boundary condition

$$v(a\omega) = u_0 - \frac{q}{|a\omega - \xi|}, \quad \omega \in S^2.$$

This boundary condition can be further expressed as

$$\begin{aligned} v(a\omega) &= u_0 - \frac{q}{\sqrt{a^2 - 2ac(\omega\cdot\omega') + c^2}} \\ &= u_0 - \frac{q}{a\sqrt{1 - 2(\frac{c}{a})(\omega\cdot\omega') + (\frac{c}{a})^2}}. \end{aligned}$$

Applying the generating function for the Legendre polynomials, we obtain the spherical harmonic expansion of the boundary data as

$$v(a\omega) = u_0 - \frac{q}{a}\sum_{l=0}^{\infty}\left(\frac{c}{a}\right)^l P_l(\omega\cdot\omega').$$

Notice that this is a zonal expansion relative to the pole at ω'. It follows that the expansion for v has the form

$$v(\rho\omega) = \left(u_0 - \frac{q}{a}\right) - \frac{q}{a}\sum_{l=1}^{\infty}\left(\frac{c}{a}\right)^l\left(\frac{\rho}{a}\right)^l P_l(\omega\cdot\omega').$$

Thus the solution to the original problem is given by

$$u(\rho\omega) = \frac{q}{\sqrt{\rho^2 - 2c\rho(\omega\cdot\omega') + c^2}} + \left(u_0 - \frac{q}{a}\right) - \frac{q}{a}\sum_{l=1}^{\infty}\left(\frac{c}{a}\right)^l\left(\frac{\rho}{a}\right)^l P_l(\omega\cdot\omega').$$

We note that in the case $u_0 = 0$ (i.e., the sphere is grounded), the second and third terms can be rewritten using the generating function for the Legendre polynomials as

$$\frac{-\frac{qa}{c}}{\sqrt{\rho^2 - 2\rho\frac{a^2}{c}(\omega\cdot\omega') + \left(\frac{a^2}{c}\right)}}.$$

This corresponds to the free-space potential of a charge of magnitude $-\frac{qa}{c}$ placed at the point $\xi^* = \frac{a^2}{c^2}\xi$, i.e., the image point of ξ under inversion in the sphere $S_a(O)$. In other words, we have again derived the Green's function for the Dirichlet problem on a ball (where $q = 1$).

Exercises

8.11 Find the solution to Laplace's equation in the ball $B_a(O)$ in three dimensions for each of the following boundary conditions.

a. $u(a,\theta,\phi) = 1 + 2\cos\phi + \cos^4\phi$

b. $u(a,\theta,\phi) = \sin\phi\cos\theta + \sin^2\phi\sin 2\theta$

HINT: See Exercise 8.9.

8.12 Repeat Exercise 8.11 in the region exterior to the ball with the assumption that $u \to 0$ as $\rho \to \infty$.

8.13 Determine the potential in the ball $B_a(O)$ in three dimensions if the northern hemisphere is held at potential V and the southern hemisphere is held at $-V$.

8.14 In three dimensions solve the exterior problem for the ball $B_a(O)$: $\nabla^2 u = q\delta_\xi(x)$, $u = u_0$ on $S_a(O)$, and $u \to 0$ as $|x| \to \infty$. Here $|\xi| > a$. Physically, this solves for the potential due to an external point charge q outside a spherical conductor of radius $a > 0$ that is held at potential u_0.

8.15 Let $C_{a,b}$ be the region between concentric spheres $\rho = a$ and $\rho = b$, where $0 < a < b$. Determine the solution to $\nabla^2 u = q\delta_\xi(x)$ where $u = 0$ on both the inner and outer sphere and $a < |\xi| < b$.

8.16 Let $0 < \alpha < \pi$. Find the steady-state temperature distribution in the ball $B_a(O)$ where on the boundary sphere the temperature is held to a constant value u_0 in the region where $0 < \phi < \alpha$ and is held to zero where $\alpha < \phi < \pi$.

Appendix A

Uniform Convergence and Approximation

Our purpose in this appendix is to review facts concerning uniform convergence of sequences of functions, uniform convergence of improper integrals, and the Weierstrass approximation theorem. A good reference for further review is [12].

Definition A.1 Let $\{f_m\}$ be a sequence of functions defined on $[a, b]$. Then f_m converges uniformly on $[a, b]$ to f if given $\varepsilon > 0$ there exists an $M = M(\varepsilon)$ such that

$$|f_m(x) - f(x)| < \varepsilon \quad \text{provided} \quad m \geq M$$

for all $x \in [a, b]$.

Notice that uniform convergence implies pointwise convergence; the converse, however, is not true. For example, on the interval $[0, 1]$, let $f_n(x) = x^n$. In a pointwise sense, $f_n(x) \to 0$ for $0 \leq x < 1$ and converges to 1 if $x = 1$. However, the convergence is not uniform (draw a picture).

Here is the basic result from calculus.

Theorem A.2 (i) Let $\{f_m\}$ be a sequence of Riemann integrable functions on $[a, b]$. If $f_m \overset{\text{unif}}{\to} f$, then f is Riemann integrable and

$$\lim_{m\to\infty} \int_a^b f_m(x)\, dx = \int_a^b f(x)\, dx.$$

(ii) Let $\{f_m\}$ be a sequence of differentiable functions on $[a, b]$. If $f_m \overset{\text{unif}}{\to} f$ and $f'_m \overset{\text{unif}}{\to} g$, then f is differentiable and $f' = g$.

Our primary application of this result takes the form of integrating or differentiating an infinite series term by term. Hence it is useful to have a test for uniform convergence of infinite series. The best known is the following due to Weierstrass.

Theorem A.3 (Weierstrass M-test). Let $\{f_m(x)\}$ be a sequence of functions defined on $[a, b]$. If there exists a sequence of constants $\{M_m\}$ such that $|f_m(x)| \leq M_m$ for all $x \in [a, b]$ and $\sum_m M_m$ converges, then the series

$$\sum_m f_m(x)$$

converges uniformly on $[a, b]$.

Concerning the differentiation of integrals, the following is basic for bounded domains.

Theorem A.4 If $f(x, t)$ is a continuous function on $[a, b] \times I$, where I is an interval, and $\partial f/\partial t$ is also continuous on $D \times I$, then

$$\frac{d}{dt} \int_a^b f(x, t)dx = \int_a^b \frac{\partial f}{\partial t}(x, t)dx.$$

In our realm of application, the domain is the real line, and the integral is improper. A regularity property of the integral is needed to justify this interchange of derivative and integral; we state the definition for the real line, but the definition can be adapted for higher dimensions.

Definition A.5 Let $g(r, s)$ be defined and continuous on $\mathbb{R} \times I$, where I is an interval. Suppose the improper integral

$$G(s) = \int_{\mathbb{R}} g(r, s)dr = \lim_{R \to \infty} \int_{|r| \leq R} g(r, s)dr$$

exists for all $s \in I$. We say the improper integral is uniformly convergent on I if given $\varepsilon > 0$ there is an $R' = R'(\varepsilon)$ such that

$$\left| G(s) - \int_{|r| \leq R} g(r, s)dr \right| < \varepsilon \quad \text{provided} \quad R > R',$$

for all $s \in I$.

Notice what this says: the function defined by the improper integral is uniformly approximated by the truncated integral.

A companion to the Weierstrass M-test for series is the following.

Proposition A.6 Let $g(r, s)$ be defined and continuous on $\mathbb{R} \times I$, where I is an interval. Suppose the improper integral

$$G(s) = \int_{\mathbb{R}} g(r, s)\, dr = \lim_{R \to \infty} \int_{|r| \le R} g(r, s)\, dr$$

exists for all $s \in I$. If there exists a function $h(r)$ such that $|g(r, s)| \le h(r)$ for all $s \in I$ and the improper integral $\int_{\mathbb{R}} h(r)$ is finite, then the improper integral defining G is uniformly convergent.

Combining our results, we now have the following theorem.

Theorem A.7 Let $g(r, s)$ be defined and continuous on $\mathbb{R} \times I$, where I is an interval, and suppose $\partial g/\partial s$ exists and is continuous. If the improper integral for $G(s)$ as previously defined and the improper integral

$$\int_{\mathbb{R}} \frac{\partial g}{\partial s}(r, s) dr$$

are uniformly convergent on I, then G is differentiable and

$$G'(s) = \int_{\mathbb{R}} \frac{\partial g}{\partial s}(r, s) dr.$$

We finish this appendix with a fundamental result concerning the approximation of continuous functions by polynomials due to Weierstrass. Indeed, this theorem is the starting point of an area of mathematics known as approximation theory.

Theorem A.8 (Weierstrass Approximation Theorem) Let $f \in C[a, b]$. Then given $\varepsilon > 0$, there exists a polynomial $p(x)$ such that $|f(x) - p(x)| < \varepsilon$ for all $x \in [a, b]$.

For a proof of this result see [12].

Appendix B

Commonly Used Symbols

Symbol	Meaning
$\in$	is an element of, set membership
$\subset$	set containment
$\cap$	set intersection
$\cup$	set union
∂D	boundary of the set D
∂_i	partial derivative, $\frac{\partial}{\partial x_i}$
$C(D)$	continuous functions on D
$C^m(D)$	class of functions whose derivatives up to order m are continuous
$L^2(D)$	square integrable functions on D
$\boldsymbol{\nabla}$	gradient (of functions), divergence (of vector fields)
bold face	vector quantities
∇^2	Laplace operator
∇^2_{sp}	Laplace operator on sphere
$\cdot$	dot product between vectors
$\mathbb{R}$	real line
$\mathbb{R}^d$	d-dimensional Euclidean space, usually $d = 2$ or 3
$\mathbb{C}$	complex numbers
S^{d-1}	unit sphere in $\mathbb{R}^d$

Appendix C

Solutions to Selected Exercises

Chapter 1

1.5. $-ku_x(0,t) = H[T(t) - u(0,t)]$

1.7(i). **a.** The boundary conditions are $u(0,t) = T_1$ and $u(L,t) = T_2$. **b.** The steady-state solution is $U(x) = T_1 + \frac{T_2-T_1}{L}x$. **c.** The outward flux of the steady state at the left end is $kU'(0) = k(T_2-T_1)/L$ and at the right end is $-kU'(L) = -k(T_2 - T_1)/L$.

1.8. Same conditions as for a heat flow problem.

1.10. $-T\frac{\partial u}{\partial x}(L,t) = ku(L,t)$

1.14. $u(x,t) = \frac{1}{2}\left[\operatorname{erf}\left(\frac{x}{2\alpha\sqrt{t}}\right) + 1\right]$

1.16. The conserved quantity in this case becomes $\frac{1}{2}\int_0^L [u_t^2(x,t)+c^2u_x^2(x,t)]dx + \frac{1}{2}\frac{k}{\rho}u^2(L,t)$. (This uses Exercise 1.10.)

Chapter 2

2.2. **b.** Two first integrals are given by $\xi(x,y,z) = x$ and $\eta(x,y,z) = y^2 + z^2$. Therefore the integral curves are circles $y^2 + z^2 = c$ lying in planes $x = d$ (here c and d are constants).

2.3. **a.** $ye^{-x^2} = c$

2.4. **a.** $u(x,y) = \frac{1}{2}x^2 \sin y + g(y)$

2.5. **a.** $u(x,y) = g(2x - 3y)$

2.6. **a.** $u(x,y) = g(bx - ay)e^{(-c/a)x}$

2.11. **a.** Hyperbolic **b.** Elliptic

2.13. **a.** Hyperbolic in the region $|x| > 1$, $y \in \mathbb{R}$; elliptic when $|x| < 1$, $y \in \mathbb{R}$

2.17. **a.** $\mathbf{F} = (0, 0, xy)$

Chapter 3

3.12. $u(r) = B + aA\log(r/b)$

3.15. $u_n(r,\theta) = r^{-|n|}e^{in\theta}$, $n \in \mathbb{Z}$

3.16. $r^2g'' + rg' + (\lambda r^2 - n^2)g = 0$

3.25. **a.** $\frac{1}{2}\frac{\partial}{\partial t}\int_D [u_t^2 + k|\boldsymbol{\nabla} u|^{2.}]dV = \int_{\partial D} ku_t \boldsymbol{\nabla} u \cdot \mathbf{n}_o \, dS$

Chapter 4

4.5. **a.** $(\partial_1 u, \phi) = \int_{\mathbb{R}} \phi(0, x_2)dx_2$; $\partial_2 u = 0$ **b.** $\partial_{12}u = 0$

4.15. $u(x,t) = \frac{1}{2}\left[\operatorname{erf}\left(\frac{x-1}{2\alpha\sqrt{t}}\right) - \operatorname{erf}\left(\frac{x-3}{2\alpha\sqrt{t}}\right)\right]$

4.16. Set $v(x,t) = \mathcal{W}_t(x,t)$. Verify that v satisfies the wave equation. Note $v(x,0) = \mathcal{W}_t(x,0) = \delta(x)$. To demonstrate the other initial condition, use the explict form for $\mathcal{W}$.

4.18. $\mathcal{S}_n(x,y,t) = \mathcal{S}(x-y,t) + \mathcal{S}(x+y,t)$

4.21. **a.** $G(x,\xi) = \frac{1}{\pi}\begin{cases} (\xi - \pi)x, & 0 \le x \le \xi \\ \xi(x - \pi), & \xi \le x \le \pi \end{cases}$

4.22. **a.** $G(x,\xi) = \dfrac{1}{\mu \sinh \mu\pi}\begin{cases} \sinh \mu(\xi - \pi)\sinh \mu x, & 0 \le x \le \xi \\ \sinh \mu\xi \sinh \mu(x - \pi), & \xi \le x \le \pi \end{cases}$

Chapter 5

5.2. $u_n(x,t) = e^{-(\alpha n)^2 t}\cos nx$, $n = 0, 1, \ldots$

5.3. $u_n(x,t) = e^{-\left(\alpha\left(n+\frac{1}{2}\right)\right)^2 t}\sin\left(n + \frac{1}{2}\right)x$, $n = 0, 1, \ldots$

5.4. **a.** $u_n(x,t) = \cos nct \sin nx$ and $u_n(x,t) = \sin nct \sin nx$, $n = 1, 2, \ldots$

5.8. **a.** $u_n(x,t) = \sin\left(n + \frac{1}{2}\right) x \cosh\left(n + \frac{1}{2}\right) x$ and $u_n(x,t) = \sin\left(n + \frac{1}{2}\right) x \sinh\left(n + \frac{1}{2}\right) x$, $n = 0, 1, \ldots$

5.12. Use the fact that $\|f + g\|_2^2 = \|f\|_2^2 + \|g\|_2^2 + 2(f, g)$.

5.16. **b.** $\widehat{f_c}(n) = \frac{2}{\pi n^2}((-1)^n - 1)$, $\widehat{f_c}(0) = \pi$, $\widehat{f_s}(n) = 0$

5.17. **a.** The complex coefficients are $\widehat{f}(n) = \frac{(-1)^n \sinh a\pi}{\pi(a - in)}$.

5.38. $G(x, \xi) = \frac{2}{\pi} \sum_{n=0}^{\infty} \frac{\sin\left(n+\frac{1}{2}\right)\xi \sin\left(n+\frac{1}{2}\right)x}{\left(n+\frac{1}{2}\right)^2}$

Chapter 6

6.2. **a.** $u(x,t) = \sum_{n=0}^{\infty} \widehat{f}(n) e^{-\left(\alpha\left(n+\frac{1}{2}\right)\frac{\pi}{L}\right)^2 t} \sin\left(n + \frac{1}{2}\right) \frac{\pi x}{L}$; $\widehat{f}(n) = \frac{2}{L} \int_0^L f(y) \sin\left(n + \frac{1}{2}\right) \frac{\pi y}{L}\, dy$

6.3. **a.** $u(x,t) = \sum_{n=0}^{\infty} \widehat{f}(n) e^{-\left(n+\frac{1}{2}\right)^2 t} \cos\left(n + \frac{1}{2}\right) x$

6.5. The steady-state solution is $U(x) = \frac{hT}{1+hL} x$. The solution is given by $u(x,t) = U(x) + v(x,t)$ where $v(x,t)$ has series representation given by

$$v(x,t) = \sum_{n=1}^{\infty} \widehat{g}(n) e^{-(\alpha\lambda_n)^2 t} \sin \lambda_n x,$$

where the sequence $\{\lambda_n\}$ are the solutions of the equation $\tan \lambda L = -\frac{\lambda}{h}$ and $\widehat{g}(n)$ is the nth Fourier coefficient of $g(x) = -U(x)$ with respect to the orthogonal set $\{\sin \lambda_n x\}$.

6.17. **a.** The Fourier sine coefficients of f are given by $\widehat{f_s}(n) = \frac{2hL^2 \sin \frac{na\pi}{L}}{a(L-a)n^2\pi^2}$.

6.20. $\mathcal{W}(x, \xi, t) = \frac{2}{L} \sum_{n+1}^{\infty} \sin \frac{n\pi ct}{L} \sin \frac{n\pi\xi}{L} \sin \frac{n\pi x}{L}$

6.28. **a.** $X_{k,l,m}(x) = \sin kx_1 \sin lx_2 \sin mx_3$, $k, l, m = 1, 2, \ldots$ **b.** $X_{l,k,m}(x) = \cos kx_1 \cos lx_2 \sin mx_3$, $k, l = 0, 1, 2, \ldots$, $m = 1, 2, \ldots$

6.32. The double Fourier sine coefficients are given by $\left(\frac{2}{\pi}\right)^2 \left(\frac{1-(-1)^k}{k}\right) \left(\frac{1-(-1)^l}{l}\right)$, $k, l = 1, 2, \ldots$

Chapter 7

7.3. **a.** $u(x_1, x_2) = \frac{3\pi}{2} - \tan^{-1}\left(\frac{x_1}{x_2}\right)$

7.7. $(p_1, p_2, -p_3)$

7.9. For $d = 3$,

$$M^r f(x_1, x_2, x_3) = \frac{1}{4\pi} \int_0^{2\pi} \int_0^{\pi} f(x_1 + r\cos\theta \sin\phi, x_2 + r\sin\theta\sin\phi, x_3 + r\cos\phi)\sin\phi \, d\phi \, d\theta$$

7.10. $\frac{\pi}{8}$

7.15. $u(\rho) = \frac{1}{6}\rho^2 + \frac{ab}{6(a+b)\rho} - \frac{a^2b+a+b}{6a^2(a+b)}$

7.18. $u(x_1, x_2) = \frac{2}{\pi}\tan^{-1}\left(\frac{x_2-1}{x_1-1}\right)$

7.21. $u(r,\theta) = \frac{\sqrt{1+h}-1}{a} r^2 \sin 2\theta$

7.27. **a.** $u(\rho) = \frac{1}{6}\rho^2$, which corresponds with $u(x) = \frac{1}{6}|x|^2$, a smooth function in the disk. **b.** $u(\rho) = \frac{1}{12}\rho^3$, corresponding to $u(x) = \frac{1}{12}|x|^3$.

7.31. Apply inversion to the Poisson kernel for the disk.

7.33. Let $\xi = (\xi_1, \xi_2, \xi_3)$ be in the upper hemisphere and define $\xi' = (\xi_1, \xi_2, -\xi_3)$. Then the Green's function is given by

$$G(x,\xi) = K(x-\xi) - K(x-\xi^*) - K(x-\xi') + K(x-\xi'^*)$$

7.36. $u(x_1, x_2) = \frac{1}{\pi x_2}\left(\tan^{-1}\left(\frac{\pi - x_1}{x_2}\right) + \tan^{-1}\left(\frac{\pi + x_1}{x_2}\right)\right)$

7.38. $u(r,\theta) = \sum_{|n|=1}^{\infty} \frac{\widehat{g}(n)}{|n| a^{|n|-1}} r^{|n|} e^{in\theta}$

7.42. $u(r,\theta) = \sum_{|n|<\infty} \widehat{f}(n) \left(\frac{a}{r}\right)^{|n|} e^{in\theta}$

Chapter 8

8.6. The Fourier–Legendre coefficients are $\widehat{f}_0 = \frac{1}{4}$, $\widehat{f}_1 = \frac{1}{2}$,

$$\widehat{f}_l = \frac{1}{4}\left[P_{l-1}(0) - \frac{1}{2l+3}P_l(0) - \frac{2l+2}{2l+3}P_{l+1}(0)\right], \quad l \geq 2.$$

8.7. For $|x| < |\xi|$, $G(x,\xi) = -\frac{1}{4\pi|\xi|} \sum_{l=0}^{\infty} \left(1 - \left(\frac{|\xi|}{a}\right)^{2l+1}\right) \left(\frac{|x|}{|\xi|}\right)^l P_l(\cos\Theta)$

Bibliography

[1] S. Antman, "The equations for large vibrations of strings," *Amer. Math. Monthly* 87 (1980) 359-370.

[2] E. T. Bell, *Men of Mathematics,* Simon & Schuster (1937).

[3] G. Birkhoff, G. Rota, *Ordinary Differential Equations,* 2nd edition, Blaisdell Publ. (1969).

[4] F. Brauer, J. A. Nohel, *The Qualitative Theory of Ordinary Differential Equations,* W.A. Benjamin (1969).

[5] E. A. Coddington, N. Levinson, *Theory of Ordinary Differential Equations,* McGraw-Hill (1955).

[6] W. Dunham, *The Calculus Gallery: Masterpieces from Newton to Lebesgué,* Princeton Univ. Press (2005).

[7] H. Dym, H. P. McKean, *Fourier Series and Integrals,* Academic Press (1972).

[8] L. Eyges, *The Classical Electromagnetic Field,* Addison-Wesley (1972).

[9] G. Folland, *Introduction to Partial Differential Equations,* Princeton Univ. Press (1976).

[10] G. Folland, *Fourier Analysis and Its Applications,* Brooks/Cole (1992).

[11] P. R. Garbedian, *Partial Differential Equations,* Wiley (1964).

[12] R. R. Goldberg, *Methods of Real Analysis,* Blaisdell Publ. (1964).

[13] George Green, *An Essay on the Application of Mathematical Analysis to the Theories of Electricity and Magnetism,* Nottingham (1828). Reprinted in 1850 in three parts: *Journal fur die reine und angewandtc Mathematik* Vol. 39, 1 (1850) pp. 73–89, Vol. 44, 4 (1852) pp. 356–374, and Vol. 47, 3 (1854) pp. 161–221. Currently available online at http://arxiv.org/abs/0807.0088.

[14] R. Haberman, *Elementary Applied Partial Differential Equations*, 3rd edition, Prentice-Hall (1998).

[15] T. Hawkins, *Lebesgué's Theory of Integrations: Its Origins and Development*, Chelsea Publ. (1975).

[16] H. Hochstadt, *The Functions of Mathematical Physics,* Dover Publ. (1986).

[17] R. D. Jackson, *Classical Electrodynamics*, 2nd edition, Wiley (1975).

[18] F. John, *Plane Waves and Spherical Means Applied to Partial Differential Equations,* Springer-Verlag (1981).

[19] O. D. Kellog, *Foundations of Potential Theory*, Dover Publ. (1953).

[20] N. J. Lebedev, *Special Functions and Their Applications*, Dover Publ. (1972).

[21] J. E. Marsden, A. J. Tromba, *Vector Calculus*, 5th edition, Freeman Publ. (2003).

[22] R. C. McOwen, *Partial Differential Equations, Methods and Applications,* 2nd edition, Prentice-Hall (2003).

[23] M. A. Pinsky, *Partial Differential Equations and Boundary Value Problems with Applications*, 3rd edition, McGraw-Hill (1998).

[24] M. A. Pinsky, *Introduction to Fourier Analysis and Wavelets*, Brooks/Cole (2002).

[25] H. Poincaré, *The Value of Science,* Dover Publ. (1958) (English translation of the original 1905 work published in French).

[26] M. Protter, H. Weinberger, *Maximum Principles in Partial Differential Equations*, Prentice-Hall (1967).

[27] C. S. Rees, S. M. Shah, Č.V. Stanojević, *Theory and Applications of Fourier Analysis,* Marcel Dekker (1981).

[28] I. Rubinstein, L. Rubinstein, *Partial Differential Equations in Classical Mathematical Physics*, Cambridge Univ. Press (1998).

[29] W. Strauss, *Partial Differential Equations, An Introduction*, Wiley (1992).

[30] R. S. Strichartz, *Guide to Distribution Theory and Fourier Transforms,* CRC Press (1994).

[31] C. A. Taylor, *The Physics of Musical Sounds*, Elsevier (1965).

[32] M. Taylor, *Partial Differential Equations, Basic Theory*, Springer-Verlag (1996).

[33] H. F. Weinberger, *A First Course in Partial Differential Equations*, Blaisdell (1965).

[34] R. Wheeden, A. Zygmund, *Measure and Integral*, Marcel Dekker (1981).

[35] E. C. Zachmanoglou, D. W. Thoe, *Introduction to Partial Differential Equations with Applications*, Dover Publ. (1975).

[36] A.C. Zemanian, *Distribution Theory and Transform Analysis*, Dover Publ. (1987).

[37] A. Zygmund, *Trigonometric Series,* Cambridge Univ. Press (1959).

Index